Optical Sensors and Microsystems

Optical Sensors and Microsystems

New Concepts, Materials, Technologies

Edited by

Sergio Martellucci
The University of Rome "Tor Vergata"
Rome, Italy

Arthur N. Chester
Hughes Research Laboratories, Inc.
Malibu, California

and

Anna Grazia Mignani
Institute for Research on Electromagnetic Waves "Nello Carrara"
Florence, Italy

SPRINGER SCIENCE+BUSINESS MEDIA, LLC

Library of Congress Cataloging-in-Publication Data

Course of the International School of Quantum Electronics on Optical Sensors and Microsystems: New Concepts, Materials, Technologies (22nd: 1997: Erice, Italy)
Optical sensors and microsystems: new concepts, materials, technologies/edited by Sergio Martellucci, Arthur N. Chester and Anna Grazia Mignani.
p. cm.
Includes bibliographical references and index.
ISBN 978-1-4757-7279-1 ISBN 978-0-306-47099-8 (eBook)
DOI 10.1007/978-0-306-47099-8
1. Optoelectronic devices. 2. Optical detectors. 3. Microelectronics. 4. Micromechanics. I. Martellucci, S. II. Chester, A. N. III. Mignani, Anna Grazia. IV. Title.

TA1750 .C69 2000
681'.25—dc21

00-023898

Proceedings of the 22nd Course of the International School of Quantum Electronics on Optical Sensors and Microsystems: New Concepts, Materials, Technologies, held November 27–December 2, 1997, in Erice, Sicily

ISBN 978-1-4757-7279-1

Originally published by Kluwer Academic/Plenum Publishers, New York in 2000
Softcover reprint of the hardcover 1st edition 2000

http://www.wkap.nl/

10 9 8 7 6 5 4 3 2 1

A C.I.P. record for this book is available from the Library of Congress.

PREFACE

In recent years, fiber optical sensors and optical microsystems have moved out of the laboratory and assumed a significant role in sensing and measurement of many kinds. These optical techniques are important to a broad range of applications, including biomedicine, environmental sensing, mechanical and industrial measurement, and art preservation.

In November 1997, an international group of scientists convened in Erice, Sicily, for a Conference on the subject of "Optical Sensors and Microsystems." This Conference was the 22nd Course of the International School of Quantum Electronics, under the auspices of the "Ettore Majorana Foundation and Center for Scientific Culture" and was directed by Prof. A. Domanski of the Institute of Physics, Warsaw University of Tecknology in Warsaw, Poland, and Prof. A.G. Mignani of the "Nello Carrara" Institute of Research on Electromagnetic Waves (IROE-CNR) in Florence, Italy. This book presents the Proceedings of this Conference, providing a fundamental introduction to the topic as well as reports on recent research results.

The aim of the Conference was to bring together some of the world's acknowledged scientists who have as a common link the use of optoelectronics instrumentation, techniques and procedures related to the fields of optical sensors and microsystems. Most of the lecturers attended all the lectures and devoted their spare hours to stimulating discussions. We would like to thank them all for their admirable contributions. The Conference also took advantage of a very active audience; most of the participants were active researchers in the field and contributed with discussions and seminars. Some of these seminars are also included in these Proceedings.

The Conference was an important opportunity to discuss the latest developments and emerging perspectives on the use of new concepts, materials, technologies for optical sensors and microsystems.

The Chapters in these Proceedings are not ordered exactly according to the chronology of the Conference but they give a fairly complete accounting of the Conference lectures with the exception of the informal panel discussions. The contributions presented at the Conference are written as extended, review-like papers to provide a broad and representative coverage of the fields of diffractive optics and optical microsystems. We did not modify the original manuscripts in editing this book, except to assist in uniformity of style; but we did group them according to the following Sections:

- Technology of optical sensors and microsystems: Measurement techniques (six Chapters); and, Design and fabrication techniques (two chapters).
- Major applications areas: Biomedical applications (three chapters); Environmental sensing (six chapters); Art preservation (three chapters); and Mechanical measurements (three chapters).

These Proceedings update and augment the material contained in the previous ISQE volumes, "Integrated Optics: Physics and Applications," S. Martellucci and A.N. Chester, Eds., NATO ASI Series B, Vol. 91 (Plenum, 1983), "Advances in Integrated Optics," S. Martellucci, A.N. Chester and M. Bertolotti, Eds. (Plenum, 1994), and "Diffractive Optics and Optical Microsystems" S. Martellucci and A.N. Chester, Eds. (Plenum, 1997). For some closely related technology, to the topical Section devoted to "Fiber Sensors," the reader may also wish to consult the ISQE volume, "Optical Fiber Sensors," A.N. Chester, S. Martellucci and A.M. Scheggi, Eds., NATO ASI Series E, Vol. 132 (Nijhof, 1987).

We are grateful to Profs. A. Domansky and A.G. Mignani for their able organization and direction of the Course, to our editor at Plenum Press London, Joanna Lawrence, for outstanding professional support. We also greatly appreciate the expert help from our assistants Carol Harris and Margaret Hayashi, and the support of Paolo Di Maggio for much of the computer processing work. This International School of Quantum Electronics (A.N. Chester and S.Martellucci, Directors) is being held under the auspices of the "Ettore Majorana" Foundation and Center for Scientific Culture, Erice, Italy. We acknowledge with gratitude the cooperation of the Quantum Elecronics and Plasma Physics Research Group of the Italian Research Council (GNEQP – CNR) and support from the Italian Ministry of Education, the Italian Ministry of University and Scientific Research, and the Sicilian Regional Government.

Sergio Martellucci
Professor of Physics
University of Rome "Tor Vergata"
Rome, Italy

Arthur N. Chester
Chairman and President
Hughes Research Laboratories, Inc.
Malibu, California

Anna Grazia Mignani
Director of Research
Institute for Research on Electromagnetic Waves "Nello Carrara"
Florence, Italy

CONTENTS

I. Technology

II. Applications

TECHNOLOGY

ADVANCED OPTOELECTRONICS IN OPTICAL FIBRE SENSORS

B. Culshaw

University of Strathclyde
Department of Electronic & Electrical Engineering
204 George Street
Glasgow G1 1XW, UK

1. INTRODUCTION

Optical fibre sensing is, in common with most other sensor technologies, highly specialised, niche oriented, and prone to long research and development times. The first fibre sensors date back to the mid 1960s and in the intervening 30 or more years a few real life applications coupled to fully engineered systems have emerged.[1] These have been characterised by long development times into niche markets. The most successful has probably been the optical fibre gyroscope first reported in 1977 and emerging as a production instrument in the mid 1990s.[2] Other fully engineered sensor systems have emerged in distributed temperature measurement, especially systems using Raman backscatter[3] and in a few biomedical applications where small size and electrical passivity are very important. There are other applications in civil engineering measurement and safety and security systems[4] which are also emerging as market competitive sensors.

Many of these sensors involve elegant optoelectronic design exemplified in the gyroscope and in the Raman and Brillouin[5] distributed temperature and strain measurement systems. Others are conceptually relatively simple, for example the chemically sensitive optrode engineers a flexible light path between a source and chemically active area.[6] The chemically active medium modulates the incoming light - usually by changing its spectrum - and the engineered light path returns this modulated light for processing.

The aim of this paper is to highlight the "advanced" optoelectronic tools which offer the potential for expanding the performance of optical fibre sensing systems either into new sensing domains or into a more comprehensive and effective development on existing sensing concepts. Arguably the fibre gyroscope and the Raman and Brillouin non-linear systems - all of which are currently commercially competitive within their niches - exploit advanced concepts in non-linear and guided wave optics. It is then reasonable to extrapolate that other "advanced" optoelectronic concepts which are currently being explored could contribute to future systems. Such advanced concepts could include lasers and amplifiers especially in all fibre format, non-linear phenomena other than Raman and Brillouin scatter,

Optical Sensors and Microsystems: New Concepts, Materials, Technologies
Edited by Martellucci et al., Kluwer Academic / Plenum Publishers, New York, 2000.

non-reciprocal processes, near field concepts, more variations on integrated optics, coherent detection, optical preamplifiers and a host of other possibilities. The remainder of this paper looks very quickly at some of these concepts and speculates on potential applications within a sensor context.

2. LASERS AND AMPLIFIERS

The fibre laser amplifier has within a decade or so progressed from concept to telecommunications product* and its role in future telecommunications networks is critical and inevitable. The technical problems in realising low noise high gain amplifiers in the erbium (1.55μm) band are now largely solved with issues such as gain flattening for dense wavelength division multiplexed systems now occupying the device engineers.[7]

In the sensor context, the impact of fibre lasers and amplifiers has been relatively modest. They have successfully been incorporated into Brillouin backscatter distributed strain monitoring systems both as power output stages and input pre-amplifiers. In the more general sensor context, fibre lasers and amplifiers offer technical possibilities which have yet to be fully exploited within the sensor community.

Fibre lasers, for example, have been configured in broadband tunable architectures offering relatively high power levels directly into the fibre (up to 10mW CW or in the region of 1W peak). Such lasers can be configured in an all fibre architecture (Figure 1.) which is electro-optically tunable and can be frequency locked with line widths in the kHz region.

The limited path length range over which an all fibre architecture can be tuned does imply a limited tuning range without mode hopping, though the electro-optic tunable filter can, of course, scan the entire tuning range. The all fibre architecture is convenient and rugged and adequate for many measurement applications. If continuous tuning without mode hopping is necessary then a much longer optical path tuning must be incorporated (Figure 2.) thereby necessitating precision mechanical adjustment.

In both the all fibre and the long air path architectures, the simplest gain medium is probably an erbium fibre though this obviously restricts the available gain bandwidth to within the 1.55μm area. Other gain blocks can be inserted with due attention to the interface and of these the most flexible are undoubtedly semiconductor amplifying media.[8]

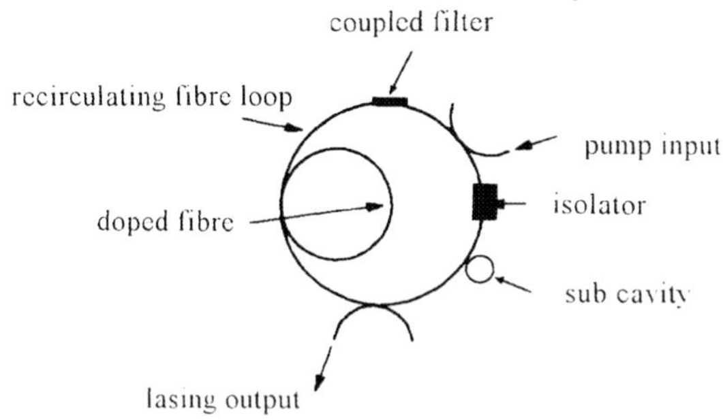

Figure 1. Tunable single mode all fibre laser architecture.

* Though we should remember that the fibre amplifier was first mooted by Snitzer over a quarter of a century ago!

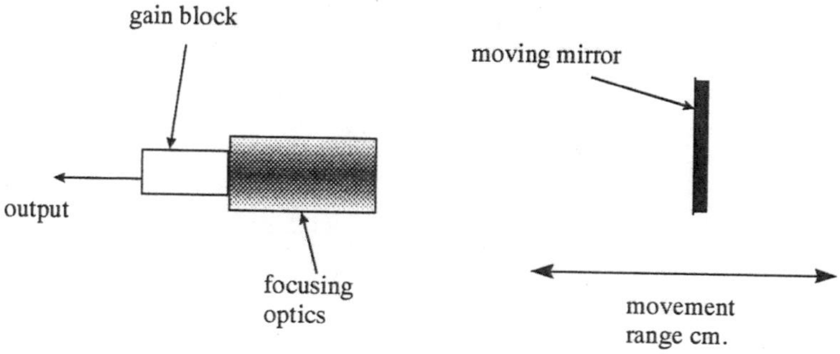

Figure 2. Continuously tunable laser architecture using long path mechanical tuning to eliminate mode hopping.

In the sensor context the fibre laser architectures are potentially attractive sources for spectroscopy since, for example the erbium gain curve addresses several interesting and relevant absorption bands including H_2S, some nitrogen oxides, SO_2 and CO (Figure 3.). Controlled wavelength switching and frequency locking within the gain band can be achieved using the all fibre architecture. Other potential applications include the use of tunable lasers for network testing in the telecommunications context (Figure 4.) typically as a source in an optical time domain reflectometer for wavelength division multiplexed systems and as probes for precision tuned and non-linear processes including, for example, multiple photon spectroscopy and soliton propagation.

The fibre amplifier - nothing more than the laser without the feedback - is now a very well characterised optoelectronic tool. The design rules which dictate noise levels, saturation power levels, pump powers, interaction lengths and related parameters are well established and can even be made available as software packages. In sensors and sensor systems, amplifiers - either using fibre or semiconductor gain media - offer potential applications in detection pre-amplification (where very low noise figures can be achieved) as power output stages especially in relatively high peak power pulsed systems and as signal boosters in sensor chains[9] (Figure 5.). In this last context, remote amplification can enhance the performance of multiplexed networks by an order of magnitude though obviously at the cost of system complexity.

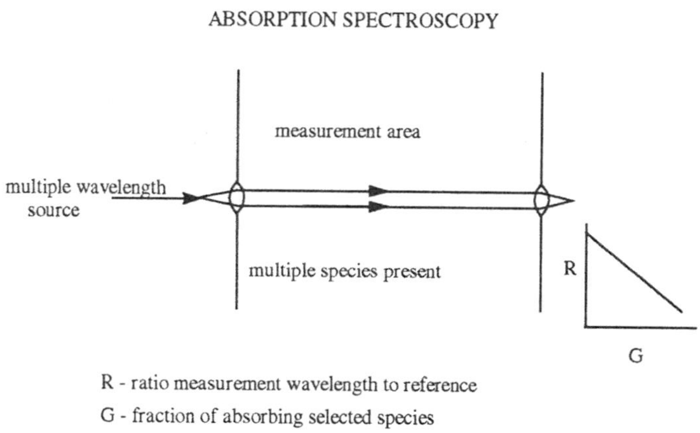

Figure 3. Multiple wavelength laser source for addressing multiple species in gas spectroscopy.

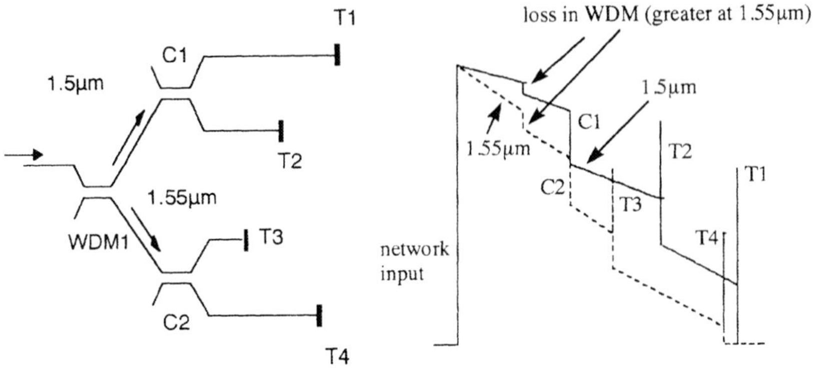

Figure 4. Tunable source OTDR for network fault testing and diagnosis.

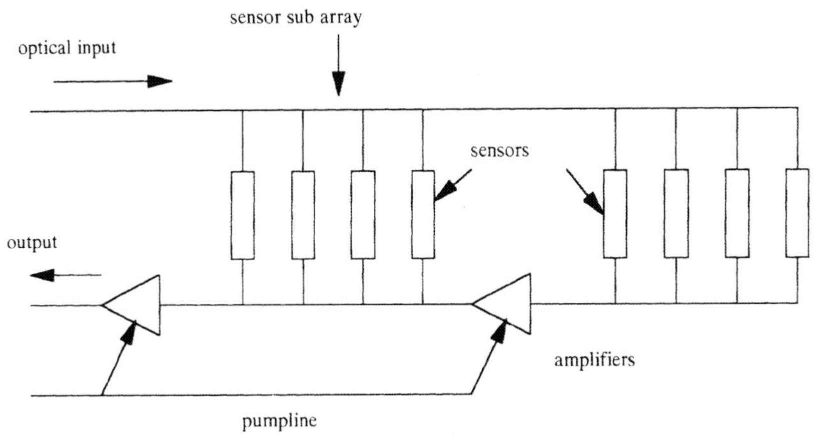

Figure 5. Schematic of one architecture for amplified sensor array. The location of the amplifiers can be varied to optimise noise performance.

Fibre lasers and amplifiers are emerging slowly into the sensor community. Their role is evolving and gradually being recognised by sensor engineers. Currently the cost of fibre lasers and amplifiers (or semiconductor lasers and amplifiers) is typically in the $20,000 area and so rather expensive for most sensors. However, there is no intrinsic reason why the cost should not be reduced significantly for a laser or amplifier system specifically tailored to a defined application and there is also the anticipation that increased volume in telecommunications will contribute to further cost reduction in the basic components. There will, I am sure, be other examples of exploitation of fibre amplifiers and tunable single mode lasers in the sensor context initially - as is customary - at the laboratory phase but with some emerging into the field.

3. NON LINEAR OPTICS IN SENSOR SYSTEMS

Non-linear optics or, equivalently multi photon optical phenomena may be very simply generically characterised as optical interactions for which some of one or more optical signals at one or more optical frequencies are converted during the interaction to other optical frequencies which are sums, differences, multiples, sub-multiples or any combination of the input frequency or frequencies. Whether the phenomena are described as non-linear or multi photon depends on whether the output signal is regarded as a distorted version of the input signal (a classical treatment) or as sums and differences of input photon energies (a quantum treatment). Alternatively when a photon interacts with a phonon analogous interactions can be identified between photon and phonon frequencies or alternatively photon and phonon energies. They are numerous standard textbooks on the subject.[10]

In the optical fibre sensing context, non-linear optics may be either exploited as a measurand sensitive interaction mechanism or as an optical signal processing technique to assist in configuring an optical signal used to probe a sensor or sensor network.

Measurement sensitive non-linear interactions all have the intrinsic advantage that the output light is at a different optical frequency than the input light. This greatly facilitates the detection process since it is relatively straightforward to shield the detector from stray interrogating light whilst monitoring the measurand induced signal as a different frequency. To date the principal interest in optical fibre sensing using non-linear phenomena has been in Raman or Brillouin scattering systems (Figure 6.) where the input probe light interacts with an optical phonon (Raman) or an acoustic phonon (Brillouin) to produce a difference frequency. Raman backscatter forms the basis of a commercially interesting distributed temperature sensor in which the ratio of the amplitude of the Stokes (downshifted) and mirror imaged anti-Stokes (upshifted) beams is a direct measure of temperature and of temperature alone. Therefore, when incorporated into a non-linear optical time domain reflectometer (Figure 7.) a measurement of this ratio can map a temperature as a function of position along the fibre.

The Brillouin system is used predominantly for strain measurement. In stimulated Brillouin scatter in single mode fibres, the acoustic phonon, typically at frequencies in the 10GHz region, produces backscattered radiation at a very specific frequency determined by the acoustic velocity in the fibre and the wavelength of the incoming probe beam. The acoustic phonon sets up a grating in the fibre so that the backscatter occurs when the grating wavelength (i.e. the acoustic wavelength) is exactly one half that of the incoming optical radiation.

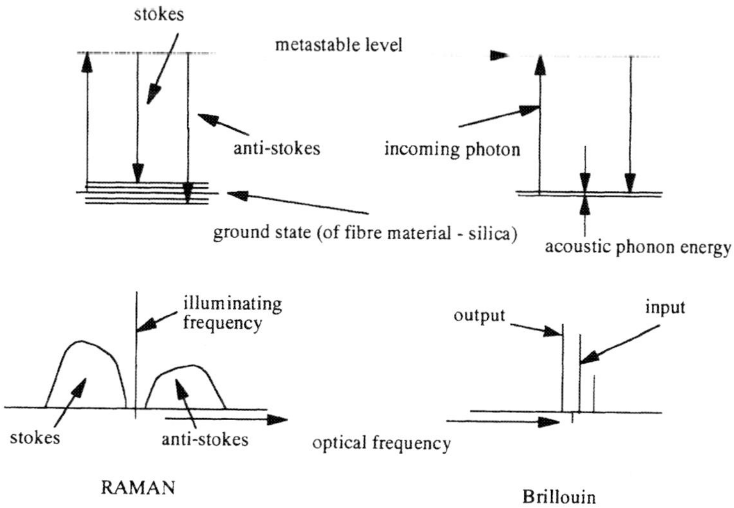

Figure 6. Non-linear scattering processes in optical fibre.

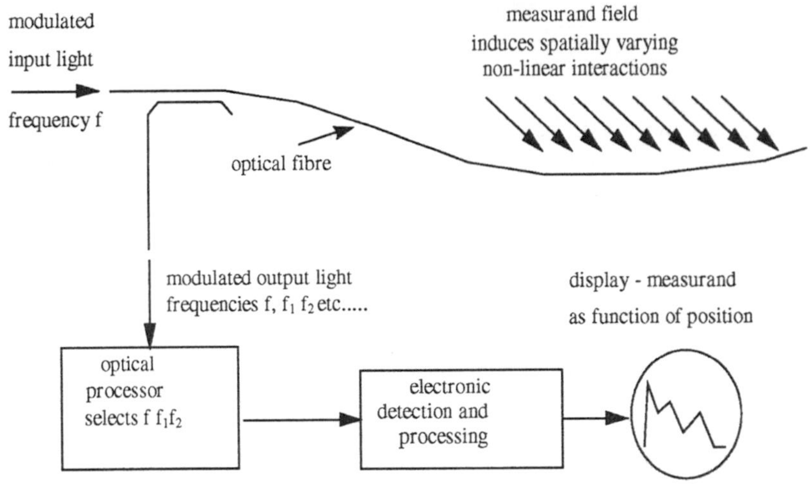

Figure 7. Use of non-linear phenomena for distributed sensing - general features.

This relationship sets the backscatter frequency. The acoustic velocity depends upon the local modulus of the material (which in turn is temperature dependent) and the local mechanical tension. Consequently if the temperature is known or can be inferred, distributed strain measurements become possible. The benefits of Brillouin systems are primarily the very long range over which the Brillouin interaction can take place, extending strain measurement systems to tens or even hundreds of kilometres with effective gauge lengths in the 1-100m region depending on the application. For both the Brillouin and the Raman systems, the sensing fibre is completely standard and needs no modifications whatsoever. Both systems are relatively expensive but have unique performance characteristics to address specific niche applications. The Brillouin system still requires temperature compensation and research is ongoing concerning methods whereby this may be realised.[11]

Other non-linear phenomena of interest include fluorescence and multi photon spectroscopy both of which may be used as probes for well defined chemical species. Fluorescence spectroscopy is well established and extensively used in chemical sensing. It is incorporated into many optrode designs. In contrast multi photon spectroscopy has received relatively little attention though, with the ability to generate a wide range of optical frequencies and to scan these frequencies, possibly using the techniques outlined in section 2, a new range of interesting probes for chemical species especially in gaseous form, becomes possible.

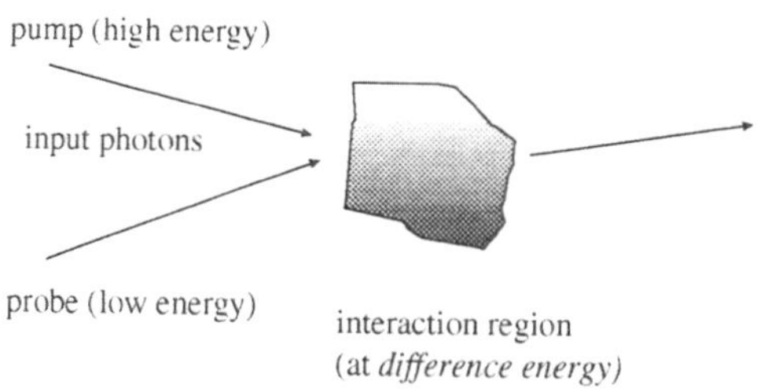

Figure 8. Principle of multiphoton spectroscopy.

The basic idea (Figure 8.) is that a high energy pulse beam and a low energy (swept frequency) probe beam interact within the medium of interest at the difference frequency. If the medium has absorption bands at this difference frequency, then the detecting probe beam will produce an absorption at the probe frequency. If the probe beam is frequency modulated then the benefits of FM spectroscopy and lock-in detection can be applied to enhance system sensitivity to absorption. Scanning the probe beam through a range of 10-100nm can enable the system to address a range of species. The fractional bandwidth at the probe frequency is also much less than that at the absorption wavelength, significantly simplifying the tuning process.

Multi-photon spectroscopy has historically been difficult since very precise overlap of the probe and pump beams is essential. However, addressing the interaction region through single mode optical fibre automatically ensures this overlap. The other interesting feature is that the interaction takes place at the difference frequency between the beams propagating in the fibre, so that in principle, this technique can address remote spectroscopic measurements at wavelengths for which the fibre is opaque but at which it transmits a suitable pair of pump:probe frequencies. Additionally, in principle, the technique can be time gated to realise distributed measurements in a forward rather than backscattered mode thereby increasing the detector signal levels. Paradoxically the concepts have yet to receive thorough evaluation as a sensing tool.[12]

The optical Kerr effect has been explored as a measurand sensitive interaction mechanism but with little success.[13] However, the Kerr effect does have one very important role in optical fibre systems - when coupled to the anomalous dispersion observed in the 1.5μm band, the Kerr effect can produce solitons. These are very short optical pulses which propagate without dispersion with pulse lengths typically in the 10picosecond or less region corresponding to optical pulses of the order of 1mm in length. Solitons have never been seriously explored as probe impulses for optical fibre measurement systems though, of course, precision short pulse OTDR is available.[14] Solitons which can, in principle, be generated with sub picosecond duration could further enhance the resolution of precision OTDR systems but the interesting potential is in the use of solitons as probes for very high speed spectroscopy, possibly, who knows, using multi photon probing techniques.

The role of non-linear optics in sensing is potentially far reaching but aside from Brillouin and Raman scatter systems, optical fibre sensor technology has but little exploited this potential. The specialised spectroscopist is accustomed to using non-linear interactions but currently the technology is specialised and difficult - maybe the optical fibre engineer could change all that.

4. STRANGE AND DIFFERENT PROPAGATION PHENOMENA

There are countless esoteric optical phenomena ranging from the Zeeman effect to the relatively simplistic such as spatial filtering. In the sensor context most of these have a potential role but to go through the whole range of possibilities would be not only impossible but also incredibly tedious - so we shall look at three unrelated fibre phenomena simply to illustrate the diversity.

Non-reciprocal phenomena manifests themselves in countless areas of optical sensing and probably most elegantly in the gyroscope.[15] Lorenz reciprocity states quite simply that non-reciprocal effects can take place in systems which propagate more than one mode in two directions, in systems in which there are inherently non-reciprocal propagation phenomena, e.g. the Faraday effect for circularly polarised light, in systems which are not time invariant and in systems which are non-linear.

- Established components such as isolators (and circulators)

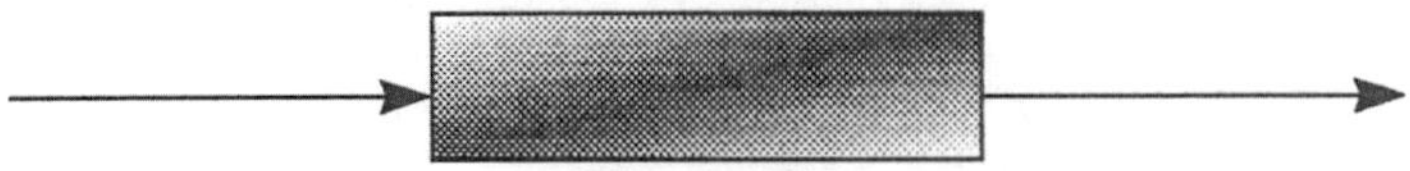

- New components such as non reciprocal mirrors

Figure 9. Non reciprocal effects.

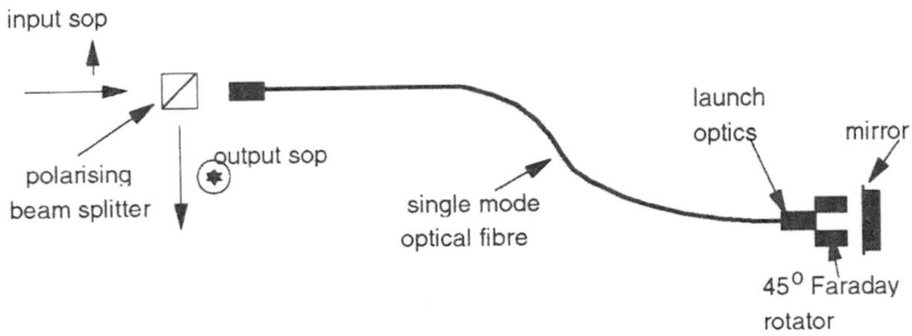

Figure 10. Non reciprocal mirrors.

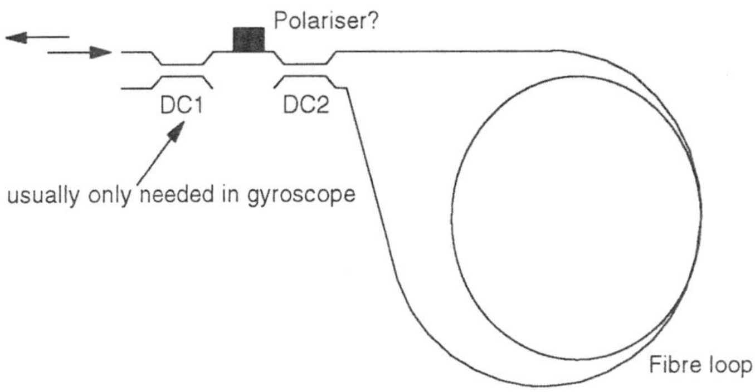

Figure 11. Loop mirrors - also known as gyroscopes.

Non-reciprocal effects can be utilised to good effect (Figure 9.) in contrast to their being a major design consideration in systems such as the gyroscope and in some architectures for optical fibre current sensing.[16]

The most common utilisation of non-reciprocal phenomena is in relatively simple optical devices such as isolators and circulators which are commercial items. However, the concepts can be extended considerably into (Figure 10.) sub systems such as the non-reciprocal mirror which can be used with a length of single mode fibre and configured to ensure that exactly the same state of polarisation arrives at the launch as the one which left.[17] This has possibilities in systems such as all fibre Michelson interferometers and has also been used to good effect in an all fibre Fourier transform infra-red spectrometer[18] used on this particular occasion to decipher Bragg gratings.

It is, of course, tempting to put the Bragg grating itself into the strange and unusual optics category in advanced optoelectronics - but the grating has reached the stage that it features in perhaps a quarter of all publications on optical fibre sensors and whilst many strange and wonderful techniques are used to decode the wavelength response, these are dealt with extensively elsewhere in this volume and in other publications.[19]

The gyroscope (Figure 11.) is a refined and well established optical system - but it can also contribute to sensors in several other guises of which the most common is probably the loop mirror (Figure 11.). This usually comprises a very short length of fibre and provided that the loop coupler is exactly tuned to 50% the loop mirror is a 100% reflector regardless of wavelength. The fact that most couplers are in themselves wavelength sensitive implies that the mirror is also wavelength sensitive. The uses for loop mirrors are many and diverse and include terminating all fibre reflection based interferometers and, when viewed from the other direction, as a monitoring process for the fabrication of 3dB optical fibre couplers. The loop mirror has the virtue of simplicity and is an elegant but (for the reasons explained above) wavelength sensitive 100% reflection system.

Near field devices show promise for subwavelength resolution and can be configured as chemical or physical sensors responding to an optical evanescent field.[20]

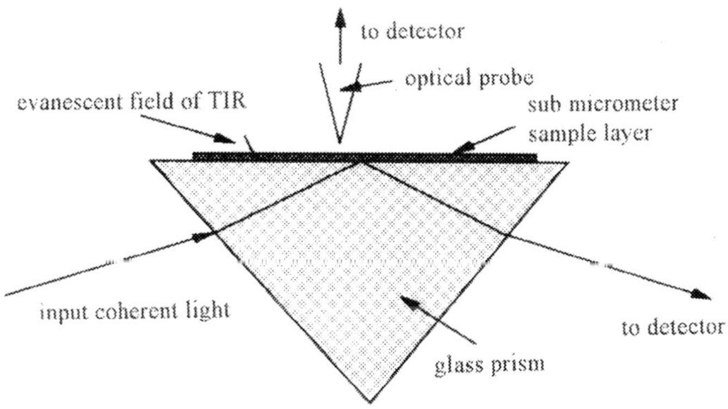

Figure 12. Near field devices.

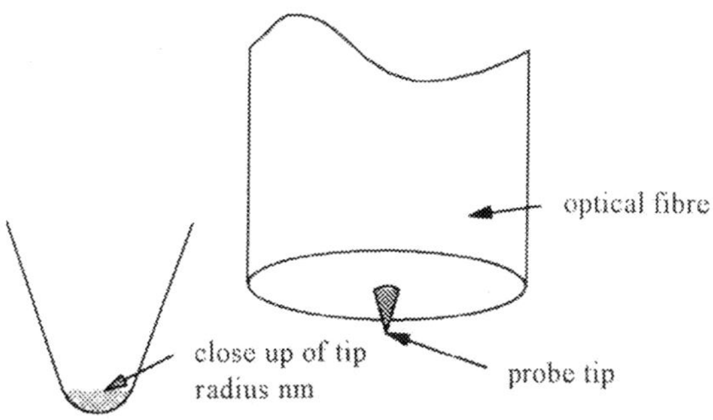

Figure 13. Tapered fibre probe (etched geometry).

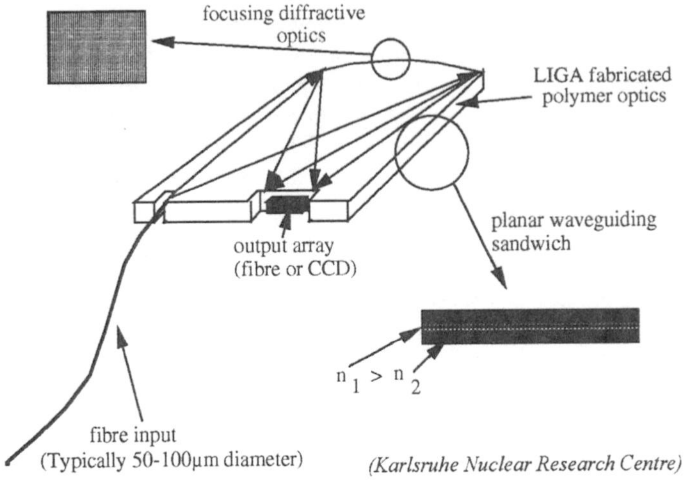

Figure 14. Monolithic spectrometer fabricated using the LIGA process.

The traditional format is shown in Figure 12. where the optical probe responds to small changes in thin film layer caused by the modifications imposed by this layer on the evanescent field from total internal reflection.

The spatial resolution of such probes depends upon the probe tip dimensions which can be readily formed in the sub 100nm region. In order to fully exploit this geometry as a probe the tapered region should be metal coated to prevent spreading of the evanescent field within the guiding region. The probe tip may be coated with intermediate chemistry to provide probes capable of resolving spatial variations in biological cellular chemistry for parameters such as pH and oxygen concentration (Figure 13.).[13]

5. INTEGRATED OPTICS AND MICROMACHINING

Integrated optics like fibre sensors has been with us for around three decades.[22] Lithium niobate integrated optics chips have a vital part to play in many optical fibre gyroscope architectures and in some Brillouin systems. There have been a few integrated optics sensor devices usually comprising coatings on glass substrate planar waveguides but aside from that there has been very little penetration of integrated optic technology into the domain of optical fibre sensors and measurement.

The best established integrated optic technologies on lithium niobate and glass substrates usually exploiting proton exchange and ion exchange technologies respectively. A number of new substrate materials are emerging particular silica on silicon, silicon/Simox and some III V compounds. The last of these has some active device capability including lasers and detectors as well as waveguide modulators and switches but the remainder have the same basic functionality in that they can produce thin film, planar and channel guides with some very restricted possibilities for active device tuning.

Micromachining, particularly in silicon has made some impact. The simple V-groove is a straightforward and most effective optical fibre alignment system whilst some micromachined based optical fibre sensors have made inroads into measurements such as pressure and temperature. More recently the micromachined optical bench has emerged.[23,24]

LIGA is yet another interesting technological opportunity. This offers the possibility of micromachined systems with very deep parallel sides and is basically a moulding process using precision masters prepared using particle beams. The problem with LIGA is essentially access to the technology since its implementation requires a nuclear reactor. The concept, which originated in Germany[25] can produce some interesting guided wave structures of which the single process step moulded spectrometers shown in Figure 14. is an example.

6. OPTICAL SIGNAL DETECTION SYSTEMS

Most sensor engineers use the very simple PIN diode direct detection approach to receiver design. Those in spectral analysis and sometime those in white light interferometry will use a CCD array to give some spatial information on the signals of interest. With careful design to maximise signal to noise ratios, these detection techniques usually suffice though the implicit choice of silicon almost invariably implies that our sensor system operates in the visible and very near infra-red up to a wavelength of a little over 1μm. For some systems, especially long range distributed sensors, operation at longer wavelengths is desirable even essential so that 1.3 and 1.55μm band detection systems borrowed the communications industry are frequently exploited - though at a very significant increase in component cost.

If we venture from the relatively simple use of silicon based detector electronics , then the gains are usually in signal detection sensitivity and/or signal detection concept flexibility but the loss is in the inherent and frequently very significant increase in system cost. But there are a lot of options:

- coherent detection based on both heterodyne and homodyne concepts can increase sensitivity and modify the basic input optical power to output electrical signal response from square law to linear. Heterodyne systems also offer the possibility of detecting outside the 1/f noise region which plagues most detector circuits. Heterodyne systems based on Bragg cell modulators have been used to enhance the performance of many detection interferometers particularly those used in ultrasonic have detection exploiting surface displacement phenomena.
- Pre-amplifiers usually based on optical fibre gain blocks can produce sensitivities comparable to photomultipliers with modest noise penalties. Again for most applications we are restricted to the 1.55μm communication band but have very long sensor systems. Preamplification will have a role to play.
- Time gating the source to act as a detector in suitable time delay systems for example gyroscopes and distributed sensors has been proposed and demonstrated as a very low cost transmitter/receiver architecture
- Interferometry from multiple spatial mode optical fields is by definition very difficult but adaptive holograms based upon photorefractive materials can perform a spatial recombination function on the multimode field to bring it into spatial coherence with the reference. This technique whilst proposed using BSO about 20 years ago remains to be demonstrated practically but the emergence of more responsive photorefractive materials, such as non-linear polymers and multiple quantum well devices may eventually produce a more practical demonstration system.

7. CONCLUSIONS

In the sensor context, all these ideas are simply tools which can be added to all the possibilities which are available to design sensors and sensor systems. Using these tools to best effect requires insight and specialist knowledge - so that in the final reckoning the sensor engineer needs to be aware that these tools exist, needs to be aware of their potential in a practical environment and, most important of all, needs to be aware of how to obtain the necessary skills to exploit these tools.

But we should always remember that the scope is immense and that by incorporating advanced optoelectronics into the design and realisation of optical sensor systems, very significant performance improvements are often possible. From time to time, (e.g. reference 26) some energetic journal editor presents a selective potted history. It is useful - even important - to ensure that the sensing and instrumentation community is aware of these broader developments.

REFERENCES

1. B. Culshaw & J.P. Dakin (eds) *Optical Fibre Sensors* Vols. I, II, II, IV, Artech House, Norwood (1988, 1989, 1997, 1997).
2. Fibre Optic Gyroscopes, Proc.20th Anniversary Conference Denver, Colorado (1996) (Proceedings published by SPIE, Bellingham, Washington).
3. J.P. Dakin, D.JPratt, G.W. Bibby and J.N. Ross, "Distributed optical fibre Raman temperature sensor using a semiconductor light source and detector", *Electron. Letts*, Vol 21 pp 569-70 (1985).
4. D. Inaudi, Field testing and application of fibre optic displacement sensors in civil structures" *Proceedings OFS(12)*, Williamsburg Virginia, p596 (published by Optical Society of America) (1997).
5. T. Horiguchi, Brillouin scatter distributed strain sensing, in ref (1) Volume IV, chapter 14 (1997).
6. O.S. Wolfbeis, *Fibre Optic Chemical Sensors and Biosensors"*, CRC Press (1991).
7. A. Bjarklev *Optical Fibre Amplifiers: Design and Systems Applications*, Artech House, London and Norwood (1994).
8. I. Andonovic, N. Langford and B. Culshaw, Optical fibre lasers and amplifiers in ref 1 Vol III, Chapter 4 (1997).
9. G.J. Cowie, Optical fibre sources, amplifiers and special fibres for application in multiplexed and distributed sensor systems, *Proc SPIE 1586*, p130-139, Boston (1991).
10. G.P. Agraval, *Non Linear Optical Fibres*, Academic Press, New York (1989).
11. T.R. Parker, M. Farhadiroushan, R. Feced, F.A. Handarek and A.J. Rogers, Simultaneous strain and temperature sensing suing noise initiated Brillouin scattering, *Proc OFS(12)*, p344, Williamsburg, October (1997).
12. S. Ezekial, New opportunities in fibre optic sensors, *Proc OFS(10)*, p3, SPIE. Vol2360, Glasgow (1994).
13. A.J. Rogers, Non-linear distributed optical fibre sensing, *Proc SPIE*, Vol1787, p50, Boston (1992).
14. B. Garside, Advances in high speed OTDR techniques, in ref 1 vol.III, chapter 5.
15. R.A. Bergh, H.C. Lefèvre and H.J. Shaw, An overview of optical fibre gyroscopes, *IEEE J Lightwave Technology*, Vol. LT2,2, pp91-107 (1984).
16. G.W. Day, M.N. Deeter and A.J. Rose, Faraday effect sensors: a review of recent progress, *Advances in Optical Fibre Sensors*, SPIE Vol.PM07, pp11-26 (1992).
17. M.A. Davis and A.D. Kersey, Fibre Fourier transform spectrometer for decoding Bragg grating sensors, *Proc OFS(10)*,Glasgow, SPIE Vol. 2360, p167 (1994).
18. M. Martinelli, Time reversal for the polarization state in optical fibre circuits, *Proc OFS(10)*,Glasgow, SPIE Vol. 2360, p312 (1994).
19. *Proceedings Topical meeting on Bragg gratings, photosensitivity and poling in glass fibres and waveguides,* (Optical Society of America) Williamsburg, October 1997.
20. D.W. Pohl and D. Courjon (eds) *Near Field Optics*, Klewer, Dordrecht (1993).
21. S. McCulloch and D. Uttamchandani, A fibre optic micro-optrode for dissolved oxygen measurements, *Proc OFS(12)*, SPIE Vol.2360, p428, Williamsburg, October (1997).
22. R.Th. Kersten, Integrated optics in sensing in Ref.1, Vol.I (1988).
23. B. Culshaw, Fibre optic sensors: integration with micromachined devices, *Sensors and Actuators*, 46-47, p463-469, (1995).
24. M.C. Wu, Optical MEMS for sensing applications, *OFS(12)*, P7, Williamsburg, October (1997).
25. W. Ehrfeld, F. Gotz, D. Munchmerger, W. Schelb and D. Schmidt, LIGA process: sensor construction via X-ray lithography, *IEEE Sensor and Actuator Workshop (1988)* reprinted in R S Muller and R T Howe (eds) *Microsensors*, IEEE, New York, pp. 141-144 (1992).
26. *Proceedings IEEE*, Vol85, 11, Special Issue on Optoelectronics Technology (1997).

INTERFEROMETRIC DISTANCE SENSORS

U. Minoni, L. Rovati, and F. Docchio

Università degli Studi di Brescia
Via Branze, 38
25123 Brescia, Ialy

1. INTRODUCTION

Optical interferometry is a well-known technique successfully used to perform displacement measurements in the industrial framework. In general, commercial devices use incremental interferometry: to measure a displacement it is necessary to fix a reference position and to carefully displace a suitable reflector (or some other optical component of the interferometer) from that position to the final one. This must be done avoiding any optical misalignment that could result in a loss of the interference signal and therefore of the measure. An incremental interferometric displacement measurement always implies a precise device to move the reflector, and a counter to keep trace of the number of periods of interference fringes that the photodetector has detected during the movement. The main advantage of this measurement technique is a very low measurement uncertainty (down to some fractions of λ in commercial meters) coupled to an extended range (up to tens of meters). In general, therefore, an interferometric system is the right choice when both high accuracy and a long measuring range are required.

A number of industrial needs cannot be successfully met by incremental interferometers, especially in the fields of dimensional gauging of surfaces, robot monitoring and characterization, autonomous vehicle controls, etc.. In fact in these fields the quantity to be measured is distance and not displacement. Taking as an example the problem of industrial robot calibration, the ideal measuring tool should be able to freely follow the robot end-effector, to measure the actual position along an arbitrary trajectory. An absolute (i.e. non-incremental) distance meter pointing to the robot manipulator could accomplish this task. Interferometric sensors have great potential in performing non-incremental measurement at a very high degree of accuracy.

The following paragraphs describe the available interferometric techniques to perform non-incremental measurements outlining the principle of operation and the applicability to industrial cases. At the end a comparison of the investigated techniques is given.

2. INTERFEROMETRIC DISTANCE MEASUREMENTS

Interferometry can be implemented either by using a high coherence light source (i.e laser) or by using a broad-band or low-coherence light source (i.e. lamp, LED, SLD, etc.). Both approaches can be successfully used to perform distance measurements.

Optical Sensors and Microsystems: New Concepts, Materials, Technologies
Edited by Martellucci et al., Kluwer Academic / Plenum Publishers, New York, 2000.

The possibility to measure distances with high-coherence interferometric techniques has been demonstrated by several authors.[1-12] The basic idea is to feed an interferometer with two or more optical wavelengths to produce a synthetic or equivalent wavelength, much longer than that of the original beams. The distance measure is extracted by gauging the phase of a signal having a period equal to that of the equivalent wavelength, thus extending the range of non-ambiguity (NAR) by some orders of magnitude (up to tens of cm) with respect to that of incremental interferometry (a fraction of the optical wavelength of the laser). The most effective approaches are based on double-heterodyne interferometry (DHI) [14 - 21] that makes use of a detection scheme based on optical heterodyning of two slightly different optical frequencies.

As an alternative to multiple wavelength interferometry, techniques based on optical frequency modulation of the laser source are very promising due to the potential for simple and robust systems.[22-32] Among the various possibilities, frequency-modulated continuous-wave interferometry (FMCW) presents the simplest optical layout which is essentially based on a Michelson-type interferometer where a tunable single frequency laser is modulated to obtain a beat signal at the photodetector.

Absolute distance measurements can also be performed on the basis of low-coherence or white light interferometry (WLI). In this case the measuring range is shorter than that of DHI or FMCW while the resolution can be better than λ. Possible light sources for WLI are i) lamps, ii) LEDs, iii) super-luminescent diodes (SLD) and iv) multi-mode laser diodes. The most popular, compact and reliable solid-state sources for WLI are SLDs. In WLI systems, the measurement resolution is practically related only to the coherence characteristics of the excitation light, since optical interferometry is used to convert spectral properties of the excitation light in space resolution. Therefore, over the last years strenuous efforts were made to develop high-power, broad-band semiconductor sources.[33]

A number of possible configurations[33-38] to implement a WLI system have been proposed. A very promising scheme is based on the use of a dispersive element in the interferometer, this element enabling the implementation of a configuration where no moving parts are needed.[37] This technique can also be combined with a FMCW scheme by using a single laser diode operating above the threshold current for the FMCW mode and below the threshold current for the DWL mode.[38]

2.1 Displacement measurements

To better understand the techniques used for non-incremental measurements, this section will remind the basic principles of incremental interferometry from which the absolute techniques can be derived.

An optical interferometer (see Figure 1.) is a device that can be used to measure the optical path difference (OPD) between its two arms. In general the optical length of one arm is kept constant (reference arm) whilst the length of the other one (signal arm or measuring arm) is changed by displacing the end reflector of the interferometer. Figure 2. shows the intensity signal detected by a photodiode as a function of the interferometer OPD.

The periodicity of the signal is evident, in fact the intensity function is a cosine function of OPD which can be written as

$$\mathrm{OPD} = \lambda \cdot \left(n + \frac{\varphi}{2\pi} \right) \qquad (1)$$

where λ is the laser wavelength, φ is the phase of the detected signal and n is an integer number taking into account for the number of complete signal periods from the position where the interferometer is perfectly balanced. Being impractical to find the zero-OPD

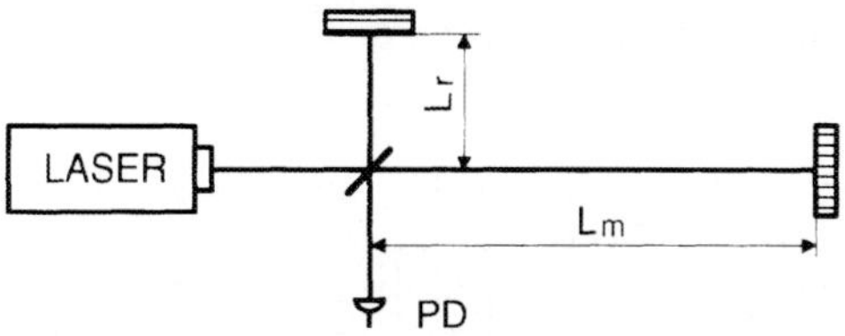

Figure 1. Simple interferometric setup for displacement measurements. PD: photo-detector; L_r optical length of the reference arm; L_m: optical length of the measuring or signal arm.

condition in an interferometer illuminated with a high coherence laser source, actually an incremental interferometer is used to measure displacements from a conventional position which is assumed as the "zero" or reference point. The displacement is calculated as a variation ΔOPD of OPD from the reference point:

$$\Delta OPD = \lambda \cdot \left(\Delta n + \frac{\Delta \varphi}{2\pi} \right) \tag{2}$$

It is evident that the displacement measurement is correctly performed only if the interferometer is well aligned during the entire movement of the signal reflector and if the laser beams are not interrupted during the movement, that is the measurement is incremental in nature. By using a direct detection as shown in Figure 1., the measurement of phase φ is very sensitive to eventual changes in the intensity of the laser source and to variations of the fringe visibility due to beam divergence as well as to small misalignments of the interferometer during the signal reflector movement. To greatly reduce the effect of this source of errors, it is common to use a detection scheme based on optical heterodyning. Figure 3. shows the basic principle of the technique. The optical beam illuminating the interferometer presents two orthogonally oriented, linearly polarized components. One has the base optical frequency equal to ν, the other one if frequency shifted by f. The two components are separated by means of polarizing beam splitter PBS, in such way each arm of the interferometer is crossed by a single optical frequency. Quarter-wave retarding plates λ/4 are inserted to rotate, after double crossing, the polarization of both beams by 90 deg. The returning beams are therefore directed towards photodetector PD in front of which a polarizer oriented at 45 deg. has been inserted to enable the interference between the two components at different frequency.
The photodetector signal has the form:

$$I(t) = I_{dc} + I_{ac} \cos\left(2\pi \cdot f \cdot t + 2\pi \frac{OPD}{\lambda} \right) \tag{3}$$

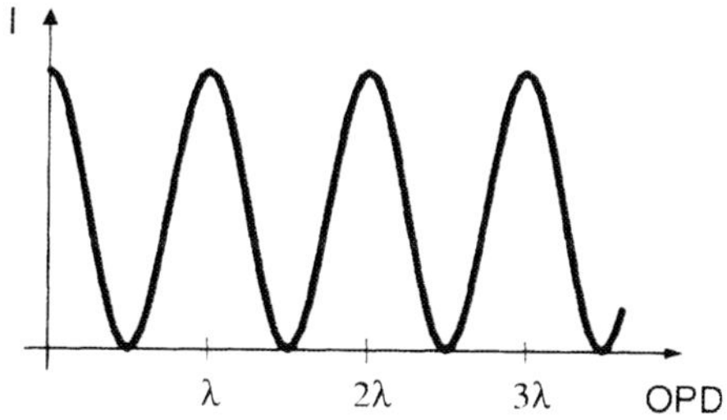

Figure 2. Intensity signal at photo-detector PD of Figure 1. as a function of the interferometer optical path difference.

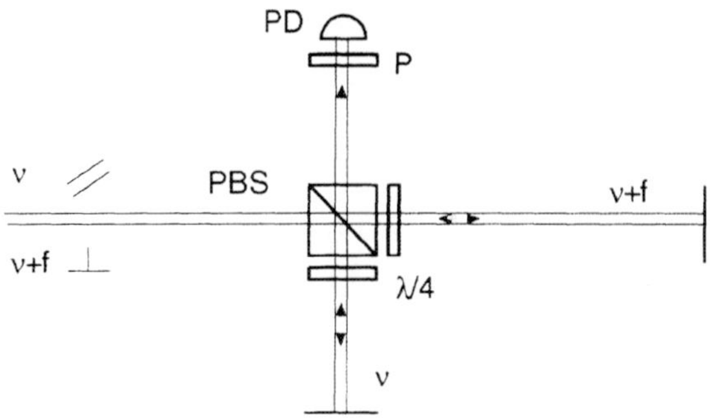

Figure 3. Optical heterodyne detection. ν is the base optical frequency, f is the optical frequency shift applied to the orthogonal polarization component. PBS: polarizing beam-splitter; P: polarizer at 45°; PD: photo-detector; λ/4: quarter wave retarding plates.

where the dc component I_{dc} does not carry any information about OPD, in fact the useful information is moved into the phase term of the ac component at frequency f. Figure 4. shows a possible layout for the implementation of optical heterodyning.

The heterodyne frequency f is obtained by means of an acousto-optical modulator which applies the frequency shift to one polarization component of the original laser beam. In order to perform the phase measurement a reference signal is generated by mixing the two optical frequencies before entering the interferometer, the reference signal is obtained at the exit of photodetector PDr.

3. DOUBLE WAVELENGTH INTERFEROMETRY

Basically double wavelength interferometry is based on the idea to perform two independent measurements of the OPD of the interferometer by using two different optical wavelengths and than to compare the measurement results. Applying Eq. (1) to both wavelengths λ_1 and λ_2 we can write:

$$OPD = \left(n_1 + \frac{\varphi_1}{2\pi}\right) \cdot \lambda_1 = \left(n_2 + \frac{\varphi_2}{2\pi}\right) \cdot \lambda_2 \tag{4}$$

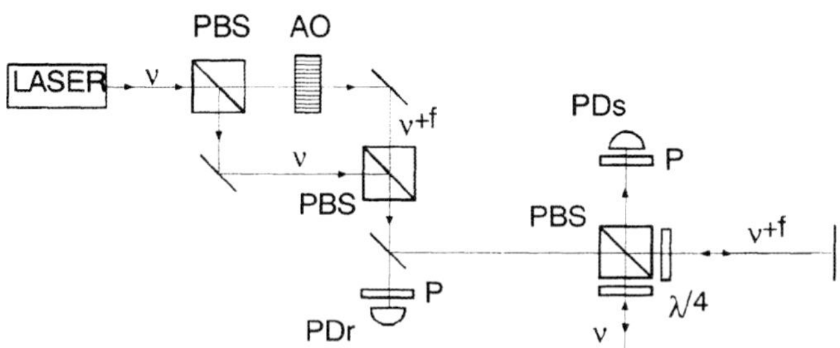

Figure 4. Example of optical setup for heterodyne detection. AO: acousto-optical frequency shifter; ν: optical frequency at the exit of the laser; ν+f: optical frequency at the exit of the modulator; PDr: photo-detector for the generation of the reference signal; PDs: interferometer photo-detector; PBS: polarizing beam-splitter; P: polarizer; λ/4: quarter wave retarding plates.

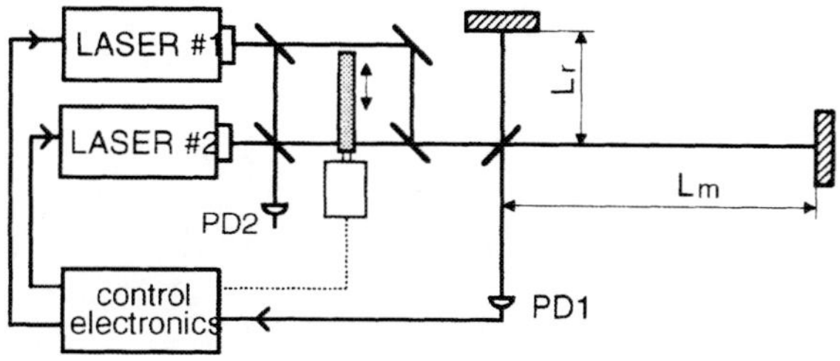

Figure 5. Principle of operation of a dual wavelength interferometer. PD1: interferometer photo-detector; PD2: reference photo-detector.

If the laser sources can be tuned, we can, for each measurement, adjust the emitted wavelength in order to make terms φ_1 and φ_2 equal to zero. A possible setup is shown in Figure 5., here a shutter can be driven to choose which beam has to be used to illuminate the interferometer.[9] Interference signal detected by photodetector PD1 is used to control the laser wavelength to satisfy relation given by Eq. (4).

When both lasers have been adjusted, the following relation can be written.

$$OPD = n_1 \cdot \lambda_1 = n_2 \cdot \lambda_2 \tag{5}$$

that is

$$\begin{cases} OPD = n_1 \cdot \lambda_1 \\ OPD = n_2 \cdot \lambda_2 \\ n_2 = n_1 + \Delta n \end{cases} \tag{6}$$

where Δn is the fringe order difference at the two wavelengths. The case of $\Delta n=1$ is depicted in Figure 6.

Solving the equation set (6), the OPD can be calculated as:

$$OPD = \Delta n \cdot \frac{\lambda_1 \cdot \lambda_2}{\lambda_1 - \lambda_2} = \Delta n \cdot \frac{c}{\upsilon_1 - \upsilon_2} = \Delta n \cdot \lambda_{synt} \tag{7}$$

The quantity λ_{synt} is the synthetic or equivalent wavelength of the interferometer.

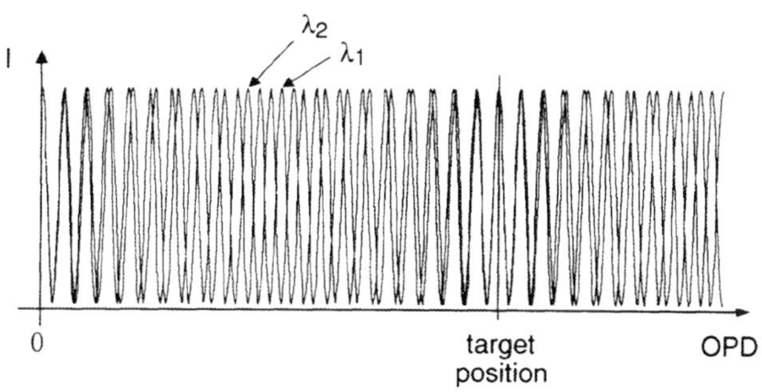

Figure 6. Interference signals as functions of OPD in a dual wavelength interferometer. Wavelengths λ_1 and λ_2 are adjusted in order to have a phase equal to zero in correspondence of the target position.

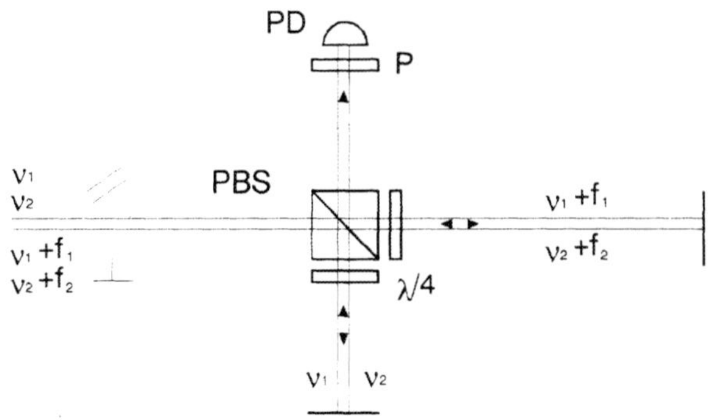

Figure 7. Optical dual-heterodyne detection. ν_1 and ν_2 are the base optical frequencies. f_1 and f_2 are the optical frequency shifts for heterodyning. PBS: polarizing beam-splitter; P: polarizer at 45°; PD: photodetector; λ/4: quarter wave retarding plates.

Combining dual wavelength interferometry with heterodyne detection it is possible to use the two wavelengths simultaneously, thus avoiding any opto-mechanical switching and shortening the measuring time. The principle of operation is shown in Figure 7. Two fundamental optical frequencies ν_1 and ν_2 together with two shifted frequencies ν_1+f_1 and ν_2+f_2 illuminate the interferometer. Taking advantage from the beam polarization, the system works by sending the two fundamental frequencies into the reference arm of the interferometer and by directing the two shifted frequencies into the signal arm of the interferometer. The intensity signal detected at photodiode PD (its bandwidth is narrower than ν_1-ν_2) is given by:[15]

$$I(t) = A_1 cos(2\pi f_1 t + \varphi_1) + A_2 cos(2\pi f_2 t + \varphi_2) \tag{8}$$

where I(t) is the sum of the two heterodyne signals at f_1 and f_2, having amplitudes A_1 and A_2 and phases φ_1 and φ_2 respectively. From Eq. (3) we have.

$$\varphi_1 = \frac{2\pi}{\lambda_1} OPD; \quad \varphi_2 = \frac{2\pi}{\lambda_2} OPD \tag{9}$$

The intensity signal (8) has the form of a carrier-suppressed amplitude-modulated signal. After amplitude demodulation with a quadratic detector followed by a low-pass filter, the superheterodyne signal is obtained:

$$I_{dem}(t) \propto cos[2\pi(f_2 - f_1)\cdot t + \Delta\varphi] \tag{10}$$

where the phase of the demodulated signal is given by:

$$\Delta\varphi = |\varphi_2 - \varphi_1| = 2\pi OPD \left| \frac{1}{\lambda_2} - \frac{1}{\lambda_1} \right| = \frac{2\pi}{\lambda_{synt}} OPD \tag{11}$$

In Eq. (11) the quantity λ_{synt} is the above defined synthetic, or equivalent, wavelength of the absolute interferometer. Therefore, the non-ambiguity range (NAR) is now moved from the optical wavelength to the synthetic one. A further extension of the NAR can be obtained[18] by changing the synthetic wavelength λ_{synt}, that is, by changing $\Delta\nu$ according to the formula:

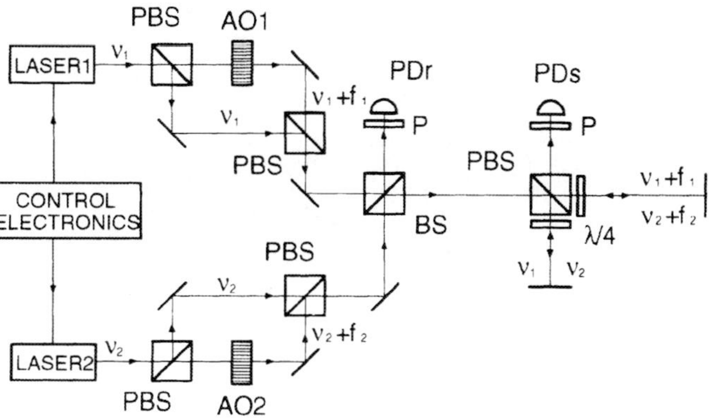

Figure 8. Double heterodyne distance sensor based on two solid state lasers which are are frequency locked. AO1 and AO2 acoustooptical frequency shifters at f_1 and f_2 respectively. PBS: polarizing beam-splitters; BS: non-polarizing beam-splitter. λ/4: quarter wave retarder; P: polarizer; PDr: photodetector for the reference signal generation; PDs: photodetector for the measuring signal generation.

$$\lambda_{synt} = \frac{\lambda_1 \lambda_2}{\lambda_2 - \lambda_1} \approx \frac{\lambda_1^2}{\Delta\lambda} = \frac{c}{\Delta\nu} \tag{12}$$

The two fundamental wavelengths λ_1 and λ_2 can be obtained from the multiple lines of a single laser or by using two lasers. In general the first solution presents a lower frequency-noise level and therefore a better stability of λ_{synt} but also presents a drawback due to the availability of only a discrete number of non-tunable synthetic wavelengths. A real implementation of a DHI sensor is shown[21] in Figure 8. In this case the synthetic wavelength is generated by using two frequency-locked diode-pumped Nd:YAG lasers.

The heterodyne frequencies are f_1= 80 MHz and f_2= 80.1 MHz yelding a super-heterodyne signal at 100 KHz. The four beams at ν_1, ν_2, ν_1+f_1, and ν_2+f_2, are superimposed by means of non-polarizing beam-splitter BS which directs a fraction of the four frequencies beam toward photo-detector PDr for the reference signal generation. The signal at the exit of reference photodiode PDr takes into account the optical path differences of the beams accumulated from the lasers up to beam splitter BS. In fact, superheterodyne signals from PDr and from PDs contain the same phase fluctuations arising from optical path changes of the optical set-up up to BS, a phase difference between the two signals results in a phase signal dependent only on OPD. The beams passing through beam-splitter BS are sent to the measuring Michelson interferometer.

After pre-amplification, the signals detected by the reference photodiode PDr and measurement photodiode PDs are sent to the electronic processing chain shown in Figure 9. In order to have signal levels suitable for subsequent amplitude demodulation, the signals enter an amplification stage based on two AGC integrated amplifiers. The demodulation stage is based on an RF detector. The low frequency demodulated signals are filtered by a sixth order Chebycheff low-pass filter with corner frequency at 100 KHz. The two output signals of the demodulation stage are squared to obtain TTL compatible signals which feed a digital phase-meter. The phase-difference between the two TTL signals gives the interferometer optical path difference measurement. Typical DHI signals are shown in Figure 10.

Figure 11. shows the principle of operation of a DHI sensor using a tunable synthetic wavelength generator[20] based on two diode-pumped Nd YAG lasers. The beat-frequency of

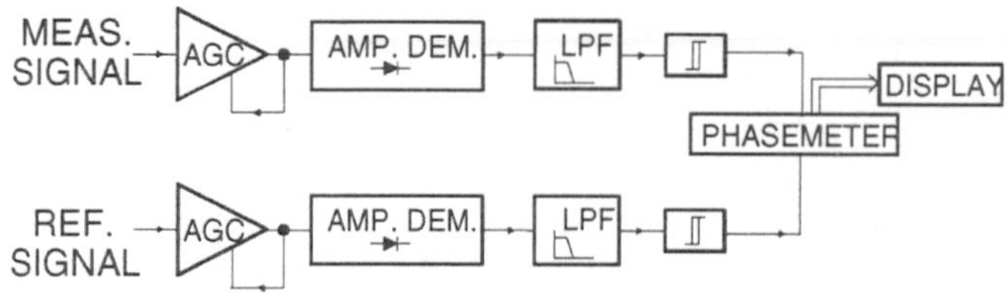

Figure 9. Signal elaboration chain for the measurement of distance using a DHI system.

the lasers is compared to that of a local oscillator. The result of the comparison is used as a feed-back signal to control the optical frequency of one of the two lasers. A small fraction of the optical signal produced by the two lasers is directed towards photo-diode PD, where a beating signal is produced. One laser works as master device, in the sense that its frequency fluctuations are followed by the other laser which therefore works as a slave. Local oscillator LO provides a reference signal f_{ref}. Both signal f_{pre} and signal f_{ref} enter a frequency comparator. Voltage output, after passing a high-voltage is used to drive the PZT controlling the optical frequency of the slave laser.

4. FREQUENCY-MODULATED INTERFEROMETRY

Frequency-modulated continuous-wave interferometry (FMCW) can be successfully used for developing non-incremental distance meters. In order to implement a reliable and compact measuring device, the use of laser diodes[26-29] as well as of solid state lasers[30-32] has been investigated. An example of a FMCW sensor[32] is shown in Figure 12.

A tunable diode-pumped Nd:YAG laser provides a Michelson interferometer with a linearly polarized beam. To avoid problems arising from the optical feedback, the laser beam passes through an optical isolator. Non-polarizing beam splitter BS divides the laser beam to form the two interferometer arms. In the measuring arm a lens is inserted to focalize the beam onto the target.

The optical frequency of the beam is controlled by applying a voltage to a piezoelectric transducer mounted onto the laser cavity. To obtain a continuous frequency modulation, a triangular signal is applied to the PZT. The beams that recombine at beam splitter BS exhibit optical frequencies slightly different, due to path difference in the unbalanced interferometer.

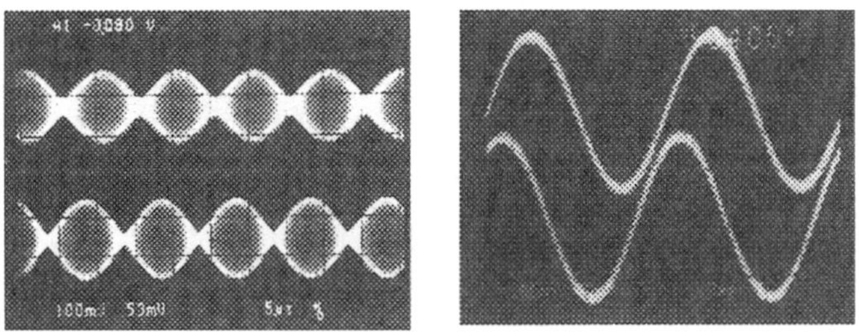

a) b)

Figure 10. Example of DHI signals. a) as detected by photo-detectors: upper trace measuring signal, lower trace reference signal. b) after amplitude demodulation: upper trace measuring signal, lower trace reference signal; the distance is given by the phase displacement between these two signals.

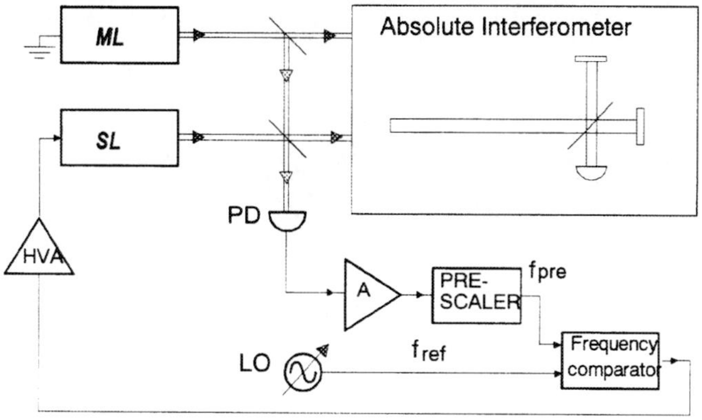

Figure 11. Distance sensor based on a tunable synthetic wavelength generator. ML: master laser; SL: slave laser; LO: local oscillator; PD: photodetector; A: amplifier; HVA: high voltage amplifier and PZT driver.

Called ν_0 the optical frequency emitted by the laser when no voltage is applied to the PZT ($t=t_0$), the optical frequency at a generic instant t is given by:

$$\nu(t) = \nu_0 + K_M \cdot V_{pzt}(t) = \nu_0 + K_M \cdot \left(V_{pp} \cdot Tri(t - t_0)\right) \tag{13}$$

where K_M is the modulation coefficient (it can be assumed to be a constant), Tri is a unit amplitude triangular function with a period equal to P_T, and V_{pp} is the p-p value of the voltage applied to the PZT.

The beams that recombine at beam splitter BS produce a beat signal having an instantaneous frequency f_{PD} (see Figure 13.) given by:

$$f_{PD} = \nu(t) - \nu(t - \Delta t) \tag{14}$$

where

$$\Delta t = \frac{OPD}{c} \tag{15}$$

The discontinuities of f_{PD} can be neglected and Eq. (14) can be rewritten as:

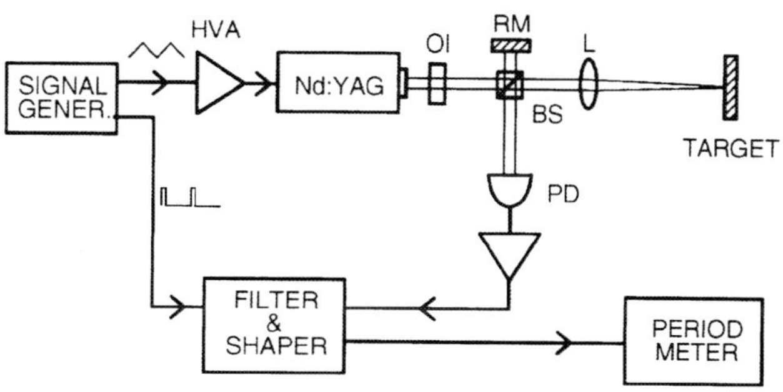

Figure 12. Layout of a FMCW distance meter. Nd:YAG: diode-pumped Nd:YAG tunable laser; HVA: high voltage amplifier driving the laser PZT; OI: optical isolator; RM: reference arm mirror; BS: non-polarizing beam-splitter; L: focusing lens; PD photodetector.

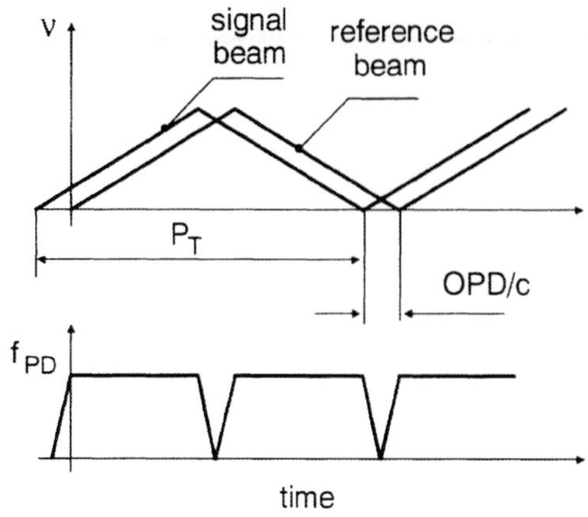

Figure 13. Principle of operation of a FMCW sensor. F_{PD}: beat frequency at the interferometer photodetector.

$$f_{PD} = \frac{2 \cdot K_M \cdot V_{pp}}{c \cdot P_T} \cdot OPD \qquad (16)$$

By measuring the frequency f_{PD} we can calculate OPD as:

$$OPD = f_{PD} \frac{P_T \cdot c}{2 \cdot K_M \cdot V_{pp}} = K \cdot f_{PD} \qquad (17)$$

The FMCW technique can be successfully applied to long range distance measurements by using a tunable diode-pumped Nd:YAG laser or to shorter ranges by using current-modulated laser diodes. Accuracy up to 10^{-4} of the measured distance can be obtained.

5. WHITE LIGHT INTERFEROMETRY

White light interferometry (WLI) is widely used in optoelectronic instrumentation for industrial as well as biomedical applications;[37] perhaps the best-known application of WLI is the optical coherence tomography[39] (OCT), used as diagnostic tool in ophthalmology.

The performance of WLI systems are largely limited by source features, therefore in the following section the main characteristics of the optical sources suitable for WLI will be presented. Therewith, we briefly introduce the three main classes of WLI systems that are: i) mechanically scanned, ii) electronically scanned and iii) dispersive white light interferometers.

5.1 Optical sources for white light interferometry.

The resolution of the WLI system is practically related only to the coherence characteristics of the excitation light, since optical interferometry is used to convert spectral properties of the excitation light in space resolution. The best resolution ΔL theoretically achievable using standard WLI can be related to the spectral width $\Delta\lambda$ of the excitation light according the following equation:[39]

Table 1. Characteristics of the light sources suitable for WLI.

Source type	Spectral width (nm)	Resolution (μm)	Mean lifetime (Hours)	Power (mW)	Cost (USD)
Lamps	200÷500	0.2÷0.5	500÷2000	10^3÷10^6	10÷>10^3
LED	20÷80	1÷8	10^4÷10^5	0.1÷10^2	10÷>10^2
SLD	10÷30	10÷30	10^3÷10^4	1÷5	100÷10^3
LD	10^{-4}÷10	30÷>10^5	10^4÷10^5	1÷10^3	5÷>10^3

$$\Delta L = \frac{2 \cdot \ln 2}{\pi} \frac{\lambda^2}{\Delta\lambda} \qquad (18)$$

where λ is the optical wavelength.

Continuous emission spectrum lamps are the classical light sources for WLI. These devices exhibit the widest emission spectrum and therefore the maximum resolution of the interferometric system. However they are normally not suitable for compact and reliable systems since they suffer from poor beam quality, low lifetime, and big size. Most popular, compact and reliable solid-state sources for WLI are SLDs. These devices have a structure similar to that of laser diodes, but the lasing effect is avoided by introducing losses in the cavity. Different technological processes have been proposed to realize SLDs. However, all these processes are based on laser diode cavities, e.g. an anti-reflection coating on the exit face of the laser. Due to the additional technological steps and to the small production volumes these devices are about ten times more expensive then laser diodes at the same emitted optical power. SLDs exhibit better beam quality and higher optical power than LEDs, and for these reasons they are more suitable to be used for WLI. On the other hand, low-cost, multi-quantum well visible laser diodes when polarized under threshold and kept at 37-40 °C exhibit spontaneous emission: therefore these devices could be efficiently used as low-cost sources for WLI when a low optical power is required.

Typical spectral width, achievable resolution, mean lifetime, size and cost of sources suitable for WLI are summarized in Table 1.

5.2 Mechanically scanned white light interferometer

In mechanical scanning white light interferometers the interference signal is obtained by scanning the position of a reference mirror.[34]

The distance measurement is performed by detecting the maximum of the interference signal envelope. The common optical setup, shown in Figure 14., is based on a Michelson interferometer with a variable reference arm. Reference mirror M2 is moved by using a mechanical or piezoelectric scanning system (MP). This element mainly limits the dynamic range of the interferometer, whereas the resolution is usually limited by the coherence length of the optical source.

Mechanical scanning WLI is suitable for wide measurement ranges, but normally it suffers from poor accuracy and a limited speed of operation due to the mechanical scanning systems used to move the reference mirror. The waveform processing procedure is shown in Figure 15. First, the current signal from the photo-detector is converted into a voltage signal by a transimpedance amplifier. This enters a full-wave active rectifier followed by an envelope detection circuit. The RC time constant of the envelope detector is set to match the reference mirror velocity since this velocity determines the main spectral component of the

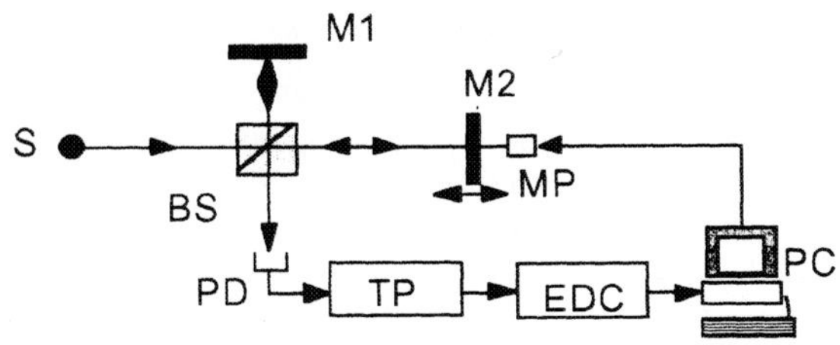

Figure 14. Standard optical setup for mechanical scanning WLI measurements. S. optical source; M1: measuring reflector; M2 reference reflector; BS: beam splitter; PD: photodetector; TP: transimpedance amplifier; EDC: envelope detection circuit.

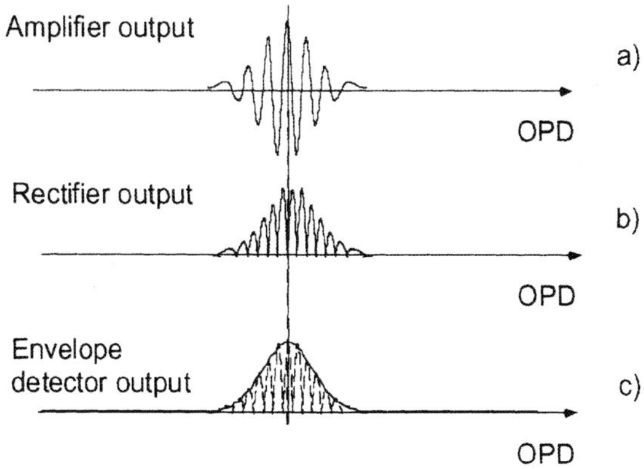

Figure 15. Waveform processing of the interference signal obtained by a mechanical scanning WLI. a) signal at the output of the transimpedance amplifier of Figure 14. b) and c) second and third step of the elaboration process.

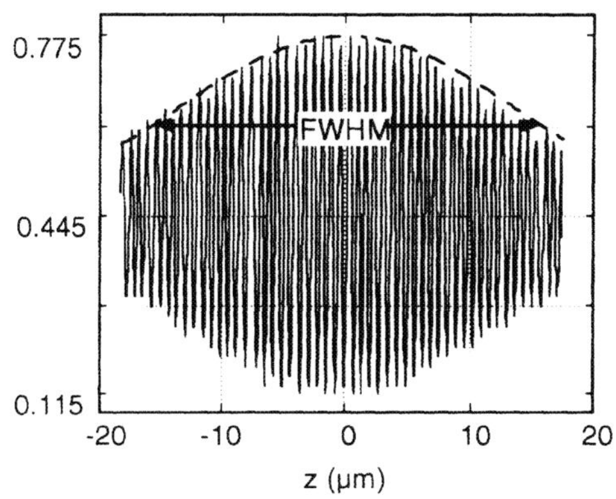

Figure 16. Typical interference signal due to a step variation of the diffraction index in the target. Z: displacement of the reference mirror from the zero-OPD condition.

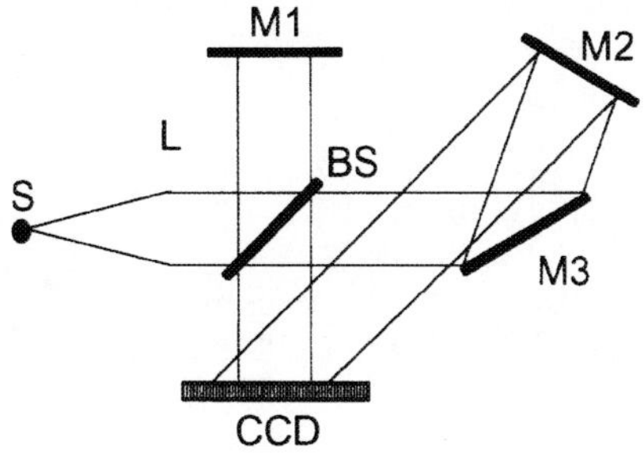

Figure 17. Typical optical setup of an electronic scanning WLI. S: optical source; L: collimating lens; BS: beam splitter; M1: measuring arm reflector; M2 and M3: reference arm mirrors.

interference signal. A personal computer generates a start TTL signal for the translation stage and begins the acquisition of the processed signal. Figure 16. shows a typical interferogram in the neighborhood of the balanced condition.

5.3 Electronic-scanning white light interferometry

In an electronic-scanning white light interferometer (ESWLI) the interference signal is converted into spatial fringes in a Moiré fashion by overlapping the light beams from the two arms at an angle on a CCD receiver, so that the output from the CCD array is equivalent to the output resulting from a spatial scan.[35]

The main advantage of electronic-scanning WLI systems is in the absence of any mechanical or piezoelectric scanning device. However, ESWLI systems normally suffer from limited dynamic range. A typical optical setup of an electronic-scanning white light interferometer is reported in Figure 17. Typical interferograms obtained for three different OPDs are reported in Figure 18. Dynamic range R and resolution ΔR of an electronic scanning white light interferometer can be calculated by using the following equations:[35]

$$R = \frac{\lambda \cdot M}{N}$$

$$\Delta R = \frac{\lambda}{2 \cdot N} \tag{19}$$

where λ is the optical wavelength, M and N are the total number of pixels of the CCD and the minimum admissible number of pixels for each interference period respectively.

5.4 Dispersive white light interferometry

As for the electronic scanning white light interferometer, in the dispersive white light

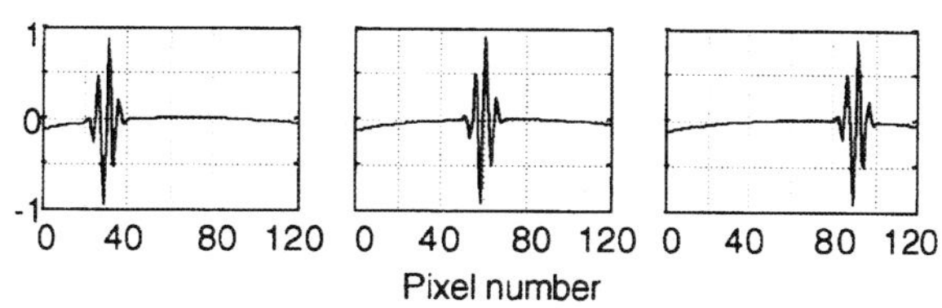

Figure 18. Example of ESWLI interferograms for three different positions of the target.

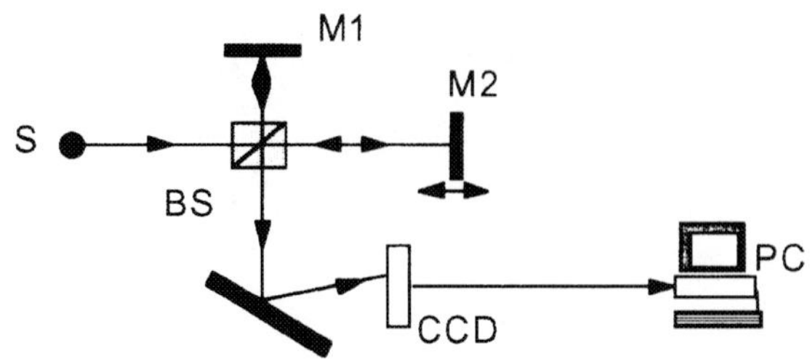

Figure 19. Typical optical setup of a dispersive WLI. S: optical source; BS: beam splitter; M1: measuring arm mirror; M2: reference arm mirror.

interferometer (DWLI) no moving parts are present.[36] The main component in such a system is the dispersion grating that spatially separates the spectral components of the broad band excitation light.

The standard optical setup is shown in Figure 19. It is based on a Michelson interferometer where the optical output is processed by dispersion grating DG and linear camera CCD. The collected signal is the product of the emission spectral shape of the low coherence source and a sinusoidal interference function. The information concerning the absolute distance of a reflective interface in the target is contained in the frequency $\delta\phi/\delta\nu$ of the sinusoidal modulation signal. According to Shnell et al.,[36,37] the interference phase of the dispersed output light varies linearly with the optical frequency ν as:

$$\phi(\nu) = \frac{4\pi}{c} \cdot OPD \cdot \nu \qquad (20)$$

where OPD is the absolute optical path difference of the interferometer, ν the optical frequency and c the speed of light.

Typical waveform processing to calculate the OPD is shown in Figure 20. First, the signal at the output of the CCD camera (Figure 20a.) is acquired by using an A/D board plugged into a personal computer, than the FFT algorithm is then used to elaborate this signal in order to obtain a shifted gaussian-like shape (Figure 20b.). The frequency shift is proportional to the absolute OPD.

The maximum measuring distance is limited by the spectral resolution $\Delta\nu_g$ of the diffraction grating. The distance resolution can be theoretically better than $\lambda/8$. Practically the finite fringe visibility and the limited number of acquired periods limit the resolution to λ, i.e. about 20 times less than the resolution achievable from mechanical scanning white light

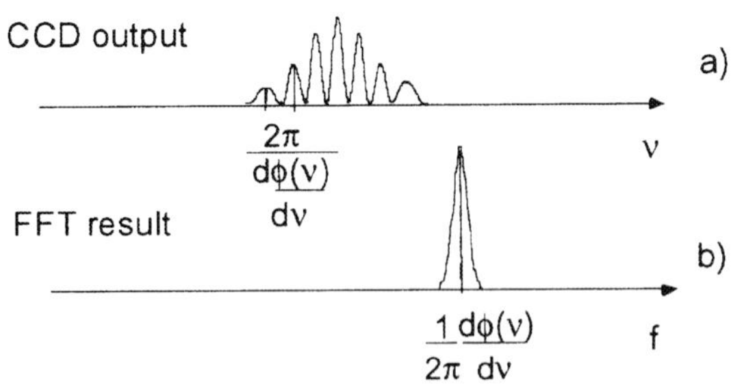

Figure 20. Waveform processing of the CCD signal due to step variation of the diffraction index in the target.

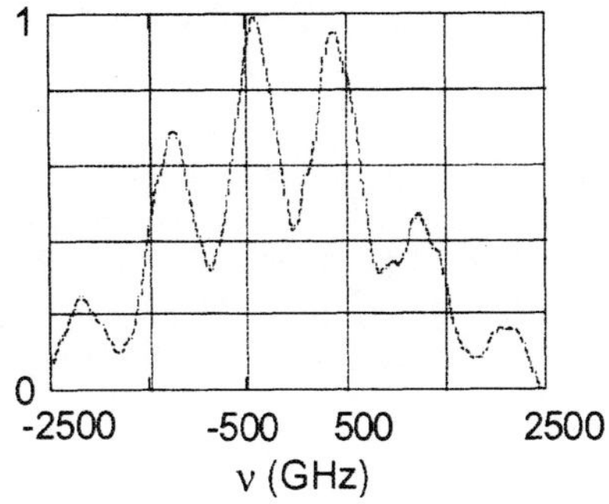

Figure 21. Typical interferogram from a dispersive white light interferometer.

interferometers. Figure 21. shows a typical interferogram obtained when OPD=200μm.

Dispersive white light interferometry can be combined with continuous wave, frequency modulation interferometry to implement a distance meter operating with both a very high resolution and a long range.[38]

6. COMPARISON OF THE DIFFERENT TECHNIQUES

A common property of the measuring techniques so far presented is the ability to perform non-incremental measurements. However the techniques are not interchangeable; for a given application, each method can present advantages and disadvantages with respect to the others. In order to attempt a classification from the application point of view, we can concentrate the analysis to a very few parameters as summarized in Table 2. First we have to compare the different laser sources that can be used to implement a particular measuring technique. In general for compact systems only semiconductor devices and solid state lasers can be considered, gas lasers are generally classified as cumbersome devices.

Portable meters can be exclusively based on laser diodes, LEDs, and SLDs. As the metrological point of view, measuring range and resolution (it is assumed that the total

Table 2. Comparison of the techniques.

Tech.	Sources	Range	Resolution	Complexity	Commercial immediate	Availability near future
DHI	gas lasers, solid state lasers, diode lasers	from tens of cm up to tens of m	0.01-0.001 % FS	Very high	No	No
FMCW	solid state lasers, diode lasers	from few mm up to few meters	0.1-0.01 %FS	Medium	No	Yes
MSWLI	lamps, LEDs, SLDs, diode lasers	limited by mechanics	1-50um	low	Yes	
ESWLI	lamps, LEDs, SLDs, diode lasers	30-200 um	0.5-2 um	medium	No	Yes
DWLI	lamps, LEDs, SLDs, diode lasers	200-2000 um	0.1-1 um	high	No	Yes

measurement uncertainty is equal to some small multiple of the resolution) have to be evaluated, from the table it is evident that the different techniques present different performances. In particular, DHI and FMCW systems are suitable for long measuring ranges, WLI systems are suitable for short ranges.

As to the industrial applicability, column *Complexity* of Table 2 gives an estimation of the effectiveness of the technique. The DHI technique is rated *very high*, this means that this technique can be successfully used only if other methods do not solve a given measuring problem. In contrast MSWLI is rated *low* in complexity, as a consequence some commercially available devices have been already developed.

REFERENCES

1. C. Polhemus. Two-wavelength interferometry, *Appl. Opt.* 12:2071 (1973).
2. C. R. Tilford, Analytical procedure for determining lengths from fractional fringes, *Appl. Opt.* 16:1857 (1977).
3. G. L. Bourdet, A. G. Orszag, Absolute distance measurements by CO_2 laser multiwavelength interferometry, *Appl. Opt.* 18:225 (1979).
4. C. W. Gillard, N. E. Buholz, Progress on absolute distance interferometry, *Opt. Eng.* 22:348 (1983).
5. H. Matsumoto, Synthetic milimeter-wave signal generation for length measurement, *Appl. Opt.* 23:973 (1984).
6. H. Matsumoto, Synthetic interferometric distance measuring system using a CO_2 laser, *Appl. Opt.* 25:493 (1986).
7. C. J. Walsh, Limit to multiwavelength interferometry imposed by frequency instability of the source, *Appl. Opt.* 26:29 (1987).
8. C. J. Walsh, Measurements of absolute distances to 25 m by multiwavelength CO_2 laser interferometry, *Appl. Opt.* 26:1680 (1987).
9. F. Bien, M. Camac, H. J. Caulfield, S. Ezekiel, Absolute distance measurements by variable wavelength interferometry, *Appl. Opt.* 20:400 (1981).
10. N. A. Massie, H. J. Cauldfield, Absolute distance interferometry, in: *Proc of SPIE* 816:149 (1987).
11. U. Minoni, G. Sansoni, F. Docchio, E. Paganini, U. Perini, G. ReGarbagnati, Interferometria laser per misure industriali e sue applicazioni, *Fisica e Tecnologia* 12:113 (1989).
12. A. F. Fercher, H. Z. Hu, U. Vry, Rough surface interferometry with a two-wavelength heterodyne speckle interferometer, *Appl. Opt.* 24:2181 (1985).
13. A. F. Fercher, Two-wavelength speckle interferometric technique for rough surface contour measurement, *Opt. Eng.* 25:623 (1986).
14. R. Dändliker, R. Thalmann, D. Prongué, Two-wavelength laser interferometry using superheterodyne detection, in: *Proc. SPIE* 813:9 (1987).
15. R. Dändliker, R. Thalmann, D. Prongué, Two-wavelength laser interferometry using superheterodyne detection, *Opt. Lett.*, 13, 339-341 (1988).
16. Z. Södnik, E. Fischer, T. Ittner, H. J. Tiziani, Two-wavelength double heterodyne interferometry using a matched grating technique, *Appl. Opt.* 30:3139 (1991).
17. F. Docchio, U. Perini, H. Tiziani, A combined distance and surface profile measurement system for industrial applications: a European project, *Meas. Sci. Technol.* 5:807 (1994).
18. E. Gelmini, U. Minoni, F. Docchio, Tunable, double-wavelength heterodyne detection interferometer for absolute-distance measurements, *Opt. Lett.* 19:213 (1994).
19. F. Docchio, U. Minoni, E. Gelmini, G. Sansoni, Optical distance meter based on dual-wavelength interferometry, in: *Proc. of IMEKO-XIII*, 3:2017, Torino (1994).
20. E. Gelmini, U. Minoni, F. Docchio, Simple digital technique for efficient frequency locking of diode-pumped Nd:YAG lasers, *Elect. Lett.* 30:47 (1994).
21. E. Gelmini, U. Minoni, F. Docchio, A tunable, double-wavelength heterodyne detection interferometer with frequency locked diode-pumped Nd:YAG sources for absolute measurements, *Rev. Sci. Instrum.* 66:8, 4073 (1995).
22. H. Kikuta, K. Iwata, R. Nagata, Distance measurement by wavelength shift of laser diode light, *Appl. Opt.* 25:2976 (1986).
23. H. Kikuta, K. Iwata, R. Nagata, Absolute distance measurement by wavelength shift interferometry with a diode laser: some systematic error sources, *Appl. Opt.* 26:1654 (1987).

24. U. Minoni, E. Gelmini, F. Docchio, Tunable laser sources for absolute optical interferometry, in: *Proc. of SPIE* 2349:131, (1994).
25. G. Beheim, K. Fritsch, Remote displacement measurements using a laser diode. *Electr. Lett.* 21:93 (1985).
26. G. Economou, R. C. Youngquist, D. E. N. Davies, Limitation and noise in interferometric systems using frequency ramped single mode diode lasers, *J. Lightw. Technol.* LT-4:1601 (1986).
27. T. Kubota, M. Nara, T. Yoshino, Interferometer for measuring displacement and distance, *Opt. Lett.* 12:310 (1987).
28. M. Imai, K. Kawakita, Optical-heterodyne displacement measurement using a frequency ramped laser diode, *Opt. Comm.* 78:113 (1990).
29. A. Dieckmann, M. C. Amann, Phase noise limited accuracy of distance measurements in a frequency-modulated continous-wave LIDAR with a tunable twin-guide laser diode, *Opt. Eng.* 34:896 (1995).
30. Y. Zhu, H. Matsumoto, T. O'ishi, Arm length measurement of an interferometer using the optical-frequency-scanning technique, *Appl. Opt.* 30:3561 (1991).
31. W. Yongjun, Z. Yang, Dacheng, New method of large-scale absolute distance measurement, in: *Proc. of SPIE* 2066:162 (1993).
32. U. Minoni, G. Scotti, F. Docchio, Wide range distance meter based on frequency modulation of a Nd:YAG laser, *Opt. Eng.* 35:1949 (1996).
33. G. A. Alphonse, D. B. Gilbert, M. G. Harvey, M. Ettenberg, High power superluminescent diodes, *IEEE J. Quantum Electron.*, 24:2454 (1988).
34. D. Leslie, P. de Groot, High speed noncontact profiler based on scanning white-light interferometry, *Appl. Opt.* 33:7334 (1994).
35. S. Chen, A.W. Palmer, K.T.V. Grattan, B.T. Meggitt, S. Martin, Study of electronically-scanned optical-fibre white-light Fizeau interferometer, *Elect. Lett.* 27:1032 (1991).
36. U. Schnell, E. Zimmermann, R. Dändliker, Absolute distance measurement with synchronously sampled white-light channelled spectrum interferometry, *Pure Appl. Opt* 4:643 (1994).
37. U. Schnell, R. Dändlinker, S. Gray, Dispersive white-light interferometry for absolute distance measurements with dielectric multilayer systems on the target, *Opt. Lett.* 21:528 (1996).
38. L. Rovati, U. Minoni, F. Docchio, Dispersive white light combined with a frequency-modulated continuous-wave interferometer for high-resolution absolute measurements of distance, *Opt. Lett.* 22:850 (1997).
39. D. Huang, E.A. Swanson, C. P. Lin, J. S. Shuman, W. G. Stinson, Optical coherence tomography, *Science*, 254:1178 (1991).

OPTICAL TOMOGRAPHY: TECHNIQUES AND APPLICATIONS

A.W. Domański

Warsaw University of Technology
Institute of Physics
Koszykowa 75
00-662 Warszawa, Poland

1. INTRODUCTION

Tomography is a method of two-dimensional imaging based on analysis many one-dimensional images (Figure 1.). From theoretical point of view tomography requires several calculations. Starting point of the calculations is to obtain mathematical description of several one-dimensional transmittance images of object. It may be made based on Radon transform.[1]

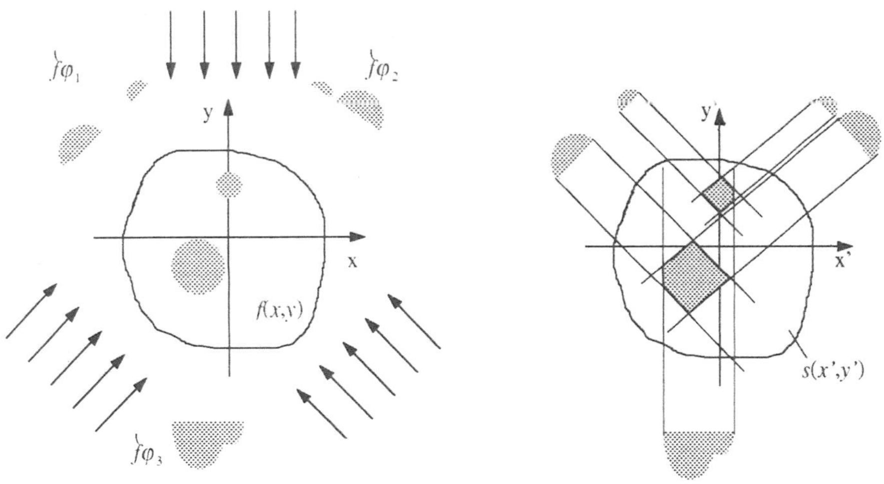

Figure 1. Transmitation images $\breve{f}\varphi$ of real object f(x,y) allow to calculate function s(x',y') describing images of the object.

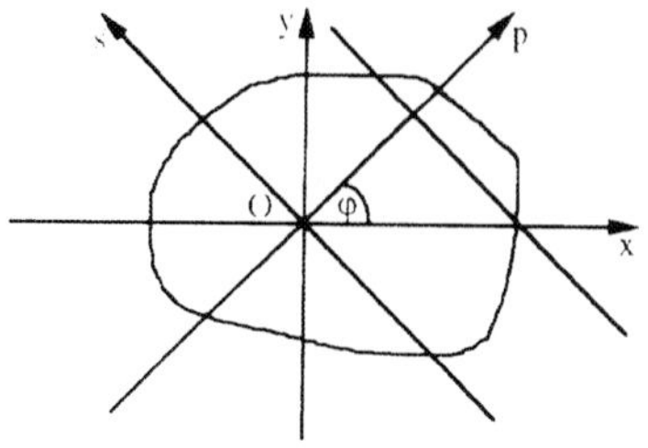

Figure 2. Geometrical parameters in Radon transform.

In general Radon transform is done by formula as follows:

$$\check{f} = \int_L f(x,y)ds \quad (1)$$

where $f(x,y)$ is object function. For two dimensional case the formula has more precisely view:

$$\check{f}(p,\varphi) = \int_{-\infty}^{\infty} f(p\cos\varphi - s\sin\varphi, p\sin\varphi + s\cos\varphi)ds \quad (2)$$

where geometrical parameters are shown in Figure 2. Substituting

$$p = (\mathbf{x},\xi) = x\cos\varphi + y\sin\varphi \quad (3)$$

we can get general case:

$$\check{f}(p,\xi) = \int f(\mathbf{x})\delta(p-(\xi,\mathbf{x}))d\mathbf{x} \quad (4)$$

Radon has proved that his transform is analogue to the two-dimensional Fourier transform.[1]

Next step in calculations is to apply Fast Fourier Transform (instead of Fourier transform due to simplification of calculations) for each one-dimensional function. Results of calculations are collected in matrix of complex components.

In order to obtain the cross section of the two-dimensional object reverse two-dimensional Fast Fourier Transform must be used. All the calculations are terrible time consuming and without computers it was impossible to have useful tomographs. There are many types of computer assisted tomography (CAT) dependent on radiation and method used to create of one-dimensional images. The best known is X-ray computer tomography.

X-ray computer tomography was introduced after almost eighty years of the first medical application of X-ray imaging.[2,3] An idea of the tomography is based on computer analysis of the large number electronic pictures obtained during rotation of X-ray tube and detector array around the human body.

Quality of images given by modern X-ray tomographs is very high although X-ray effective dose during one computer tomography examination is almost twice as big as natural background radiation per year.[4] Hence some attempts of application of light instead of X-rays have been carried out for several years. Application of light leads to optical tomography. Up-to now results of optical imaging are not so good as X-ray tomography mainly due to high scattering of light inside human tissue and inside others media with similar optical parameters.

Other possible types of computer assisted tomography very similar to X-ray type are:

- nuclear radiation (α, β, γ),
- electron beams,
- ultrasounds.

Magnetic resonans imaging is a little different types of CAT which requires an analysis of relaxation time of nuclear magnetic resonans inside human body.

2. OPTICAL TOMOGRAPHY

Computer assisted tomography with help of optical waves for create of one-dimensional images seams to be very attractive. From theoretical and experimental point of view there are no problems with optical tomography imaging in open space. Unfortunately optical tomography should be mainly applied for visualization of objects inside high scattering medium like for example human body.

Light transmitted in high scattering medium like living tissue does not allow to obtain an image of different objects which are inside human body. We can observe only low quality shadow image which is insufficient for typical computer tomography. Some research of Alfano and coworkers[5] gives a view for the problem. Only ballistic photons and partially snake-like photons can give a good image which is disturbed by photonic noise caused by scattered photons (Figure 3.). A fraction of photons which are useful for creation of image is extremely small sometimes much less then one percent. In order to omit a problem with interference scattered photons many methods are introduced. The methods may be collected in a few groups named as follows.

- time - resolved imaging;
- fluorescence imaging;
- coherence imaging;
- direct transillumination imaging.

3. TIME-RESOLVED IMAGING

Time-resolved imaging is one of the possible method of separation of direct transmitted i.e. ballistic and snake-like photons and scattered photons (Figure 4.). It may be realized by various techniques of gated detection like by use Kerr gate,[6] streak camera,[7] stimulated Raman scattering.[8]

The time-resolved method requires very short pulses (100 fs) of light illuminating the object and high speed photodetection techniques (response time from 100 fs to 10 ps). It is very costly but laboratory results are very promising.[9]

4. FLUORESCENCE IMAGING

There are other possibility to obtain an object image in high scattering medium. The method named fluorescence imaging is based on comparison laser induced fluorescence image of different kind of living tissue.[10]

The method utilizes results of laser fluorescence spectra analysis. In order to increase the spectra differences some special dopants are introduced to the human body. Hence the method may be associated with photodynamic therapy because both methods require the agents which are selectively retained in human tissue. Due to effects of accumulation of the agents in the skin and up-to-now low quality of such imaging the method seems to be inconvenient for optical computer tomography.

5. COHERENCE IMAGING

The best known method of coherence imaging is holography. The holography allows to create of three-dimensional image based on hologram i.e. interferogram in which amplitude and phase of wave object is registered. The holography requires illumination of object by light with good coherence and without disturbing of amplitude and phase of wave diffracted on the object as well as reference wave helpful in creation of interferogram. Hence the holography which is very promising technique of imaging in open space is useless in optical imaging in high scattering medium.

Optical coherence imaging in light scattering medium is based on utilization of phase correlation of ballistic as well as snake like photons with reference photons.[11,12]

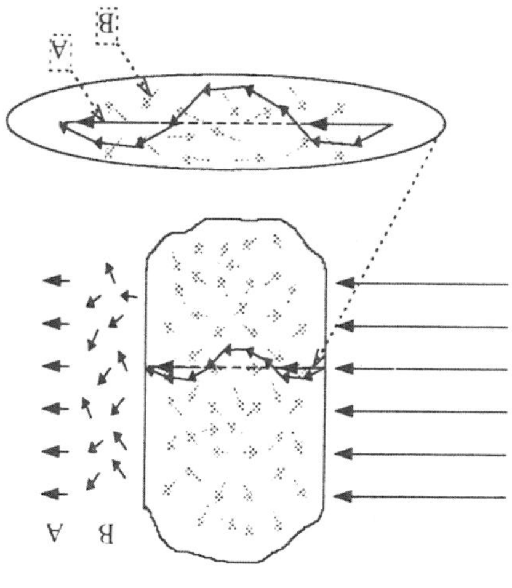

Figure 3. Ways of three types photons in tissue: ballistic (A), snake-like and diffuse (B).

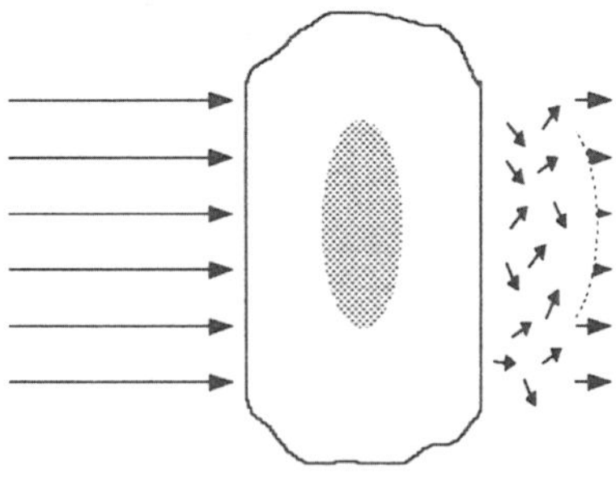

Figure 4. Imaging in high scattering medium is disturbed by photonic noise created by large number diffuse photons.

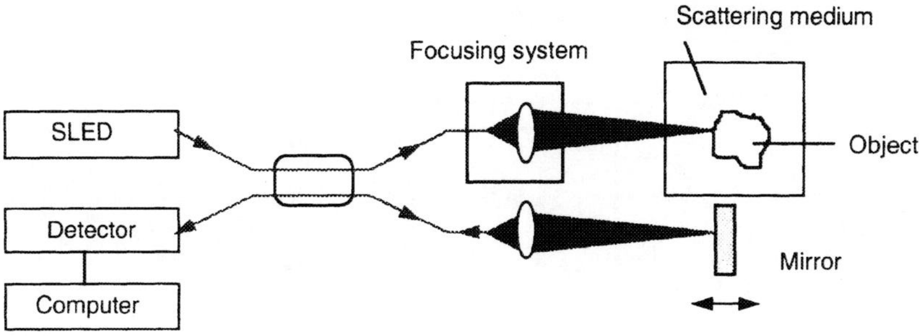

Figure 5. Diagram of optical coherence imaging.

Scattered photons have too long optical path in the medium, longer than coherence length of light used in experiments. Hence the scattered photons create a noncoherent photonic noise which can be eliminated. The method may be applied for microscopy imaging of objects in high scattering media.[13] An idea is based on application of an optical fiber Michelson interferometer with superluminescent light emitting diode (SLED) as a light source (Figure 5.). Coherence of light emitted by SLED is not very high and it allows to obtain interference pattern only when optical path of light in the object arm and reference arm with mirror is almost the same.

Reflectivity of object in scattering medium influences on intensity of interference pattern. It allows to create a reflectivity map of object with help of moving reference mirror. Additional scanning in two dimensions gives multiple maps of reflectivity of different slices of object. Hence such imaging allows to build up optical coherence tomograph.

6. DIRECT TRANSILLUMINATION IMAGING

Direct transillumination method of imaging was introduced more than fifty years ago for breast cancer diagnose with not very good results.[14] Development of light sources like gas lasers and recently visible laser diodes and sensitivity of detectors gives the researchers new possibility to build up equipment for direct transillumination imaging.

There are at least two groups of researchers who built up and clinically tested a two wavelength optoelectronic mammoscopy for breast cancer imaging.[15,16] One group from Hammamatsu used He-Ne gas laser and infrared laser diode to create shadow images. Results of clinical tests of the mammoscope are very good and some electronic pictures of tumor are better than mammography films used as a standard. Second group from Warsaw University of Technology built up the mammoscope by use only two kinds of laser diodes (Figure 6.). The mammoscope was tested in Warsaw Institute of Oncology and some results after additional image processing were very promising[17] and similar to the results of the Japanese group.[18]

7. OPTICAL TOMOGRAPHY SCANNERS

Obtained in different way images of objects should be transformed by computer in similar way as X-ray images are processed. There are some researchers who try to do that.

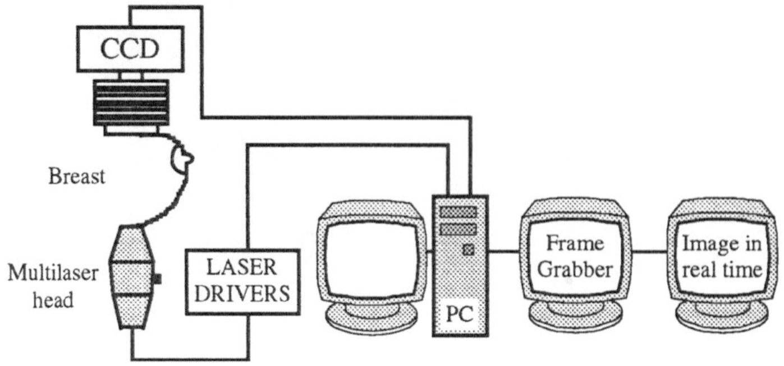

Figure 6. Schematic drawing of the laser diode mammoscope.

They apply direct transillumination method as a very similar to the X-ray computer tomography configuration.A titanium - doped sapphire laser or solid state laser pumped by an argon - ion laser as a light source may be used. Several hundreds of avalanche photodiodes help to measure transmission in sequence, synchronous with sweep of the fun. Such optical tomography scanner was built up by Grable and co-workers.[19]

Optical coherence tomography also allows to built up a scanner but up to now it is limited only for imaging of objects in not very high scattering medium like inside eye or layers of skin close to surface.

8. CONCLUSIONS

Optical computer tomography requires high quality of image and up-to-now no method of optical imaging presented in the paper are sufficient for that. Nevertheless we know about optical imaging in high scattering media much more than few years ago and some results are promising.

Medical problems of optical imaging are close to problems of submarines in murky water and airplanes in clouds. Hence combination of military and medical requirements gives a chance to support of future research leading to real time optical computer tomography.

REFERENCES

1. J.C. Russ „The Image Proccesing Hanbook" CRC Press Inc. 1992.
2. G.N. Hounsfield, "Computerized transverse axial scanning (tomography)- Part I Description of system.", Br. J. Radiol., pp. 46-1016, 1973.
3. J. Ambrose, "Computerized transverse axial scanning (tomography)- Part II. Clinical application.", Br. J. Radiol., 46-1047, 1973.
4. W.A. Kalender, "X-Ray Computed Tomography - State of the Art.", SPIE V.IS11 Medical Optical Tomography: Functional Imaging and Monitoring , pp.10-27, 1993.
5. F. Liu, K.M. Yoo and R.R. Alfano, "Ultrafast laser-pulse transmission and imaging through biological tissues", Applied Optics, vol.32, pp.554-58, 1993.
6. L. Wang, Y. Liu, P.P. Ho and R.R. Alfano, "Ballistic imaging of biomedical samples using picosecond optical Kerr gate", Time Resolved Spectroskopy and Imaging of Tissue, B. Chance, ed., SPIE vol. 1431, pp. 97-101, 1991.
7. J.C. Hebdan, R.A. Kruger, K.S. Wong, "Time-resolved imaging through a highly scattering medium", Applied Optics vol.30, pp.788-794, 1991.
8. M.D. Duncan, R. Mahon, L.L. Tankersley, J. Reintjes, "Time-gated imaging through scattering media using simulated Raman Amplification", Opt.Lett. 16, pp.1868-1870, 1991.

9. R. Berg, S. Andersson, S. Svanberg, "Time-resolved transillumination imaging", SPIE V.IS11 Medical Optical Tomography: Functional Imaging and Monitoring, pp. 397-424, 1993.
10. S. Svanberg, "Optical Tissue Diagnostics: Fluorescence and Transilluminescence Imaging", Optical and Photonics News, Vol.3, No 10, pp.31-34, 1992.
11. J.A. Izatt, M.R. Hee, D. Huang, J.G. Fujimoto, E.A. Swanson, C.P. Lin, J.S. Schuman, C.A. Puliafito,"Optical Coherence Tomography for Medical Diagnostics", SPIE V.IS11 Medical Optical Tomography: Functional Imaging and Monitoring, pp. 450-472, 1993.
12. H. Inaba, "Coherent detection imaging for medical laser tomography", SPIE V.IS11 Medical Optical Tomography: Functional Imaging and Monitoring, pp. 317-347, 1993.
13. J.A. Izatt et al., Optical coherence tomography for biodiagnostics, Optics and Photonics News. Vol.8, No 5, pp. 41-47, 1997.
14. M. Cutler, "Transillumination as an aid to diagnosis of breast lesions", Surg.Gynee Obstet, vol. 48, pp.721-29, 1929.
15. Y. Yamashita, S. Suzuki, S. Miyaki, T. Hayakawa,"The neonate brain (NIR) and breast imaging using transillumination", Photon migration in Tissues (edited by B.Chance,Plenum, New York), pp.55-67, 1989.
16. A.W. Domański et al., "Optoelectronic Mammoscope for Breast Cancer Diagnosis", SPIE vol.2085, pp.96-103,1993.
17. A.W. Domański, J. Dziukowa, M. Karpierz, M. Świłło, E. Wesołowska, "Medical test of multilaser mammoscope for breast cancer diagnosis" in Laser in Medicine, Springler-Verlag, Berlin, pp.591-594, 1996.
18. P. Hee et al."Breast Cancer Diagnosis by Laser Transmission Photo-Scanning with Spectro-Analysis", Radiation Medicine, vol.8, pp. 1-5, 1990.
19. R.J. Grable, Optical tomography improves mammography, Laser Focus World, Vol.32, No 10, pp.113-118, 1996.

OPTICAL WAVEGUIDE REFRACTOMETERS

R. Ramponi, M. Marangoni, and R. Osellame

Dipartimento di Fisica del Politecnico di Milano
Piazzale Leonardo da Vinci, 32
20133 Milan, Italy

1. INTRODUCTION

The use of laser sources in many application fields has led to new kinds of experiments where the accurate determination of the optical properties of the media used in device fabrication is essential. On the other hand, the availability of laser sources and sophisticated optical devices has led to an increasing use of the optical characterisation as an investigation technique for many materials. Among optical properties, the refractive index n plays a role of primary importance. Many methods have been proposed to determine the refractive index of materials, and in particular of liquid samples.[1-7] The propagation of electromagnetic waves in any material depends on its refractive index both in the case of free propagation, when geometrical optics approximation applies, and in the case of guided configuration. Thus, measuring techniques exploiting both geometrical optics components, such as prisms, dielectric slabs, etc., and guiding structures, such as optical fibers and waveguides, can be employed. To achieve higher precision, interferometric measurements can be of use.

In this chapter, some methods to determine the refractive index of liquids are described, and a new method proposed, based on a nonlinear phenomenon in optical waveguides. In particular, different types of refractometers will be presented, grouped depending on their working principle: based on geometrical optics, exploiting interferometric techniques and non conventional methods, in optical fibers and in waveguides.

The new refractometer proposed is based on second harmonic generation from nonlinear optical waveguides in the Cerenkov configuration. It exploits the dependence of the output angle of the Cerenkov beam on the refractive index of the waveguide cover (liquid sample). The main advantage of this method is that it allows the determination of the refractive index of the liquid in the infrared by means of a plain angle measurement performed on a visible beam (Cerenkov second harmonic). Moreover, the refractometer is realised in a simple planar step-index waveguide and device full-integration can be foreseen.

The possibility of exploiting the same measurement technique as a method to characterise nonlinear optical waveguides starting from liquids of known refractive index is also considered.

The rationale of the technique proposed is discussed, and preliminary experimental results are reported.

Optical Sensors and Microsystems: New Concepts, Materials, Technologies
Edited by Martellucci et al., Kluwer Academic / Plenum Publishers, New York, 2000.

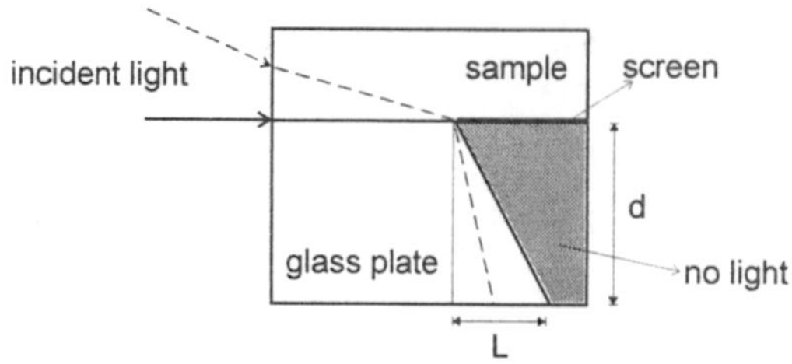

Figure 1. Abbe refractometer.

2. REFRACTOMETERS FOR LIQUIDS: AN OVERVIEW

2.1. Classical refractometers

Classical refractometers for the determination of *n* in liquids mainly exploit reflection and Snell's refraction laws in different geometrical configurations. This is the case for Abbe refractometer,[1] Pulfrich refractometer[2] and refractometers based on minimum deviation angle or on beam-displacement measurements.[3-5] In general, all these methods require precise knowledge of the refractive index and of the geometrical characteristics of the optical components used in the measurements. Actually, both Abbe refractometer and Pulfrich refractometer are known and used since long time, but recently they have been proposed in new configurations employing lasers as the light source. Indeed, on one side, the spatial and temporal coherence of laser beams allow more precise measurements; on the other side the wide use of lasers in optical experiments makes it important to perform the optical characterisation of materials with the same light sources. The working principle of the above mentioned devices is illustrated in Figures 1. to 4.

In particular, Figure 1. illustrates the scheme of a typical Abbe refractometer, that exploits critical refraction conditions, the ligth path being opposite to that of total reflection: by measuring L and d, the critical angle is obtained, and hence, through Snell's law, the sample refractive index. Only indices lower than that of the glass plate, that must be known, can be measured.

Figure 2. illustrates one possible configuration for Pulfrich refractometer: a V-shape prism filled with the liquid to be characterised is used, and the deflection angle measured; a second configuration similar to Abbe refractometer, where the critical angle is measured, has also been reported, the only difference being in the method used for the angle determination. The first configuration requires the knowledge both of the V aperture angle and of the refractive index of the prism, whereas the second one only requires the knowledge of the refractive index of the prism, but again the latter represents an upper limit for the indices to be measured.

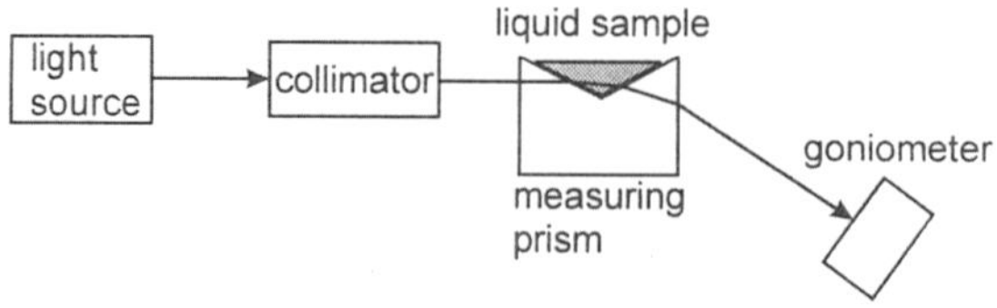

Figure 2. A possible configuration of the Pulfrich refractometer.

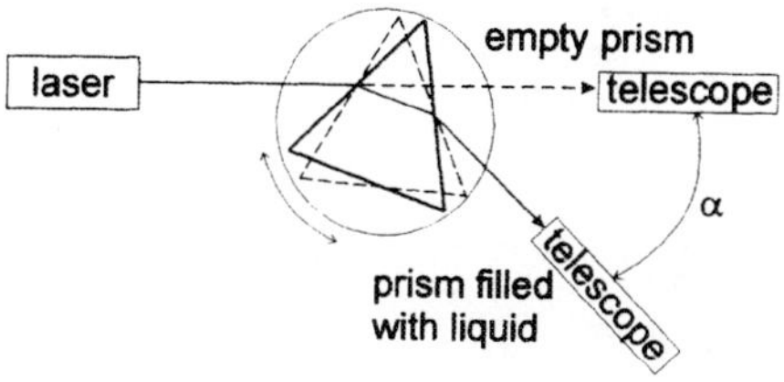

Figure 3. Scheme of the minimum deviation angle refractometer.

Figure 3. shows the scheme for the refractive index determination by means of minimum deviation angle measurement. A glass hollow prism is placed on a rotating platform. The zero position is determined when the prism is empty. Then the prism is filled with the liquid to be measured and the angular position corresponding to the minimum deviation condition is determined. Only the prism angle must be known provided that entrance and exit faces of the hollow glass prism are optically flat and each with parallel sides.

Figure 4. illustrates how to determine the refractive index of a liquid by measuring the displacement d of a laser beam passing through a rectangular hollow cell filled with the liquid sample. In the basic scheme, all geometrical parameters of the cell and the refractive index of its walls must be known with high precision. Otherwise, a reference measurement can be performed with the empty cell: in such case only the length of the internal dimension of the cell has to be known.

2.2. Interferometric refractometers

Interferometric methods are a typical choice in optical measurements when high precision is needed. The common principle of interferometric refractometers for liquids is that of having the liquid in the measurement arm, the optical path being thus affected by the refractive index of the sample, and air (or vacuum) in the reference arm. Different configurations can be exploited, such as plain Michelson interferometer[6], Mach-Zender interferometer,[7-8] Jamin or Rayleigh refractometers.[9] More complex configurations are reported, e.g. combining two Michelson interferometers,[10] or inserting a very stable stepped cell in an arm of the interferometer,[11] or using grating interferometers.[12-13]

2.3. Non conventional refractometers

Refractometers combining different phenomena or exploiting particular situations that take place in small drops of liquid have also been proposed.

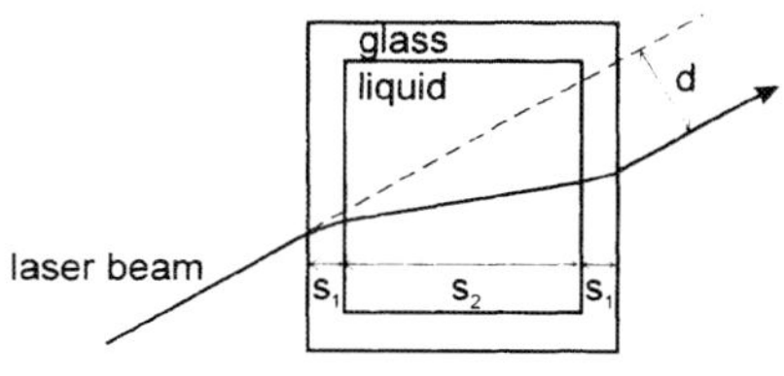

Figure 4. Beam displacement method.

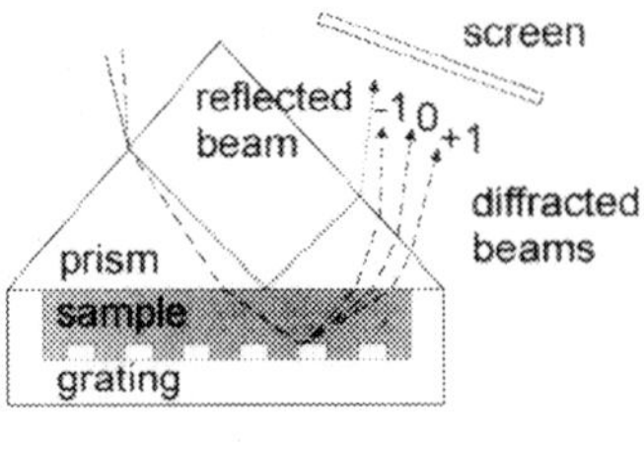

Figure 5. Diffraction grating microrefractometer.

Figure 5. shows a diffraction grating microrefractometer[14] where the liquid sample is placed on a diffraction grating and a glass prism is superimposed to it. When the input-beam incidence angle on the prism is such that light is transmitted through the liquid and hits the diffraction grating, diffracted beams will be observed on the screen. By varying the incidence angle, diffracted beams will disappear when the angle corresponding to total reflection at the prism-liquid interface is reached. The measurement of this angle, together with the knowledge of the refractive index of the prism, allows the determination of the refractive index of the liquid. The latter can not be higher than the prism index, as it always happens when total reflection is involved.

The refractive index of liquid samples has also been measured by exploiting high-order rainbows formed by a pendant drop,[15] or by studying multiple-beam Fizeau fringes (in transmission) produced when a small drop of the liquid sample is introduced in a wedge interferometer.[16]

2.4. Optical fiber refractometers

Optical fibers characteristics and the possibility of altering light propagation and coupling conditions by acting on the interfaces have been exploited in some refractometers.

When the optical fiber end is plunged in the liquid sample, the refractive index of the liquid can be determined by monitoring the back-reflected power with respect to the coupled input power, according to the well known Fresnel coefficients.[17] Since equations contain square parameters, in particular the square of the difference between the refractive index of the fiber core and that of the liquid sample, to avoid ambiguity at least two determinations must be performed, with two fibers having different core index.

Instead of exploiting back reflected power, the optical fiber end can be again dipped in

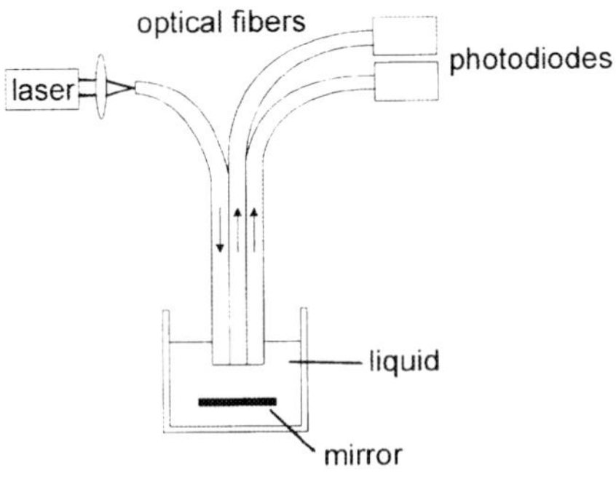

Figure 6. Optical fiber reflection refractometer.

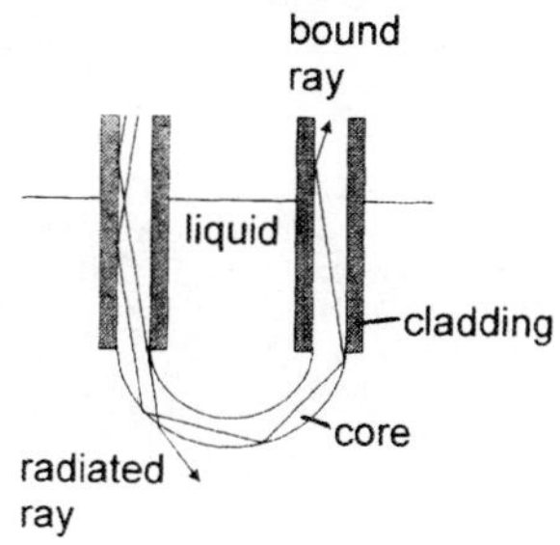

Figure 7. U-shaped fiber optic refractometer.

the liquid sample, but facing a mirror. The power collected by a second fiber placed next to the incoming one, as shown in Figure 6., is a function of the liquid refractive index.[18] Actually, it is convenient to use two parallel collecting fibers to minimise the effects of possible laser fluctuations, variations in the mirror reflectivity and bending losses.

When the cladding is removed in a region of an optical fiber, and the core exposed to the liquid sample, mode confinement conditions, and thus losses, are affected and depend on the refractive index of the liquid. Therefore, by measuring radiation losses or transmitted power in suitable geometrical conditions, the refractive index of the liquid can be measured.[19-20] A possible scheme based on this working principle is shown in Figure 7.

The cladding of the fiber can be substituted in a region with a gold layer, and the corresponding part of the fiber dipped in the liquid sample,[21] as shown in Figure 8. By adjusting the input coupling angle α, phase matching between the fiber guided mode and the plasmon mode is achieved. When this occurs, transmitted power drops dramatically. Since the phase matching condition depends on the refractive index of the surrounding liquid sample, the latter parameter can be determined by monitoring the transmitted power as a function of α.

2.5. Waveguide refractometers

The basic principles that allow the design of optical fiber refractometers can be extended to waveguides. Moreover, waveguide interferometers can be designed to achieve higher precision, as for example in the Mach-Zender interferometer proposed by Liu et al.,[22] where the reference arm is shielded, while the sensing arm is in contact with the liquid sample.

A waveguide refractometer based on surface plasmon resonance[23] similar to that realised in optical fiber has been proposed. Actually, in this case the sensor design is far

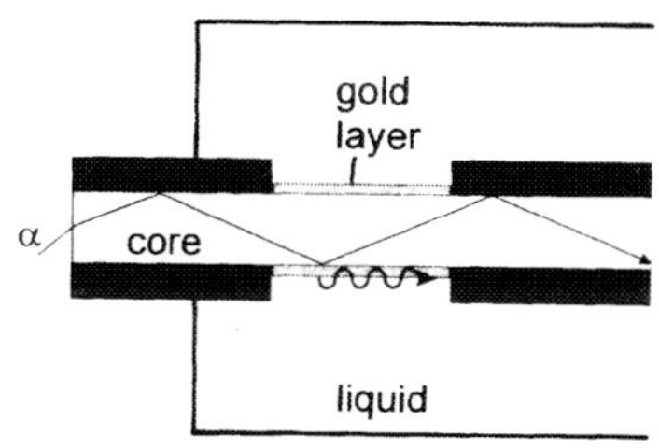

Figure 8. Fiber optic refractometer based on surface plasmon excitacion

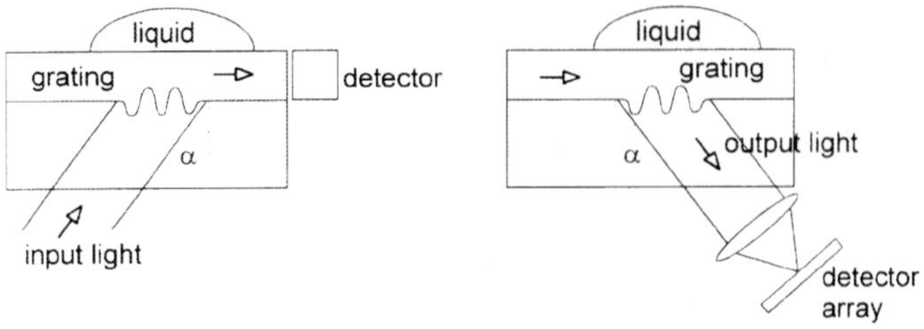

Figure 9. Waveguide refractometer exploiting input or output grating couplers.

more complicated. Indeed, in the configuration proposed a SiO_2 buffer layer, a TiO_2 tuning layer and a gold layer are placed on a K^+-Na^+ ion exchanged waveguide and a layer of the sample (not necessarily a liquid) is placed as the cover. The occurrence of phase matching between the guided mode and the surface plasmon mode, which depends on the cover refractive index, deeply affects the guide losses. Thus, by measuring the transmitted power, the sample refractive index is evaluated. The guide has to be carefully designed for superstrate refractive indices not too far from that to be measured. Moreover, preliminary calibration curve of the transmission coefficient as a function of the liquid refractive index is needed.

Coupling gratings have been exploited to design another type of waveguide optical refractometer. Guided modes propagating in a waveguide are characterised by the so-called effective refractive index (n_{eff}) defined as the ratio β/k_0, where β is the propagation constant of the mode itself, and k_0 is its wavenumber. The sensor is constituted by a planar waveguide provided with a coupling grating[24] and the liquid sample is placed as the cover, as shown in Figure 9. The value of the refractive index of the cover affects the evanescent field and thus the n_{eff} of the guided mode. The latter can be determined by measuring the coupling or out-coupling angle α. From the n_{eff}, it is then possible to calculate the refractive index of the liquid sample.

3. CERENKOV REFRACTOMETRY

As illustrated in the previous paragraph, many methods have been proposed to measure the refractive index of liquid samples. However, most of them suffer from some limitations.

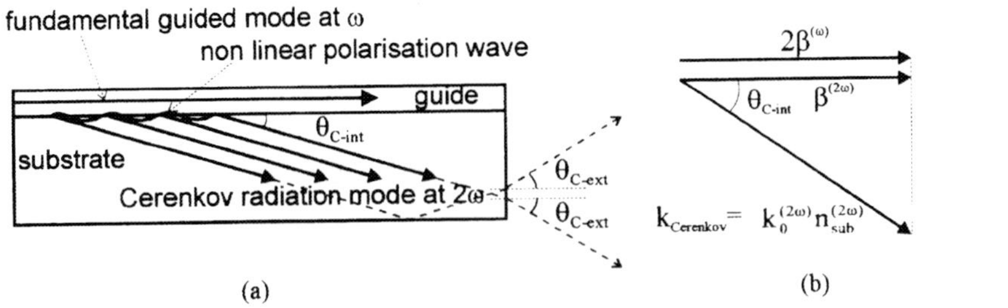

Figure 10. Cerenkov second harmonic generation scheme (a) and wave vector representation of the phase matching condition (b).

Indeed, depending on the principle and the material used, each system limits the range of n-values that can be measured. Moreover the device or system used needs to be fully characterised, in terms of optical, physical and geometrical parameters, with very high precision. It must also be taken into account that the most precise devices are also very complex. It has to be pointed out that, although the methods reported can in principle be used both in the visible and in outer ranges (for example in the infrared), they have typically been tested only in the visible, mainly due to experimental difficulties.

In this chapter, we propose a new waveguide refractometer based on second harmonic generation in the Cerenkov configuration from planar waveguides. As already stated, for a given optical waveguide, the effective refractive index of a guided mode is directly related to the cover refractive index, and thus the cover refractive index itself can be determined by measuring the n_{eff}. When dealing with second order nonlinear waveguides, the n_{eff} of a guided mode can be determined by measuring the output angle of the corresponding second harmonic (SH) radiation mode generated in the Cerenkov configuration.

It is worthwile recalling the basics of Cerenkov second harmonic generation[25] so as to better clarify the working principle of the new refractometer proposed. As shown in Figure 10a., when a guided mode at ω (fundamental) propagates within a second order nonlinear waveguide (planar in the example reported in the figure), a nonlinear polarisation wave arises at the waveguide-substrate interface and a Cerenkov radiation mode at 2ω (second harmonic) is generated. Of course, the phase matching condition $[\beta(2\omega) = 2\beta(\omega)]$ must be satisfied, but, in the configuration described, this is not a critical point since such condition is automatically fulfilled by means of a self-adjustment of the output angle of the Cerenkov mode provided that $2\beta(\omega) < k_{Cer} = k_0(2\omega)\, n_{sub}(2\omega)$, as illustrated in Figure 10b. In fact, if this is the case, the Cerenkov radiation mode will propagate in the substrate at an angle $\theta_{C\text{-}int}$ such that the projection of k_{Cer} along the propagation-plane of the waveguide equals $2\beta(\omega)$. Actually, $\beta(\omega) = k_0(\omega)\, n_{eff}(\omega)$, and since $k_0(2\omega) = 2k_0(\omega)$, the following equation is obtained

$$\theta_{C\text{-}int} = \arccos[n_{eff}(\omega)/n_{sub}(2\omega)]$$

The relationship between $\theta_{C\text{-}int}$ and the measured angle $\theta_{C\text{-}ext}$ is given by Snell's refraction law. Thus, $\theta_{C\text{-}ext}$ is a simple function of n_{eff}. This is the basis for the refractometer that we propose. In fact, when a liquid of unknown refractive index n_s is placed as the cover of the waveguide acting as the sensor, it alters the boundary conditions and thus the n_{eff} of the guided mode. The corresponding analytical relation can be obtained by the propagation theory in waveguides.[26] If the waveguide is properly designed for Cerenkov second harmonic generation, a Cerenkov radiation mode will be observed at an output angle $\theta_{C\text{-}ext}$ related to n_{eff} and thus to n_s. Therefore, if the optical parameters of the waveguide and of the substrate are known with the sufficient precision, by measuring $\theta_{C\text{-}ext}$, n_{eff} can be evaluated and finally n_s. Actually, since the setting of an absolute zero for the measurement of $\theta_{C\text{-}ext}$ is very delicate, it is more convenient to choose a reference cover (typically air) and then measure the variation of Cerenkov output angle $\Delta\theta_{C\text{-}ext}$, calculate the corresponding Δn_{eff}, then Δn_s, and thus the absolute value for n_s. The refractive index of the guiding film gives the upper limit for the n_s that can be measured.

To achieve a good sensitivity, the waveguide sensor must be properly designed. We have chosen Z-cut X-propagation proton-exchanged (PE) planar waveguides in lithium niobate ($LiNbO_3$). Indeed, these step-index waveguides exhibit a good efficiency for Cerenkov second harmonic generation, and they support only extraordinary modes. Due to the high refractive index of the guiding film, a wide range of n_s can be measured. By numerical simulations, it is possible to show that sensitivity is higher for thin waveguides with high Δn.

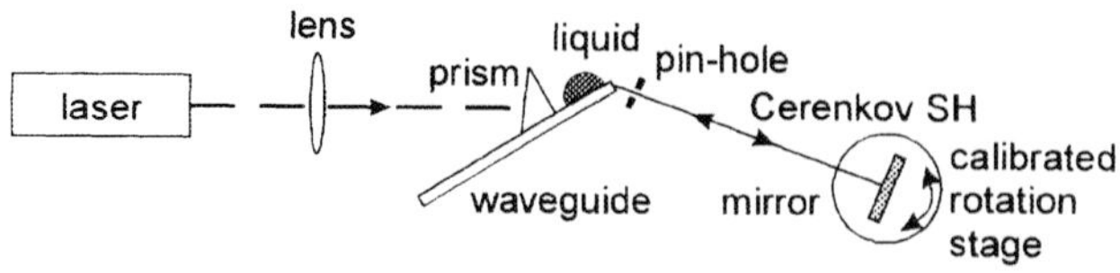

(a)

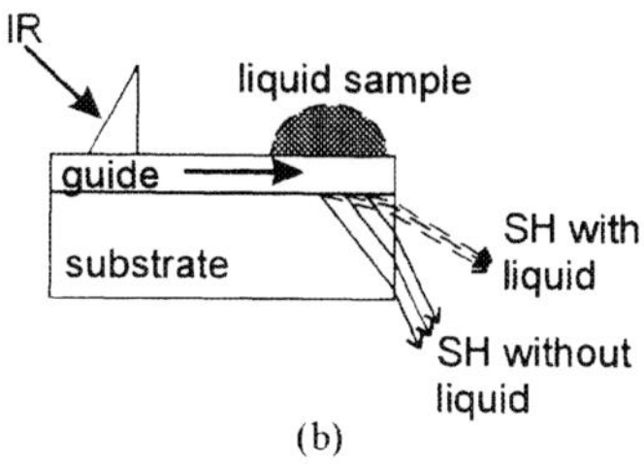

(b)

Figure 11. Experimental set-up for the Cerenkov refractometer (a). Detail of the waveguide with the liquid sample drop (b).

Figure 11. shows the experimental set-up (a) together with a detail of the sensor (b). The laser light is coupled into the waveguide by means of a rutile prism, the Cerenkov output angles with the reference-cover and with the liquid-sample are measured by finding the condition of perpendicularity respect to a mirror mounted on a high-precision rotation-stage. In the experiments we used as the laser source a Nd:YAG laser at 1.06 μm. Actually, one of the main advantages of the refractometer proposed is that it allows the determination of the sample refractive index in the near infrared by a measurement performed in the visible range. Moreover, the sensor described is quite simple: indeed, although it requires a nonlinear substrate, no complex waveguide fabrication techniques (e.g. grating realisation) are needed. The measurement is independent on the coupling technique and only a few drops of liquid are required.

The experiment reported here was performed by using a multimode waveguide as the sensor, so as to allow standard full optical characterisation of the waveguide itself. Actually higher sensitivity is expected for single-mode waveguides, due to their lower thickness, but in such case non-conventional characterisation techniques must be used.

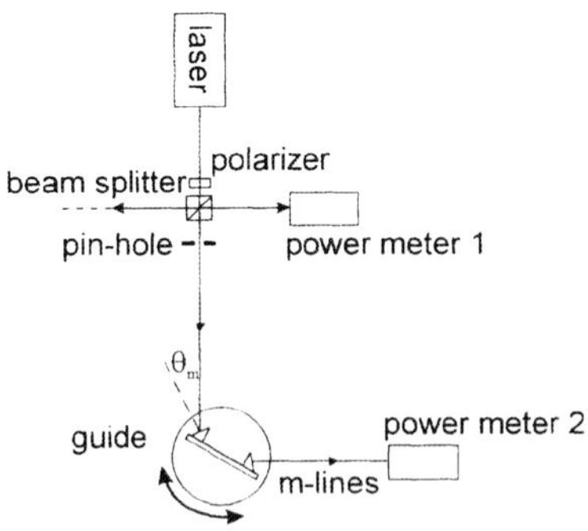

Figure 12. Waveguide characterisation set-up.

Table 1. Cerenkov refractometer: comparison between expected values and experimental results

n_s from datasheet	measured $\Delta\theta_{C\text{-ext}}$ (degrees)	n_s from the measurement
1.484	0.608	1.484
1.578	0.808	1.577
1.622	0.906	1.622
1.667	1.016	1.664

Waveguide optical depth and refractive index difference Δn were calculated by inverse methods from the n_{eff} of the different modes. Figure 12. shows the experimental set-up for conventional characterisation of planar waveguides. The laser light is coupled into the waveguide by means of a suitable prism and the guided modes are outcoupled by means of a symmetrical prism placed at the other end of the guide. The coupling angles of the different modes are measured by maximising the corresponding out-coupled m-lines and from these angles the effective refractive indices are calculated. The waveguide we used in the preliminary measurements reported here exhibits extraordinary index difference $\Delta n_e = 0.0914$ and optical depth $d_e = 1.434$ μm at the fundamental wavelength $\lambda = 1.064$ μm. The refractometer was tested on a set of liquids with known refractive index (R.P. Cargille Labs. Inc., USA). Table 1 reports the Cerenkov-angle variations obtained with the different liquids respect to air, the corresponding calculated n_s and the refractive index values from the datasheet. It is worth noting that the values reported in the datasheet at $\lambda = 1.064$ μm are extrapolated from those measured in the visible range. If keeping into account the precision of the values given in the datasheet ($\pm 1.5\times10^{-3}$) and that of our experimental values, which is of the same order, the data reported show a good agreement.

4. INVERSE USE OF CERENKOV REFRACTOMETRY

Waveguide characterisation is very delicate on the experimental point of view. Moreover, in the case of single-mode waveguides, standard optical characterisation as described above can not be performed. In the previous paragraph we have shown that, when a planar second-order nonlinear waveguide is fully characterised, it can be used as a sensor for liquid refractometry. Actually, an inverse approach is also possible. Indeed, when the waveguide optical parameters are not known, but a set of liquids with known refractive

Table 2. Experimental values of Cerenkov output angle variations corresponding to the different liquids.

i	Liquid refractive index from data-sheet (λ=1.064μm)	Cerenkov angle variation measurement $\Delta\theta_{C\text{-ext}}$ (degrees)
1	1.484	0.653
2	1.578	0.908
3	1.622	1.047
4	1.667	1.214
5	1.717	1.423
6	1.745	1.570

index is available, from the determination of $\Delta\theta_{C\text{-ext}}$ (liquid-cover versus air) the Δn_{eff} of the guided mode can be evaluated. Full characterisation of the waveguide requires the determination of both d (optical depth) and Δn (refractive index difference), that is of two parameters. Thus a single Δn_{eff} is not sufficient. But using different covers, more Δn_{eff} are determined and both d and Δn evaluated. In principle two covers are sufficient to determine d and Δn, but since variations of the two parameters may partially compensate each other, to achieve a good precision a higher number of liquids is used and data processed by means of a best fitting procedure. Table 2. reports the variations of Cerenkov output angle (in degrees) when the reference cover (air) is substituted with different liquids together with the refractive indices of the liquids as given in the datasheet. The measurements have been performed on a planar Z-cut X-propagation single-mode PE waveguide in $LiNbO_3$. Of course, each measured $\Delta\theta_{C\text{-ext}}$ can be obtained by different couples of d_e and Δn_e (in principle infinite) since variations in the two parameters can compensate. To obtain the couple of d_e and Δn_e that best describes the waveguide, a full set of d_e and Δn_e couples has been considered (chosen with a fixed step, sufficiently small) and the corresponding $\Delta\theta_{C\text{-ext}}$ has been calculated for all the liquids used. Then, best fit has been obtained by minimising the following parameter:

$$\delta = \Sigma_{i=1\text{-}6}[\Delta\theta_{C\text{-ext meas.}}(i) - \Delta\theta_{C\text{-ext calc.}}(i)]^2.$$

The resulting parameters are $d_e = 0.449$ μm and $\Delta n_e = 0.0894$ that give a $n_{eff} = 2.1572$ for the TM_0 guided mode. A full control of this result is not possible, since from standard prism characterisation the two optical parameters of the waveguide can not be obtained. However, the measured n_{eff} for TM_0 is exactly 2.1572, thus in perfect agreement with that calculated from the parameters chosen for the waveguide.

5. CONCLUSIONS

In this chapter we have illustrated a new method for the determination of the refractive index of liquid samples, based on Cerenkov second harmonic generation from suitable planar waveguides. The method allows the determination of the refractive index in the near infrared by detection in the visible. Although the sensor proposed requires a nonlinear material, it can be fabricated by standard, rather simple techniques.

The same method can be exploited in an inverse way to obtain a full optical characterisation of second-order nonlinear waveguides, even if single-mode.

We have reported preliminary experimental results that demonstrate the reliability of the method described. Suitable waveguides will be designed for the refractometer proposed, with particular attention to sensitivity optimisation. Moreover, the possibility of full integration with a semiconductor laser will be considered. Indeed, waveguides, both linear and nonlinear, are likely to play a major role in sensor applications, and in particular in refractometry, especially in the prospect of fully integrated devices.

ACKNOWLEDGMENTS. The authors wish to acknowledge the valuable scientific contributions of V. Russo and D. Conti (Integrated Optics Laboratory) of the INFM Department of Physics, Polytechnic of Milan and CEQSE-CNR. Nonlinear measurements were performed in the Solid-State Laser Laboratory of the same Department, thanks to the willingness of S. De Silvestri, V. Magni, G. Cerullo and M. Zavelani-Rossi. PE-$LiNbO_3$ waveguides were kindly provided by M. Varasi of Alenia, Rome.

The research work was partially supported by the Research Project "Nonlinear Optical Guiding Structures for All-Optical Signal Processing" of CNR-Technological Sciences and Innovation Committee.

REFERENCES

1. K. Kuhler, E. L. Dereniak, and M. Buchanan, Measurement of the index of refraction of the plastic Phenoxy PKFE, *Appl. Opt.* 30:1711 (1991)
2. E. Moreels, C. de Greef, and R. Finsy, Laser light refractometer, *Appl. Opt.* 23:3010 (1984)
3. B. W. Grange, W. H. Stevenson, and R. Viskanta, Refractive index of liquid solutions at low temperatures: an accurate measurement, *Appl. Opt.* 15:858 (1976)
4. S. Nemoto, Measurement of refractive index of liquid using laser beam displacement, *Appl. Opt.* 31:6690 (1992)
5. F. Docchio, S. Corini, M. Perini, and R. S. Kasana, A simple and reliable system for measuring the refractive index of liquids using a position-sensitive detector, *IEEE Trans. Instr. Meas.* 44:68 (1995)
6. B. Richerzhagen, Interferometer for measuring the absolute refractive index of liquid water as a function of temperature at 1.064 μm, *Appl. Opt.* 35:1650 (1996)
7. W. Lu and W. M. Worek, Two-wavelength interferometric technique for measuring the refractive index of salt-water solutions, *Appl. Opt.* 32:3992 (1993)
8. H. El Kashef, G. E. Hassan, and I. El-Ghazaly, Mach-Zehnder optical system as a sensitive measuring instrument, *Appl. Opt.* 33:3540 (1994)
9. F. A. Jenkins and H. E. White, *Fundamentals of Optics*, Mc Graw Hill Book Company, New York (1976).
10. J. M. St-Arnaud, J. Ge, J. Orbriot, T. K. Bose, and Ph. Marteau, An accurate method for refractive index measurements of liquids using two Michelson laser interferometers, *Rev. Sci. Instrum* 62:1411 (1991)
11. T. Li and X. Tan, Stepwise interferometric method of measuring the refractive index of liquid samples, *Appl. Opt.* 32:2274 (1993)
12. M. de Angelis, S. De Nicola, P. Ferraro, A. Finizioe, and G. Pierattini, A reflactive grating interferometer for measuring the refractive index of liquids, *Pure Appl. Opt.* 5:761 (1996)
13. T. H. Barnes, K. Matsumoto, T. Eiju, K. Matsuda, and N. Ooyama, Grating interferometer with extremely high stability suitable for measuring small refractive index changes, *Appl. Opt.* 30:745 (1991)
14. S. Sainov and N. Dushkina, Simple laser microrefractometer, *Appl. Opt.* 29:1406 (1990)
15. C. W. Chan and W. K. Lee, Measurement of a liquid refractive index by using high-order rainbows, *J. Opt. Soc. Am. B* 13:532 (1996)
16. S. Y. El-Zaiat and H. A. El-Hennawi, Applying multiple-beam Fizeau fringes for measuring the refractive indices of liquids, *Meas. Sci. Technol.* 7:1119 (1996)
17. M. S. Meyer and G. L. Eesley, Optical fiber refractometer, *Rev. Sci. Instrum.* 58:2047 (1987)
18. A. Suhadolnik, A. Babnik, and J. Mozina, Optical fiber reflection refractometer, *Sens. Act. B* 29:428 (1995)
19. M. Archenault, H. Gagnaire, J. P. Goure, and N. Jaffrezic-Renault, *Sens. Act. B* 5:173 (1991)
20. T. Takeo and H. Hattori, Silica glass fiber photorefractometer, *Appl. Opt.* 31:44 (1992)
21. C. Ronot-Trioli, A. Trouillet, C. Veillas, and H. Gagnaire, Monochromatic excitation of surface plasmon resonance in an optical fiber refractive index sensor, *Sens. Act. A* 54:589 (1996)
22. Y. Liu, P. Hering, and M. O. Scully, An integrated optical sensor for measuring glucose concentration, *Appl. Phys. B* 54:18 (1992)
23. M. N. Weiss, R. Srivastava, H. Groger, P. Lo, and Shu-Fang Luo, A theoretical investigation of environmental monitoring using surface plasmon resonance waveguide sensors, *Sens. Act. A* 51:211 (1996)
24. D. Clerc and W. Lukosz, Integrated optical output grating coupler as refractometer and (bio-)chemical sensor, *Sens. Act. B* 11:461 (1993)
25. N. Hashizume, T. Kondo, T. Onda, N. Ogasawara, S. Umegaki, and R. Ito, Theoretical analysis of Cerenkov-type optical second-harmonic generation in slab waveguides, *IEEE J. Quantum Electron.* 28:1798 (1992)
26. D. Marcuse, *Theory of dielectrical optical waveguides*, Academic Press, New York and London (1974)

CHARACTERIZATION OF AN OPTICAL FIBRE pH SENSOR WITH METHYL RED AS OPTICAL INDICATOR

F. Baldini, and A. Falai

Istituto di Ricerca sulle Onde Elettromagnetiche "Nello Carrara" - CNR
Via Panciatichi 64
50127, Firenze, Italy

1. INTRODUCTION

The chemical parameter most investigated with optical fibres is indubitably pH.[1] Since 1980, when the first optical fibre pH sensor was proposed,[2] many optical fibre pH sensors have been described, based in both fluorescence and absorption.[3,4] These have many advantages over the traditional pH electrodes, such as immunity from the electromagnetic field, the absence of electric contacts, and high miniaturization. These features make them very promising for several applications, especially in the biomedical field.

On the other hand, the optical approach for pH detection is characterized by certain disadvantages, such as limited working ranges and interferences coming from ionic strength. In the determination of blood acidity, for which pH sensors are already available on the market, the required pH range which has to be measured is very narrow (pH range = 7.0-7.5) and the changes in ionic strength are limited. In other fields, these two aspects are not to be underestimated. Generally pH sensors are capable of covering 2-3 pH units and this fact limits their range of application. In order to cover a broader range of pH, a combination of more than one chromophore could be used.[5,6,7] However the manufacture of the proposed probe appears too complicated to allow its possible industrialization.

The present paper is concerned with the description of an optical pH sensor which makes use of only one indicator, methyl red, covalently immobilized on controlled pore glass (CPG) and capable of covering a broad pH range.

2. OPTICAL FIBRE PROBE

Methyl red was immobilized on controlled pore glass (CPG) according to an immobilization procedure followed for other acid-base indicators and previously described.[8] The surface of the controlled pore glass (CPG) was silylated by means of a proper silane: γ-aminopropyltriethoxysilane; subsequent chemical reactions made it possible to bind methyl red

Optical Sensors and Microsystems: New Concepts, Materials, Technologies
Edited by Martellucci et al., Kluwer Academic / Plenum Publishers, New York, 2000.

covalently to the support. The treated CPG was immobilized by means of a melting process which made it possible to embed the glass support in plastic material.[9] This approach has been described previously, the only difference being the fact that the treated CPG was immobilized on a plastic reflector, and not directly on the plastic fibres. This reflector was placed, by means of a stainless steel wire, in front of two optical fibres used for the connection with the optical source and the detecting system, respectively. Fig. 1 shows a sketch of the optical fibre probe. An interesting feature of this probe is that the sensing part of the optode can easily be changed without changing anything else.

3. SPECTROPHOTOMETRIC ANALYSIS : PHOTODEGRADATION

The optical fibre probe was connected to an optical fibre spectrophotometer (Guided Wave, model 260), and the probe was dipped in appropriate buffers. The ionic strength of the buffers was kept constant and equal to 1.0 M, while the pH could range from 2 to 10.

Figure 2(a). shows absorption spectra for the different pH values, while the relationship between the absorbance A (evaluated at the absorption peak of the undissociated form of the dye $\Rightarrow \lambda$=556 nm) and pH is shown in Figure 2(b). As can be observed, the working range was quite broad, and a linear relationship could be considered from pH=3 up to pH=8.

Although the chromophore showed good sensitivity in this range, a slow but constant decrease in the absorbance values was observed when the probe was dipped in buffers. This behaviour could be ascribed either to a leakage of methyl red from the glass support, as a consequence of the break in a chemical bond of the chemical chain which bound the dye to the glass surface, or to the photodegradation of the dye. Since this effect was also observed at neutral pH values, and neutral buffers could not be considered to be chemically aggressive, the main reason for the decrease in the absorbance values was assumed to be the dye photodegradation.

This was confirmed experimentally by exposing optical probes to different lighting conditions for eighteen hours. During this period, the absorbance step (ΔA) was measured many times in correspondence with the same pH step (pH = 7.06 $\rightarrow$ pH = 4.16). Figure 3. shows the experimental results: curve (a) is related to a lighting limited to the measurement time ($\approx$5 minutes), while curve (b) is related to continuous lighting.

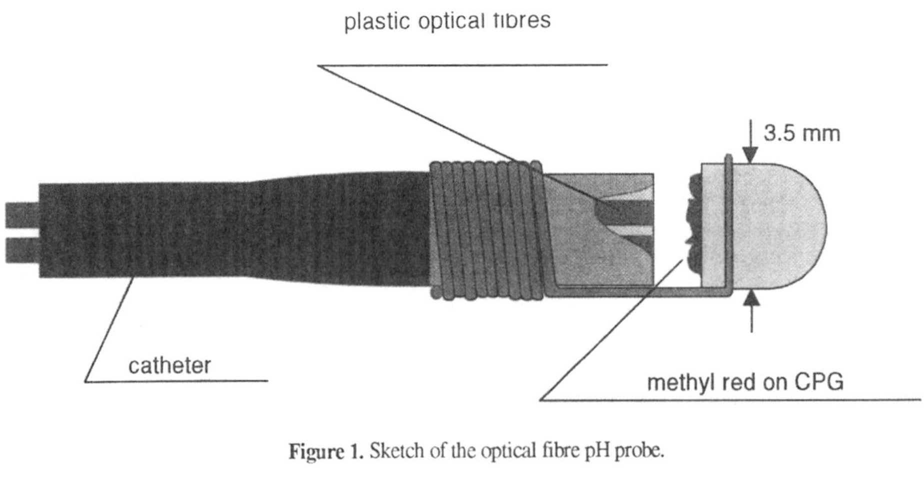

Figure 1. Sketch of the optical fibre pH probe.

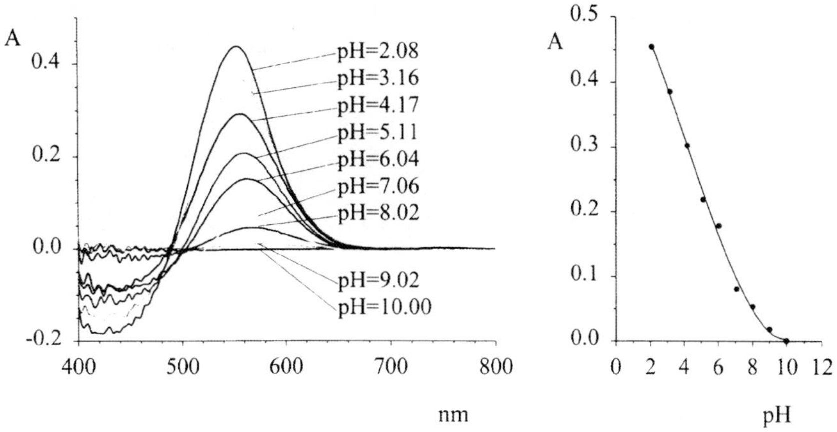

Figure 2. Absorption spectra of methyl red on CPG for different pH values (a) and absorbance (at λ=556 nm) vs pH (b).

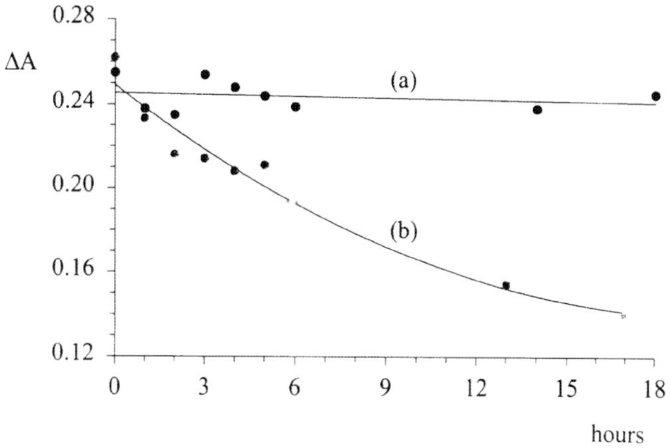

Figure 3. Absorbance step ΔA for the same pH step (pH=7.06 → pH=4.16) in the case of lighting limited to the measurement time (curve a) and in the case of continuous lighting (curve b).

The decrease in absorbance values was noticeably reduced when the light exposure time was limited, and these results confirmed that the decrease in absorbance values was due to the photodegradation of the dye, and not to its leakage from the support.

4. OPTICAL FIBRE SENSOR

The use of pulsed light-emitting diodes (LED) made it possible to noticeably decrease the time exposure of methyl red to the light. An optoelectronic unit was developed which makes use of LEDs as sources, as well as of a suitable detection system. Two LEDs were used with their emission peaks centred at 555 nm and 660 nm, for the signal and the reference, respectively.

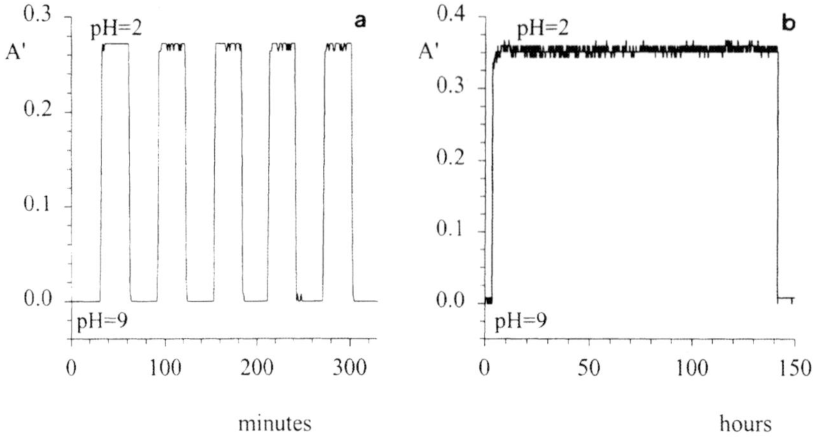

Figure 4. Stability tests of the optical fibre pH sensor: A' vs time for the same cycle between pH 9 and 2 (a) and 6-day measurement with the optical fibre costantly dipped in a buffer solution at pH 2 (b).

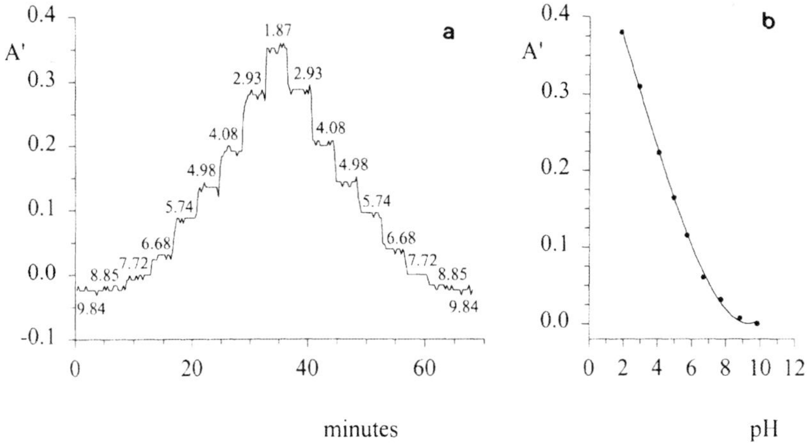

Figure 5. Response curve of the optical fibre pH sensor to different pH steps (a) and A' vs pH (b).

An electrical board equipped with a microprocessor and an internal buffer for data storage constituted the detecting system. All the stored data were transferred to a personal computer for further processing. What is calculated by the optoelectronic unit, is not the absorbance, defined at a specific wavelength, but the quantity A', defined as:

$$A' = \log\left[\frac{\int I_0(\lambda)d\lambda}{\int I_0(\lambda)d\lambda}\right]_{\lambda_{em}=555nm} - \log\left[\frac{\int I_0(\lambda)d\lambda}{\int I_0(\lambda)d\lambda}\right]_{\lambda_{em}=660nm} \qquad (1)$$

where I_o is the detected light when the probe is dipped in the buffer characterized by the lowest absorption (pH ~10), while I is the detected light with the probe dipped in a generic buffer.

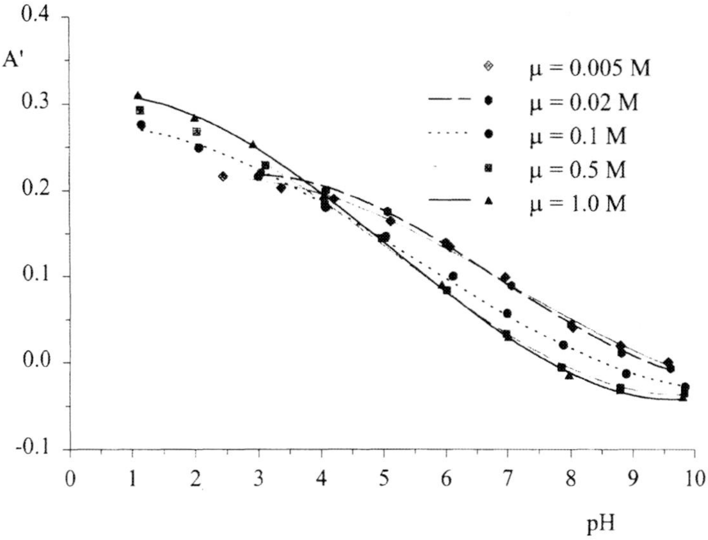

Figure 6. A' vs pH for different ionic strengths.

The integration is automatically performed by the optical detector, a preamplified photodetector, which does not carry out any discrimination of the wavelengths emitted by the LEDs.

Stability tests were performed to check the photostability of the probe. As a first step, some cycles between pH 9 and 2 were repeated (Figure 4(a).): the A' value at pH=2 was constant after five cycles. Long-term measurements were performed for several days with other probes. A typical result is shown in Figure 4(b).: the optical probe was dipped for approximately 6 days in a buffer solution at pH=2. No drift was observed, indicating that the problem of photodegradation had disappeared.

The response curve for different pH steps (between 2 and 10 pH units), obtained with the optoelectronic unit, is showed in Figure 5(a)., while Figure 5(b). shows the related A' vs pH curve.

5. IONIC STRENGTH EFFECTS

In any accurate characterization of an optical pH sensor, the effect of the ionic strength should be considered since it can heavily affect the measurement of pH.

Sensitivity, response time, working range and accuracy were investigated with buffer solutions at different ionic strengths in the 0.005÷1.0 M range. Figure 6. shows the relationship between A' and pH, for every ionic strength considered. As can be seen, both the whole range and the linear working range depends on the ionic content of the solution. A simultaneous broadening and shift towards lower pH values of the working range is observed with an increase in the ionic strength. The performed measurements also show a great dependence of sensitivity and response times on the ionic strength. Table 1. shows the linear range and the correlation coefficients for the five different ionic strengths. Sensitivity ≤0.06 pH units was obtained in the linear range of pH in any case. The linear range is quite broad if compared with the other indicators described in the literature. If a suitable algorithm is chosen, the range between 1 and 10 pH units can be covered with the same dye for an ionic strength ≥0.1 M. Better sensitivity can reasonably be obtained with an improvement in the electronic board.

Table 1. Linear ranges and related correlation coefficients for the five different ionic strengths.

ionic strength	0.005M	0.02M	0.1M	0.5M	1.0M
linear pH range	4.8 ÷ 9	5.1 ÷ 8.8	3.0 ÷ 7.9	3.1 ÷ 7.8	2.9 ÷ 8.0
r^2	0.990	0.999	0.998	0.999	0.999

Table 2. Accuracy of the pH sensor considering different ionic strength ranges.

	μ=0.005÷1.0M	μ=0.005÷0.02M	μ=0.1÷1.0M
pH range	ΔpH	ΔpH	ΔpH
2.0 ÷ 9.5	0.9	0.7	0.7
4.0 ÷ 9.5	0.9	-	0.5
4.0 ÷ 7.0	0.6	0.1	0.3

Table 3. Response time τ_{90} for different pH steps for the ionic strengths 0.005 M and 0.02 M.

μ=0.005M				μ=0.02M			
higher ⇒ lower		lower ⇒ higher		higher ⇒ lower		lower ⇒ higher	
pH step	τ_{90}	pH step	τ_{90}	pH step	τ_{90}	pH step	τ_{90}
8.80 ⇒ 8.03	712	8.03 ⇒ 8.79	464	8.81 ⇒ 8.00	496	7.97 ⇒ 8.79	208
8.03 ⇒ 6.96	976	6.96 ⇒ 8.03	528	8.00 ⇒ 7.06	368	7.04 ⇒ 7.97	128
6.96 ⇒ 6.06	528	6.02 ⇒ 6.96	208	7.06 ⇒ 5.99	208	6.00 ⇒ 7.04	64
6.06 ⇒ 5.11	560	5.11 ⇒ 6.02	104	5.99 ⇒ 5.06	112	5.07 ⇒ 6.00	80
5.11 ⇒ 4.20	352	4.20 ⇒ 5.11	30	5.06 ⇒ 4.07	80	4.07 ⇒ 5.07	48
4.20 ⇒ 3.36	18	3.35 ⇒ 4.20	32	4.07 ⇒ 3.04	32	3.03 ⇒ 4.07	32
3.36 ⇒ 2.44	16	2.44 ⇒ 3.35	20	3.04 ⇒ 2.10	32	2.09 ⇒ 3.03	48

Table 4. Response time τ_{90} for different pH steps for the ionic strengths 0.1 M and 0.5 M.

μ=0.1M				μ=0.5M			
higher ⇒ lower		lower ⇒ higher		higher ⇒ lower		lower ⇒ higher	
pH step	τ_{90}	pH step	τ_{90}	pH step	τ_{90}	pH step	τ_{90}
8.89 ⇒ 7.88	144	7.84 ⇒ 8.84	72	8.80 ⇒ 7.85	96	7.88 ⇒ 8.82	128
7.88 ⇒ 6.99	148	6.95 ⇒ 7.84	96	7.85 ⇒ 6.98	96	7.00 ⇒ 7.88	152
6.99 ⇒ 6.11	138	6.11 ⇒ 6.95	90	6.98 ⇒ 6.01	80	6.03 ⇒ 7.00	96
6.11 ⇒ 5.03	39	5.02 ⇒ 6.11	88	6.01 ⇒ 4.96	92	4.96 ⇒ 6.03	96
5.03 ⇒ 4.07	45	4.06 ⇒ 5.02	64	4.96 ⇒ 4.05	48	4.07 ⇒ 4.96	72
4.07 ⇒ 3.00	24	2.98 ⇒ 4.06	16	4.05 ⇒ 3.11	32	3.10 ⇒ 4.07	32
3.00 ⇒ 2.06	16	2.05 ⇒ 2.98	16	3.11 ⇒ 2.02	32	2.02 ⇒ 3.10	32
2.06⇒ 1.15	16	1.15 ⇒ 2.05	16	2.02 ⇒ 1.13	32	1.13 ⇒ 2.02	24

It is possible to evaluate the accuracy of the sensor when the influence of the ionic strength is not considered on the basis of calibration curves showing A' vs pH. The half of the pH change (ΔpH), which is observed with different ionic strengths keeping constant the absorbance value, gives an idea of this accuracy. Such values are reported in Table 2 for different pH ranges and with different ionic strength ranges. As can be observed from the last two columns, where ΔpH is reported for narrower ionic strength ranges, a knowledge, even a rough one, of the ionic strength of the samples whose pH has to be measured, could make it possible to obtain better accuracy.

Table 5. Response time τ_{90} for different pH steps for the ionic strength 1.0 M.

μ=1.0M			
higher ⇒ lower		lower ⇒ higher	
pH step	τ_{90}	pH step	τ_{90}
8.80 ⇒ 7.97	170	7.99 ⇒ 8.82	144
7.97 ⇒ 7.01	125	7.03 ⇒ 7.99	136
7.01 ⇒ 5.93	96	5.95 ⇒ 7.03	80
5.93 ⇒ 5.00	48	5.01 ⇒ 5.95	64
5.00 ⇒ 4.03	48	4.04 ⇒ 5.01	72
4.03 ⇒ 2.91	32	2.91 ⇒ 4.04	80
2.91 ⇒ 1.99	24	1.98 ⇒ 2.91	30
1.99 ⇒ 1.10	30	1.10 ⇒ 1.98	16

Steps of approximately 1 pH unit were considered in order to evaluate the effect of ionic strength on response times. The measured values of response time t_{90} in correspondence with each step are reported in Tables 3.-5. t_{90} is defined as 90% of the time required to reach the steady state. On the basis of the experimental results, the following comments can be made: t_{90} is less than 2.5 minutes for the higher ionic strengths (0.1-1.0 M), but it increases as the ionic strength decreases; for low ionic strengths (0.005 M and 0.02 M) and for pH>5 an increase in pH is characterized by a response time lower (two or three times) than that for a decrease in pH, while the response times in both directions are comparable for the higher ionic strengths (0.1 M÷1.0 M); response times are lower in acid solutions (pH<5) than in basic solution; the difference becomes extremely evident for the lowest ionic strength where, for example, the response time is of the order of 20 seconds for the steps 2.44⇒3.35 and viceversa and is about 12 and 8 minutes for the 8.80⇒8.03 and 8.03⇒8.79 steps, respectively.

The fact that a longer response time is observable with low ionic strengths is easily explained with the slower diffusion of hydrogen ions in samples characterized by a low concentration of ionic species.

Also, the second aspect can be explained by ascribing this behaviour to the presence of a surface charge distribution on the treated CPG. This charge distribution does not affect the response time if the ions dissolved in solution are the main source of the local electric field. On the contrary, in low ionic content solutions, the surface charge distribution acquires a dominant role in the formation of a local electric field, and is capable of lengthening the diffusion of hydrogen ions towards the optode (decrease in pH) and to make the H^+ diffusion faster away from the optode (increase in pH).

What at the moment is without explanation is the fast response at very low ionic strength and pH<5. This is very characteristic of this dye, and to our knowledge so far, these fast responses at this low ionic strength have never been reported either with optical sensor or glass electrodes.

6. CONCLUSIONS

The capability of methyl red, covalently immobilized on CPG, to cover a very broad range makes such an indicator suitable for environmental or biomedical applications in which an extended range is required. In particular, the developed optical fiber pH sensor appears suitable for gastric pH detection, for which a broad working range (1÷8 pH units) is required. The main advantages of this sensor are:

- the absence of leakage of the sensitive material from the glass support;
- the immobilization procedure of treated CPG on the plastic reflector, which is simple and greatly facilitates the construction of the probe;
- the removal of the plastic reflector, which is done very easily in order to have a new probe;
- the fast response of this probe, also at very low ionic strength for pH<5;
- the optoelectronic unit which is fed by batteries and is easy to transport.

Ionic strength affects the performance of the probe, making worse its accuracy. Therefore the behaviour of the probe is satisfactory only if ionic strength is constant as it occurs in all optical fibre pH sensors. However, since methyl red is characterized by a lower dependance on ionic strength, if compared with other acid-base indicators, also a rough knowledge of the ionic strength (the order of magnitude) can make the probe performances satisfactory.

ACKNOWLEDGMENTS. This work has been partially supported by the EC Measurement and Testing - OFOSEAM project (contract MAT1-CT93-0021).

REFERENCES

1. M.J.P. Leiner and O.S. Wolfbeis, Fiber optic pH sensors, in Fiber Optic Chemical Sensors and Biosensors, O.S. Wolfbeis, ed., CRC Press, Boca Raton, Florida, USA (1991).
2. J.I. Peterson, S.R. Goldstein, R.V. Fitzgerald and D.K. Buckhold, Fiber optic pH for physiological use, *Anal. Chem.* 52: 864 (1980).
3. M.J.P. Leiner and P. Hartmann, Theory and practice in optical pH sensing, *Sens. and Actuat.B*, 11: 281 (1993).
4. F. Baldini, Optical determination of pH with optical fibres, *Trends in Appl. Spectr.*, in press.
5. G. Boisde, F. Blanc, X. Machuron-Mandard, pH measurements with dyes co-immobilization on optodes: principles and associated instrumentation, *Intern. J. Optoelectr.*, 6: 407 (1991).
6. F. Baldini, S. Bracci and F. Cosi, An extended-range fibre-optic pH sensor, *Sens. and Actuat. A*, 37-38: 180 (1993).
7. E.J. Netto, J.I. Peterson, M. McShane and V. Hampshire, A fibre-optic broad-range pH sensor system for gastric measurements, *Sens. and Actuat. B*, 29: 157 (1995).
8. F. Baldini, M. Bacci and S. Bracci, Spectroscopic behaviour of acid-base indicators after immobilization on glass supports, *Appl. Spectr.*, 45: 26 (1991).
9. F.Baldini, S.Bracci, F.Cosi, P.Bechi and F.Pucciani, Controlled-pore glasses embedded in plastic optical fibers for gastric pH sensing purposes, *Appl. Spectr.*, 48: 549 (1994).

OPTICAL SENSORS AND MICROSYSTEMS USING LIQUID CRYSTALS

L. Sirleto, G. Abbate*, G.C. Righini, and E. Santamato*

IROE-CNR "Nello Carrara"
Optoelectronics and Photonics Department
Via Panciatichi 64
50127 Firenze, Italy

*INFM and Università degli Sudi di Napoli, Dipartimento di Scienze Fisiche
Via Cintia Monte S.Angelo
80126 Napoli, Italy

1. INTRODUCTION

A variety of physical phenomena makes liquid crystals (LC's) one of the most interesting subjects of modern fundamental science. At the same time, their unique properties, which give rise to enhanced optical anisotropy and sensitivity to external fields, can be exploited in a large number of practical applications. Even if their main field of application is concerned with displays, there are a number of other application areas where their peculiar properties may play a relevant role as well; thus, for instance, optical sensors and microsystems using LC's are gaining increasing attention.

In this paper we have tried to describe the state of the art of sensors and microsystems based on liquid crystals. Reference has been made to all the principal mesophases of thermotropic liquid crystals and also to the preliminary results obtained by using a recently developed composite material, namely polymer dispersed liquid crystals (PDLC). In order to make clear the working principle of the devices which will be mentioned in the following, a brief description of LC's principal mesophases and of their optical and electro-optical properties is presented in the next paragraph. Optical sensors and microsystems using LC's are described in the following sections, while the guided-wave optical microsystems which have been the subject of our own research are presented in the last section. In our description, we have chosen to privilege physical and intuitive aspects; however, the reader interested in entering into more detailed and more formal descriptions of these phenomena is referred to an useful list of references, which includes "classical" books, recent reviews and all the articles (at least at our knowledge) bounded to the described devices.

Optical Sensors and Microsystems: New Concepts, Materials, Technologies
Edited by Martellucci et al., Kluwer Academic / Plenum Publishers, New York, 2000

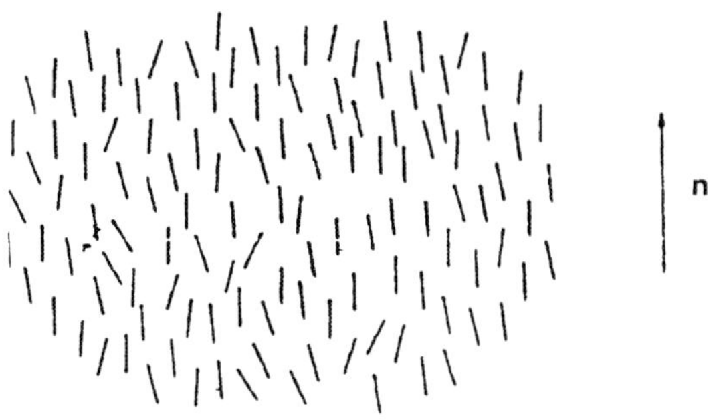

Figure 1. Schematic representation of molecular arrangement in the nematic phase.

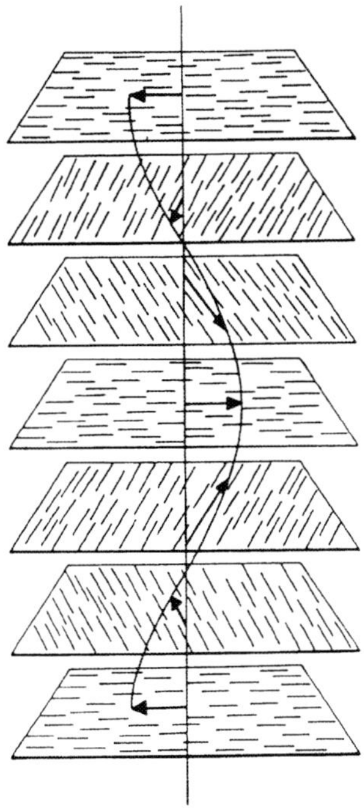

Figure 2. Schematic representation of molecular arrangement in the cholesteric phase.

2. LIQUID CRYSTALS: OPTICAL AND ELECTRO-OPTICAL PROPERTIES

Liquid crystals are composed of highly anisotropic organic molecules and possess a number of structurally different mesomorphic liquid phases in a readily accessible temperature range. In these mesophases, the macroscopic properties are intermediate between the solid state and the liquid state. The liquid crystals are fluid and still maintain some degree of molecular order.

The melting of normal compounds involves in a single process the transition between the solid state and the liquid one. In thermotropic LC's, on the contrary, as the temperature increases, the compound goes through a series of phase transitions: from solid to various liquid crystal phases, to isotropic liquid and finally to vapor phase. Therefore all the physical properties, such as refractive indices, dielectric constants, elastic constants and viscosity, depend on the temperature. Thermotropic LC's may exist in three typical distinct mesophases: nematic, cholesteric and smectic.[1,2] In the nematic phase (see Figure1.) the molecules have their centers of mass randomly distributed in the fluid but they are directionally correlated to point, on the average, along a common direction. Such a direction is described by the so-called molecular director n, which coincides also with the optical axis of the material. Nematics are uniaxial, non polar and centrosymmetric.

The nematic phase is substituted by the cholesteric phase in compounds having non centrosymmetric molecules. The cholesteric phase (see Figure 2.) is similar to the nematic one on the microscopic scale, but over distances of the order of one micron the average direction of the molecules tends to follow a characteristic helical structure, having its pitch strongly dependent on the temperature. Cholesterics are uniaxial, non polar, non centrosymmetric and show optical activity.

In the smectic phases (there are actually eight known smectic phases), the molecule's centers of mass are ordered along parallel layers (the smectic planes), and still the molecules tend to be aligned in each layer. The smectic phase is more ordered than the nematic one, and hence it is stable at lower temperature. In the smectic C (Sm-C) phase, the molecules form a layered structure with the average orientation of long axes, denoted by the smectic director n, placed at an angle θ with respect to the layer normal. The director n exhibits a continuous degeneracy in its azimuthal orientation, lying on a cone coaxial with layer normal. This tilt angle varies with temperature; it often increases with decreasing of temperature. In particular chiral smectic C (Sm-C*) is an interesting class of liquid crystal materials. The structure of Sm-C* (where the star means chirality), similar to the one of smectic C except for a helical tilt distribution from layer to layer (see Figure 3.), will be ferroelectric with a macroscopic electric dipole P locally normal to n lying in the plane of the layers. In the isotropic phase, liquid crystals behave just as other organic liquids, although they exhibit the so-called petransitional phenomena[3] in the vicinity of the phase transition temperature Tc.

Liquid crystals flow as fluids in their mesogenic phase. Thus, they have to be confined within substrates. Generally, these substrates are treated for proper surface alignment[4] to form a crystal monodomain. Four basic alignments of LC's have been developed and widely used for applications: parallel, perpendicular, hybrid, and twist alignment. In the parallel alignment, the director in the front and back substrates is parallel to the interface. Perpendicular alignment is also known as homeotropic alignment. In such case, the liquid crystal director is perpendicular to the substrate interfaces. Hybrid alignment is realized with the director parallel to one interface and perpendicular to the other. In the twist alignment, both substrates are treated similarly to those for parallel alignment except that the back substrate is twisted at an angle. Cells with twist angles greater than 90° produce the so-called super twisted nematics. In most electro-optical devices, the substrates have to be coated with an electrically conductive but optically transparent film, such as indium-tin-oxide (ITO), in order to apply an electric field.

Molecular structure plays an important role in determining the optical and electro-optical properties of liquid crystals.[5] Three types of electronic transitions are often encountered in a LC compound, namely $\sigma\rightarrow\sigma^*$ (excited states of σ-electron), $n\rightarrow\pi^*$ (excited states of π-electron), and $\pi\rightarrow\pi^*$ transitions. The $\sigma\rightarrow\sigma^*$ transitions (the λ_0-band) take place in the vacuum ultraviolet region (its central wavelength is somewhere around 120-150 nm). This band is an allowed transition and since there are many σ-electrons available in a liquid crystal molecule, its transition intensity should be very strong.

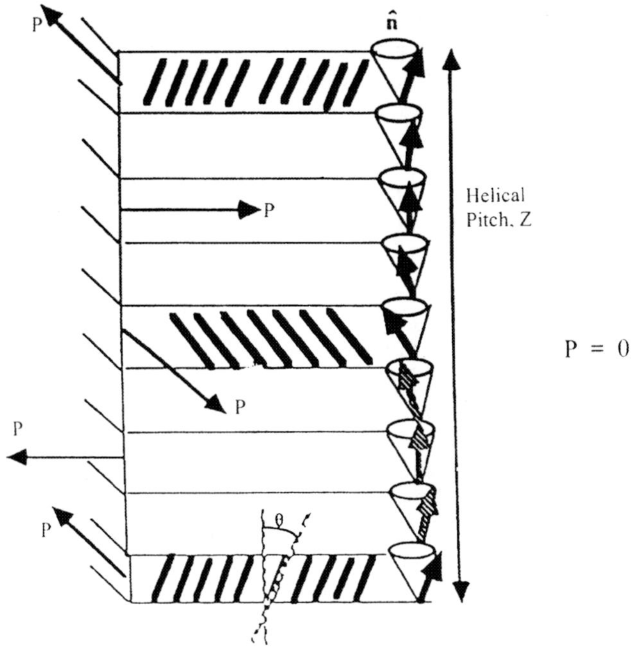

Figure 3. Schematic representation of molecular alignment in the smectic chiral (Sm-C*) phase.

The absorption of a totally saturated LC compound or mixture (only σ electrons are present) becomes negligible in the visible spectra region. As to $n \rightarrow \pi^*$ transition, the oscillator strength of such transition is not expected to be very strong, thus its contribution to the refractive index is not significant. In a conjugated liquid crystal, regarding the $\pi \rightarrow \pi^*$ transitions, it is possible to consider only the two major transition (the λ_1 and λ_2 band, with $\lambda_2 > \lambda_1$). They take place in the vacuum ultraviolet region and in particular cover the spectral range from 180 to 400 nm, depending on the conjugation length of the LC molecule. The λ_1-band of several conjugated LC molecules occurs at about 200 nm. As to the λ_2-band, its wavelength, the transition intensity and the absorption anisotropy all increase with increasing conjugation.

In the liquid crystals mesophases, at wavelengths far off resonances, e.g. in the visible spectral region, both scattering and absorption are decreasing: the scattering decreases asλ^{-4} and the absorption declines even more drastically. Consequently, scattering tends to surpass the absorption and dominates the transmission losses. On the contrary, in the isotropic state, where the scattering of liquid crystal is greatly reduced (by a factor of about one million) as compared to the nematic state, light scattering losses may be lower than the absorption.

In the IR region the absorption is generally low, even if some molecular vibrational bands appear, resulting in an increased absorption at the molecular vibrational resonances. Since IR absorption of LC's depends on the individual molecular vibrational transition frequencies, different chemical bonds exhibit different absorption bands. As to the light scattering in IR region, it is estimated to be about 3 orders of magnitude lower than that in the visible region, and its contribution can be neglected.

The refractive indices of an uniaxial LC are primarily governed by its constituents, wavelength and temperature. Wu,[6] taking into account all possible electronic transitions (λ_0, λ_1 and λ_2), has proposed an empirical model, which covers almost all LC's in the entire visible and IR regions and in the entire mesogenic phase. In the visible region, according to the three-band model, ordinary index n_o decreases as the wavelength increases and slightly increases as the temperature increases. On the other hand, extraordinary index n_e decreases gradually as the temperature or the wavelength increase. Close to the nematic-isotropic phase transition, the refractive indices changes are more pronounced, until they collapse to the value of the isotropic phase. So, in order to reduce index fluctuations at room temperature, a large mesogenic range, with a high clearing temperature (transition temperature between the nematic and isotropic phase) is highly desirable. In the IR region, due to the weak oscillator strength of vibrational transitions, the refractive indices of a LC can be considered basically independent of wavelength except at the very vicinity of molecular vibrational bands.

The useful range of LC's is quite remarkable; it spans over the UV, the visible, and the IR. Of course, in each region an appropriate LC has to be chosen. Apart the already discussed localized resonances, in general there is a broad spectral region where the liquid crystal is transparent and electro-optical effects can be usefully exploited.

As already said, the anisotropy of the macroscopic properties, due to the collective molecular alignment, is the most important LC's property. The degree of molecular alignment as well as the molecular average direction can be changed by very low external perturbations such as mechanical or thermal stresses, magnetic, electric or optical fields, giving rise to a variety of thermo-optic, electro-optic, magneto-optic and opto-optic modulation effects. The high dielectric anisotropy, the low voltage operation and low power dissipation are the main advantages of liquid crystal displays, as compared to other display technologies. Once the external field vanishes, the director may relax to the initial state or stay in its final state, depending on the type of liquid crystal and the physical mechanism involved (monostable or bistable behavior).

Several electro-optical effects in liquid crystals have been observed. These include: (i) dynamic scattering, (ii) guest-host effect, (iii) field-induced nematic-cholesteric phase change, (iv) field-induced director axis reorientation, (v) laser-addressed thermal effect, and (vi) light scattering by micro-droplets. Field induced director-axis reorientation on aligned nematic and ferroelectric liquid crystals is one of the most important electro-optic effects exploited for modulating light.

In nematics, the electric field-induced polarization of a LC molecule follows the field up to a few MHz of frequency before the dielectric relaxation takes place. The dynamic response of the LC director is obtained balancing the elastic torque (due to the elastic restoring force), the electric field-induced torques (due to electric field action) and the viscous torque. The LC director responds to the squared amplitude of the external field. Thus, electric field-induced molecular reorientation of a nematic LC does not depend on the polarity of the field. The final states of the nematic director axis can be controlled continuously by the applied field and once the external field is completely removed, the LC director relaxes back to its initial state through the elastic restoring force. The decay time of a nematic liquid crystal (NLC) device is mainly controlled by the viscosity and depends on the specific operation conditions and material employed. The typical response time of a NLC device is in the range of 1-100 ms. Many material parameters such as refractive index, dielectric constant, elastic constant, order parameter and rotational viscosity may affect the electro-optical behavior of LC devices. For instance, dielectric anisotropy and elastic constant together determine the threshold voltage. Phase or amplitude modulation of an incident light beam onto a nematic LC device has been obtained by means of a number of different physical mechanisms.

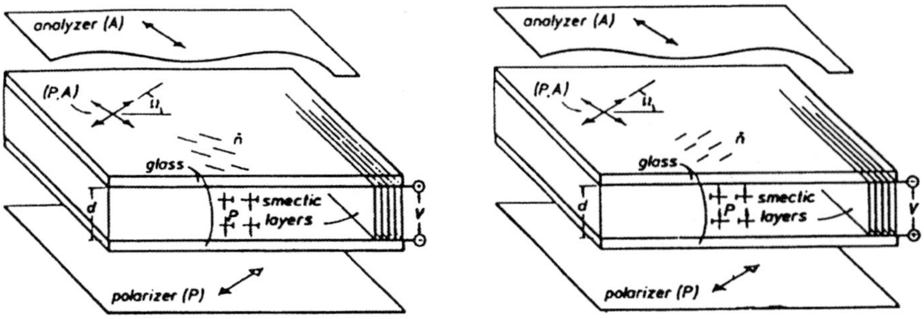

Figure 4. Surface stabilization: principle of operation of a surface-stabilized ferroelectric liquid crystal (SSFLC) device.

In the case of ferroelectric liquid crystal (FLC's) devices, two operation modes are both of fundamental and practical interest: 1) surface-stabilized FLC (SSFLC) in the smectic-C* phase,[7] and 2) soft-mode FLC (SMFLC) in the smectic-A* phase.[8]

As already said, the Sm-C* structure is ferroelectric. In principle this gives, as a result of the P•E torque, a strong linear coupling of n to the applied electric field E. However, this linear coupling is eliminated on a macroscopic scale by an additional consequence of the molecular chirality: in fact, n forms a spiral about the layer normal axis yielding an effectively antiferroelectric helical structure. It is possible to use surface interactions to suppress the antiferroelectric helix in a particular geometry.[7] The Sm-C* is confined between flat planes which are treated so that the director at a surface is constrained to lie in the plane of the surface but with no strong tendency for a particular orientation in the surface plane. Since the energy required to unwind the helix decreases as the sample thickness d, this boundary condition will suppress the helix for sufficiently thin samples ($d < p$, the helix pitch). In the absence of the helix there is the spontaneous formation of either of two surface-stabilized Sm-C* monodomains of opposite ferroelectric polarization. For ferroelectric Sm-C* these two types of monodomains will possess opposite P normal to the plates, and will be denoted as the 'UP' and 'DOWN' states. Adjacent UP and DOWN regions in a sample will be separated by well-defined domain boundaries, which may be manipulated with an applied electric field. The application of a field favoring the UP orientation produces a torque in the wall which induces a wall motion that expands the UP regions and viceversa. In the simplest geometry (see Figure 4.), the sample is placed between crossed polarisers with n parallel to the polarization direction in the down state, leading to the extinction of the light passing through the device. When an external field is applied normal to P, the ferroelectric torque P•E drives the LC director around the tilt cone (the tilt angle remains constant and the azimuthal angle varies) to the other stable state and stays there unless a second field with opposite sign is applied. In the UP state the light polarization will make an angle 2θ (twice the smectic tilt angle) with the optical axis and a fraction of the incident optical power will be transmitted. The resulting electro-optical effect exhibits a unique combination of properties: microsecond to sub-microsecond dynamics, threshold behavior, symmetric bistability and a large electro-optic response. The response time of an FLC device is generally in the range 1 to 100 μs. In order to achieve a uniform molecular rotation in a Sm-FLC cell, the layer thickness is usually limited to about 2μm. A large area, uniform alignment of an SSFLC cell is not easy to obtain.

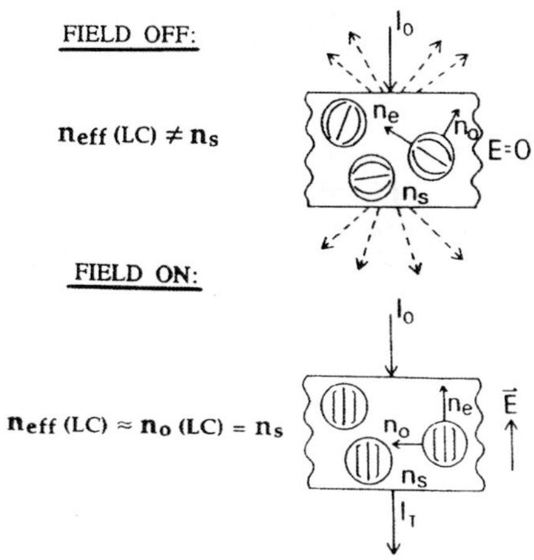

Figure 5. Light transmission characteristics of a normal mode PDLC-film

Liquid crystal may also have huge optical nonlinearity. The nonlinearity arises from the high degree of electron delocalization in the molecules, the large anisotropy of the molecular structure and the strong correlation of their motion.

Second order nonlinearity is dominated by $\chi^{(2)}$ terms and vanishes in centrosymmetric media such as nematic LC's. On the other hand, cholesterics and ferroelectric Sm-C*'s are not centrosymmetric and exhibit a non zero second-order optical nonlinearity. In general, the inversion symmetry doesn't apply at the liquid crystal interfaces and there a Second Harmonic Generation signal can be detected.

Third-order nonlinear effects may be very large in LC's. The electronic contribute to the $\chi^{(3)}$ mainly comes from individual molecules and is of the same order of magnitude as in ordinary liquids. Much more important are the effects arising from the collective motion of the molecules, i.e. from molecular reorientation. In the LC's the optical Kerr constant n_2 may be very large ($\cong 10^{-3}$ cm^2/W) due to appreciable molecular reorientation induced by the laser field. Because of their large anisotropy, in fact, the molecules tend to stay aligned along the electric field of the laser beam. The consequent refractive index change appears to be quite large, allowing the observation of nonlinear optical effects feasible even with commercial cw laser sources. However, besides a large n_2, in most practical application one requires also a fast response time and this is not achievable with LC's, whose response time is in the range of seconds.[9-13]

A class of very promising materials for optoelectronic applications, still based on LC's, is that of Polymer Dispersed Liquid Crystals (PDLC).[14] They are composite materials where a specific preparation method gives rise to random dispersion of liquid crystal droplets (which have diameter in the range 0.1 to 10 μm) into a polymeric matrix. The interest in studying PDLC's resides mainly in their peculiar combination of LC's optical properties with the mechanical properties of the polymeric matrix, yielding self standing devices with no need of external glass substrates. The electro-optical behavior of a PDLC, sketched in Figure 5., can be exploited in the following way. In a PDLC thin film the microdroplets of nematic liquid crystal are embedded in a polymer (an optically isotropic medium) of carefully selected refractive index n_s. The liquid crystal director axis varies nearly randomly from droplet to droplet in the absence of an external field. The refractive index mismatch between the liquid crystal droplets and the host polymer is then responsible for a significant light scattering in the off-state.[15,16] The degree of the off-state scattering losses depends on the size, birefringence, and concentration of the liquid crystalline droplets and on the film thickness. In this state the PDLC film appears cloudy.

By applying a sufficiently strong electric field, if the employed liquid crystal has positive dielectric anisotropy ($\Delta\varepsilon>0$), the director in the droplets is reoriented along the field direction. The effective refractive index n_{eff}(LC) of the microdroplets becomes equal to n_o, which has been chosen to be close to the polymer index n_s: the PDLC film thus appears clear. Removing the applied field, n in each droplet tends to relax back to its own preferred direction so that the film appears hazy again. This mode of operation is called the 'normal mode', but we may have a 'reverse mode' too; in the latter case, the PDLC is clear without any applied voltage and opaque when V>Vth. In some applications, such as controllable transparency windows or mirrors for automobiles, the reverse mode is preferred.

3. OPTICAL SENSORS UTILISING LIQUID CRYSTALS

Thermometry is an established application for liquid crystals. The majority of such sensors, based on the use of thermotropic liquid crystals, requires a simple direct observation of the color. LC's role is often that of a transducer in a probe head. All of the proposed schemes rely on the change in the spectral reflectivity of cholesteric liquid crystals with temperature. These devices are inexpensive, compact and portable. Other advantages include wide temperature range, chemical stability and freedom from electrical interference and noise.

To measure heat flow in large machines,[17] it was suggested to utilize liquid crystals in order to cover a whole piece of equipment and to use a video camera to record LC's color changes as it heats up and cools down. This method allows knowing which area of the equipment heats up: a detailed description can be obtained, the density of the information being limited only by the resolution of the video camera. This method is cheaper and more accurate than the traditional method of applying thermocouples to various points of the surface.

Other devices can be used for the passive detection of an accurately defined temperature. In the device proposed by Coles,[18] light is conveyed by an optical fiber into a cell where it crosses a smectic liquid crystal material, undergoing two total internal reflections before passing into the output fiber. When the temperature of the liquid crystal rises, going through its characteristic transition temperature, the amount of light received in the output fiber rises sharply. Switching temperature may range from –20 °C to 200 °C, defined to an accuracy of 0.5 °C. It is envisaged, as an example, that this device would find application in biomedical thermometry, for instance in specialized incubators where the temperature should not exceed a pre-set limit.

The Liquid Crystal Thermometry Unit[18] proposed by Wolfson is a temperature-monitoring laminar sheet consisting of a black substrate covered by a liquid crystal layer in a polymeric matrix, which is protected by an overlying clear plastic film. The color of the Bragg-reflected light from such thermochromic liquid-crystal material changes with temperature, so that it can be used as the basis of optical thermometers. Combination of materials allows monitoring of the temperature in the range of –30 °C to 250 °C. Accuracy is 0.1 °C.

LC's sensors utilizing optical fibers for remote sensing have also been constructed. The use of a reference channel to eliminate many of the problems inherent in intensity-dependent sensors has been proposed by Augousti et al.[19] They proposed a simple referenced LC fiber-optic switch. The principle of operation of the device is the following: light from two high-power light emitting diodes (LED), which are pulsed alternately, is coupled into two fibers which transmit the light to the probe head. The probe head consists of a sample of liquid crystal; the reflected light is picked up by a third fiber and transmitted to a detector. Since the amount of reflected light at each of the two wavelengths depends on the temperature of the liquid crystal and the final signal is the ratio of the two signals, most intensity-dependent errors can be substantially reduced. The response of the device, however, is non-linear and that precludes its use as a calibrated temperature measuring device; the application as a switch, on the contrary, is possible. This sensor has a response time of 5 s, within an 1 °C threshold.

Augousti et al.[20] have also proposed a temperature sensor using a hybrid optoelectronic multi-vibrating system, whose oscillatory frequency is proportional to the temperature. In this case the property of liquid crystals to be aligned by an electric field is used as the basis of transduction, for an indirect application such as temperature measurement. The main advantage of this device is its output in the frequency domain, thereby eliminating the requirement for elaborate referencing techniques. The oscillation frequency of this system depends upon the alignment response time of the liquid crystal to an applied electric field. The system has a temperature resolution of less than 0.1 °C, although the accuracy is approximately 0.5 °C in the range 0 to 50 °C. The response time is approximately 5 s.

The optical visualization of air or gas flow and the induced shear stress field is crucial to many applications in aerodynamic research. Cholesterics LC's and their shear stress and temperature-sensitive optical reflection properties have been exploited,[21] but shear introduces helix deformation through complex physical processes making the interpretation of the observed optical reflection spectra difficult. Further, the selective reflection spectrum with incident white light depends on the background surface, the incidence and viewing angles, as well as the thickness of liquid crystal coating. Preservation of the liquid crystal thickness during the experiment becomes difficult without introducing an undesired external surface. Thus, the results have been more qualitative than quantitative. In addition, the cholesteric based systems are relatively slow, typical response times being of the order of 1-100 ms. Hence rapid changes in flow patterns, with time scale in the micro and sub microsecond range cannot be recorded.

In ferroelectric liquid crystals as well, domain reorientation can be achieved by applying an external mechanical stress. The stress is a tensor interacting directly with the liquid crystal director and has the advantage of orienting the domain in either direction along the stress axis. Further, by orienting the smectic layers parallel to the sample surface, purely shear-stress-induced domain reorientation can be achieved without introducing hydrodynamic flow of the liquid crystal. These conditions are ideal for repetitive on-line measurement of the stress field. A new sensor, proposed by Parmar[22] is made up by a thin ferroelectric liquid crystal film coated on a flat glass surface exposed to the gas flow from a micro wind tunnel. The texture of the thin film consists of two stable domains separated by a wall. Gas or air flow on the liquid crystal surface induces director reorientation, resulting in optical contrast. Transmitted and reflected light intensities seen through a polarization microscope provide measurements of the flow parameters. The system response time has been estimated to be about 150 μs, and optical response is linear for applied differential pressures up to ≅800 Torr.

The study of a PDLC thin film devised to operate under externally applied shear stress has been reported.[23] In this new configuration, the liquid crystal microdroplets fill the randomly distributed micro-voids in a rigid continuous polymer matrix in such a way that the trapped droplets are fully exposed to the external flow field. This unique morphology, in which the rigid polymer matrix surrounds the droplets from all sides except that from the top, provides a direct LC-air interface. This device responds to the wind flow by a shear stress director reorientation resulting in a change of the film optical transmission when viewed under crossed polarisers. From the condition of minimum transmission in the absence of the air flow, the transmitted light intensity is an increasing function of the pressure in the laminar flow region (p ≅ 0-40 Torr).

Quantitative measurement of local and global pressure is important in several applications fields, such as in aerodynamic tests. There is a clear need for pressure sensors capable of measuring local as well as global pressures on a continuous basis, and they should be easy to mount and be cost effective. Partially exposed polymer dispersed ferroelectric liquid crystal (PEPDFLC) have been shown to respond to an external pressure.[24] When the PEPDFLC thin film is sandwiched between two transparent conducting electrodes, i.e. a glass plate and a flexible sheet (such as polyvinylidene fluoride), the switching characteristics of the thin film are a function of the pressure applied to the flexible transparent electrode and of the bias voltage. The transparent conducting film on each surface allows a bias voltage to be applied across the liquid crystal layer providing the initial alignment. Applying the pressure to the surface of the flexible electrode causes a variation in the local electric field at the site of the partially exposed droplets, which modulates the director field and the optical transmittivity of the system. Pressure in the range 0 to 1.5 psi can be measured, and the device response is linear for most of the pressure range. The electro-optic response time τ of the PEPDFLC is inversely proportional to the pressure, but τ for an FLC is also inversely proportional to the electric field, so the change in electric field is proportional to pressure.

In recent years a certain number of experiments have been undertaken with optical fibers and liquid crystals. It has been shown, both experimentally and theoretically, that a monomode tapered fiber is sensitive to any change in the external medium (index of refraction, temperature, mechanical constraints).[25,26] The effect is due to the interference of cladding modes that can propagate in the tapered region. Increasing the taper elongation renders it more sensitive to variation of n_{ex}. Using LC's as cladding materials makes this variation very easy to obtain. Obviously, one would like to have LC's with both ordinary no and extraordinary indices n_e lower than 1.46, which corresponds to the cladding index of the commercial optical fibers. Thus, Veilleux et al.[27] used a lyotropic LC, due to its low refractive indices; they proved that this material as a cladding of a highly tapered monomode fiber could be exploited for temperature sensing. Their sensor has 0.03 °C sensitivity, but the temperature range of operation is very short. The overall dimensions of this sensor are relatively large compared to a thermistor, but it has the advantage of immunity to electromagnetic noise.

Using a tapered monomode fiber surrounded by a liquid crystal external cladding, Warenghem et al. showed how the losses, induced by bending or tapering a fiber, can be controlled by the electrically induced reorientation of the LC. The tapered part of the fiber was inserted between two glass conductive plates separated by Mylar spacers and filled with a nematic liquid crystal (NLC). The initial alignment of the NLC molecules was planar, parallel to the fiber axis. The used NLC had a positive dielectric anisotropy, leading to homeotropic alignment under electric field. After injection of light, directional losses appeared, in the form of a sharp beam irradiated outside of the fiber taper. The angle between the leaky beam and the fiber axis is given by Snell's law and depends on the refractive index of the NLC seen by the light. In the planar alignment configuration, both TE and TM polarisations see the ordinary index no, leading to a specific angle between the fiber and the leaky beam. By applying the electric field, with the reorientation of NLC, TM polarization sees always the ordinary index while TE polarization sees the extraordinary one, leading to two weaker beams respectively corresponding to the previous angle and a new angle.[28]

Few years ago, an interesting LC device was suggested for fault check of electronic circuits.[29] In that device, the chip is covered with a NLC layer and a transparent and conductive plate. As a particular wire is addressed, a potential exists between that line and the upper plate, which reorients locally the NLC: as a result, the addressed wire becomes visible provided polarisers are properly installed. In the case of any trouble in the circuitry, the fault spot is immediately visualized and this control can be done on a quite large scale. The main used property in this case is the well known Fréedericksz transition.

It was experimentally demonstrated that liquid crystals can play the role of 'streak cameras' and visualize traces of leakage of optical radiation from microscopic areas of fiber-optics components such as tapers, couplers, switches, as well as from inhomogeneous regions and imperfections. Extended regions of fibers can be simultaneously monitored and distribution of strain and temperature in long fibers can be visualised.[30] The technique, called LC leakography, exploited the following LC properties: 1) extraordinarily large cross-section of light scattering due to thermal fluctuations of the optic axis; 2) negligibly small absorption of NLC throughout the visible and near infrared wavelengths. Thus the scattered light is visible during propagation in the NLC over considerable distances, and there are no unwanted consequences like absorption-induced heating; 3) the liquid nature of NLC. This allows objects of complex geometrical form or very big size to be embedded in the NLC.

The arrangement used in order to demonstrate the principle of LC leakography has been the following: a tapered fiber was inserted in a capillary filled by NLC. A heating coil was placed around one side of the capillary to induce a temperature gradient. Thus, the NLC was heated up to the isotropic state closer to the coil region, and the nematic phase was maintained in the region far further from the coil. They observed the leaky beam propagating through the nematic medium, undergoing reflection on the internal surface of the capillary and disappearing at the nematic-isotropic interface. Actually, since the scattering of light is negligibly small in the isotropic region, the wave propagating in it is no more visible. Thus, the light scattering phenomenon, conventionally regarded as the main obstacle for the use of LC in fiber optics, can find unexpected and important application: a thin layer of NLC, which can be easily "dropped" on any surface due to its small viscosity, plays the role of a 'streak camera' for the light leaking. The method allows coverage and check of extended fiber regions with varying geometrical shape, which is hardly possible by other methods.

4. OPTICAL MICROSYSTEMS UTILISING LIQUID CRYSTALS

The potential of Integrated Optics (IO) extends well beyond the field of communications. Planarity makes IO naturally compatible and likely to be associated with other technologies, such as silicon based micro-mechanics and micro-electronics, and it opens the way to the design and fabrication of even complex microsystems. These microsystems, being compact, rugged, suitable to efficient packaging, and capable of novel optical functions that cannot be performed in volume optics, may find a number of appealing applications, especially in optical signal processing and optical sensing.

Thin dielectric films have been used extensively for confining and guiding light; most of these films are amorphous and thereby act as optically isotropic media. Increasing emphasis, however, has been placed on films that are single-crystal in nature and optically uniaxial in character, because they generally possess larger electro-optical, magneto-optical or acousto-optical figures of merit: these characteristics are fundamental to the development of controllable devices, such as IO modulators and switches. In materials like GaAs, $LiNbO_3$, or poled polymers, the operation mechanism often is the linear (Pockels) electro-optic effect. The main problem usually is related to the relatively large switching voltage which may be required.

Liquid crystals thin films, which are transparent and suitable for light guiding, are birefringent and exhibit large electro-optic coefficients (at least an order of magnitude larger than those of other typical materials). The ease of manufacture, the low cost, and the possibility of integration with silicon circuitry technology make the investigation of LC integrated devices attractive. Moreover, being fluid materials, LC's can be easily inserted into microscopic devices or mixed with other substances to form new composite thin films having peculiar optical and thermo-mechanical properties.

Besides the electro-optic effect, other physical mechanisms can be exploited in LC IO devices. Liquid crystal molecules have a magnetic moment, and domains may be oriented through application of magnetic field. Magnetic fields, however, are generally less convenient to apply than electric fields and there is no reason to believe that they would be faster in response time.[31] Acoustic streaming has been demonstrated in liquid crystals[32] and some attention has been devoted to acousto-optic effect. Reliable large arrays of LC's have been manufactured at low cost for information displays. LC materials have also been integrated with silicon circuitry to produce monolithic Spatial Light Modulators with large numbers of pixels.[33]

There are, however, some limitations to the use of LC devices for integrated optics. Liquid crystals have slow response time (typically milliseconds with a few devices responding more rapidly than tens of microseconds), because the effects require displacements or rotation of molecules and thus suffer from viscous force, representing interactions with neighboring molecules. Another problem arises from the relatively large scattering losses of liquid crystal. Giallorenzi et al. showed that the scattered light from a nematic waveguide is affected by the presence of waveguide boundaries. They observed large scattering losses (about 20 dB/cm) in absence of an applied field; the presence of an applied electric field reduces the losses by at least an order of magnitude.[34] Green and Madden[35] found that nematic liquid crystals under strong two-dimensional confinement exhibit far less losses (1.5 dB/cm) caused by scattering.

Sosnowski[36] reported on experimental studies using a NLC as a waveguide cladding material, exploiting its uniaxial properties to realize a mode filter. Channin[37] and Sheridan[38] et al. successfully demonstrated waveguiding in the liquid crystal medium itself. They also demonstrated that electronically controlled amplitude switches could be fabricated in these guides (or in the cladding) using dynamic scattering and electro-hydrodynamic scattering phenomena.

The high propagation losses of a liquid crystal can be eluded by employing it as an overlayer. Okamura[39] et al. demonstrated two types of electro-optically controlled leaky anisotropic waveguides with NLC overlayers: by applying an electric field, a guiding structure could be changed to a leaky one or viceversa. They obtained switching operation in the ms range with extinction ratio of 20 dB, applying a 30 Vp-p voltage. Later on, they achieved similar results applying only a few volts (4 Vp-p): this reduction in the voltage[40] was obtained by removing a buffer layer previously put in between the guiding film and the electrode.

Another suggested device used a rectangular-core glass waveguide embedded in a glass substrate, with a covering layer made of an NLC film: the molecular alignment was tuned acting on the electric force and the surface anchoring force. Both the theoretical analysis and the experiment showed that in the ON state some components of hybrid modes are well-guided and other components are cut off; in the OFF state all components are cut off.[41]

Recent results on ferroelectric liquid crystals seem to be more promising. Giallorenzi et al.[42] studied losses in homeotropic smectic-A LC waveguides and they found losses (~2 dB) an order of magnitude lower than in nematics, because the smectic-A phase has higher degree of ordering. Lo et al.[43] examined a homeotropic smectic-A cylindrical waveguide and also measured low losses (1.7 dB/cm). FLC's in the surface-stabilized geometry (SSFLC) should exhibit low losses in a waveguide configuration, because the smectic-C phase is more ordered than the smectic-A phase and the alignment of the molecules is bound along the surface. In addition, even in bulk FLC devices transmission losses caused by scattering have been measured to be low.

There have been a few studies investigating waveguide switches that use FLC's. Clark and Handschy[44] proposed to control the optical coupling between two ion-exchanged waveguides by applying a suitable voltage to a SSFLC film layered in between them. The device exhibited a response time of 200 μs and a 40:1 switching ratio. The FLC is oriented so that the wave of TE mode evanescent into the FLC sees a large refractive index for one voltage polarity, and a smaller one for the opposite polarity. TE light incident in the first waveguide remains confined there for the small-index FLC state, but it is radiated away for the large-index FLC state. In this basic geometry, the core refractive index must lie between the highest and lowest indices available in the FLC: the combination of a typical available FLC material (CS-1014) and of a waveguide fabricated by K-Ag ion-exchange in a standard microscope slide proved to be successful.

Ozaki et al.[45] proposed a polymer-SSFLC composite for fast optical switching in a waveguide. The SSFLC layer is placed between one conducting glass plate and a polymer waveguide onto another conducting glass plate. In the ON state, light was guided only in the polymer waveguide. In the OFF state, light was guided in both the polymer and the FLC material. Losses were due to the large discontinuity between the thickness of the polymer waveguide and the polymer-FLC waveguide. The contrast ratio was about 40 and the optical rise and fall times were very short, both about few microseconds.

Placing the LC material in the cover is not desirable in general, because the interaction of the LC with the modes in the waveguide is reduced. Clearly, larger effects can be realized if the LC material is used as the guiding film itself. Walker et al.[46] explored the possibility of using FLC's as an active waveguide, and analyzed several modulator configurations, namely a cut-off modulator, a deflection modulator, and an input coupler. In order to properly study these structures, a mode matching technique was developed: the numerical simulations showed that FLC modulators have many desirable performance characteristics. In the deflection modulator, large deflection angle and low switching losses are attainable. In the cut-off modulator large radiation losses are foreseen. An experimental realization of the cut-off modulator proposed by Walker has been recently demonstrated by our group.[47]

5. OUR EXPERIMENTAL WORK

With the aim of developing electro-optical switches using LC's in integrated optics, we started to focus our attention onto the optimization of the device design, on one side, and onto the fabrication technology of the glass waveguides including the LC cell, on the other side. We investigated in particular the problem of an efficient coupling of the light into and out from a waveguide, with special attention to the modal coupling between the different sections of the device.[48] A typical device structure, in fact, is made of three sections in a glass substrate: the first and the last ones are glass waveguides, while the middle one is a liquid crystal waveguide, which may operate either in a linear or in a nonlinear optical regime.

Problems to be solved in the manufacture of LC integrated optical devices include the following ones: a) choice of the most suitable waveguide fabrication method; b) design of the guiding structure and of the LC cell which assure the most efficient interaction; c) design and realization of the transparent conductive electrodes (usually ITO thin films). The guiding structure, sketched in Figure 6., was designed and realized in order to study the nonlinear light propagation in a waveguide filled with an NLC material. First a planar dielectric waveguide is fabricated, then a rectangular cell having the same depth as the waveguide is etched in it. The cell is filled with nematic LC and covered with a glass plate. The cell's bottom and the walls are previously treated to obtain the wanted molecular alignment. The realization of electro-optical devices further requires the deposition of thin transparent electrodes. To avoid the need for high voltages and to ensure a high uniformity of the control electric field, the electrodes have to be integrated into the device. ITO (Indium Tin Oxide) electrodes are laid between the substrate and the guiding film. The choice of the deposition technique is critical: we must not alter neither the ITO film characteristics nor the guided modes pattern of the multi-layer structure.

The waveguide sections may be fabricated by ion-exchange in the glass substrate; however, in order to simplify the process, by taking advantage of ITO-covered substrates already available, it was preferred to fabricate the waveguides by using deposition techniques such as RF sputtering and sol-gel. The former is a well-known and widely used technique, which allows the control of the film refractive index as well as of the deposition rate by acting on the sputtering gas and the partial pressure; the control of the waveguide thickness is easy and accurate by acting on the deposition time. The sol-gel method has gained considerable attention in recent years. This attention is justified by the versatility of the process, which allows the production of glasses with largely different physico-chemical properties just by changing the starting solution composition. In principle it allows a fine tuning of the refractive index of the produced film over a very large range (e.g. using silica-titania binary glasses, refractive index may be varied from 1.6 up to 2.3).[49] Another fundamental advantage of the sol-gel process is the relatively low fabrication temperature. The guiding films were deposited on top of the ITO layer, and were characterized by the standard dark-line technique and by loss measurements.[50] The final multilayer structure, quite obviously, may be significantly affected by ITO properties: this can be ascribed mainly to its refractive index, which is high (effect on the waveguide modal structure) and complex (effect on waveguide loss).

The comparison of modal properties of sol-gel layers onto soda-lime and ITO-coated glasses (20 nm-thick ITO coating) confirmed that, by using this deposition technique, the presence of ITO has negligible effects on the modal structure. Loss measurements indicated a sufficiently good quality of sol-gel waveguides onto uncoated substrates (with loss below 0.8 dB/cm), while higher losses (up to 2 dB/cm) were caused by the presence of the ITO layer, partly due to its intrinsic absorption, and partly due to its effect on the confinement of the guided optical field.

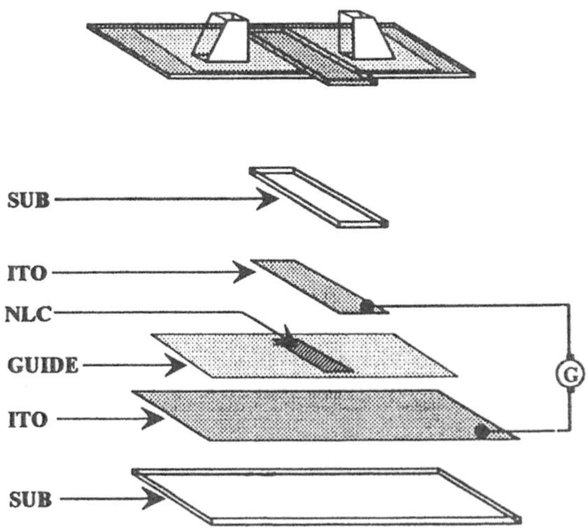

Figure 6. Schematic representation of the fabricated integrated optical LC electro-optical switch. SUB: substrate and cover; ITO: indium tin oxide; NLC: nematic liquid crystal; G: voltage generator.

As to the LC cell fabrication, a photolithographic and etching procedure was used. The required pattern was imprinted in a photoresist layer deposited on the top of the waveguide, and the subsequent etching permitted to transfer the pattern inside the waveguide. In order to limit losses due to light scattering in the LC, the propagation length in the LC cell was reduced to a minimum. The etching process is particularly critical because one needs cell's walls of the lowest possible roughness, to avoid additional scattering of light at the boundaries, as well as to maintain a good alignment of LC molecules. We used a chemical etching procedure, and good results have been obtained by using a fluoridric acid solution. It guarantees a good etching rate and relatively low-roughness surfaces, with the further advantage that ITO is not attacked by this solution and therefore it can constitute an easy end-point of the process. Tests have been made also by using a reactive-ion-etching (RIE) process, but no significant advantages have been noticed.

With this device an integrated electro-optical switch based on a planar nematic liquid crystal waveguide was realized and tested. The electro-optical behavior and the response times were studied for different configurations. The parameters of the glass waveguide were optimized to match the refractive index of the NLC and this resulted in a quite high transmittivity of the device in the ON state. We achieved peak performances of 35% transmission and response time of 200 μs. Adding a constant bias voltage the performances were significantly improved: transmission up to 40% and response time less than 100 μs are promising results in view of new integrated electro-optical switches. Further investigations, not only on this device but also on fundamental aspects of liquid crystal physics, are suggested by this study.[51]

Employing the same device, we also realized an electro-optic switch based on a planar, ferroelectric liquid crystal waveguide.[47] The FLC was oriented in the bookshelf geometry; as in this case the FLC parameters and those of the glass waveguide were not optimized, only a rather low transmittivity was obtained. The contrast ratio was at best 5:1, while the rise and fall times were ~120μs. These results could be considered as a first experimental demonstration of the FLC cut-off modulator proposed by Walker.[46] A proper device optimization, which is now in progress, could lead to improve the time response and transmittivity performance by more than one order of magnitude.

6. CONCLUSIONS

An overview of the basic properties of liquid crystals has been presented as an introduction to the subject of design and fabrication processes of optoelectronic LC devices.

Even besides the main application area of LC's, namely displays, there has been a large and fruitful R&D activity aimed at the development of novel or advanced devices exploiting the peculiar properties of these materials. A number of sensors and optical modulators or switches have been demonstrated, that exhibit interesting characteristics.

Integration of liquid crystals in guided-wave structures leads to compact, efficient and reliable devices: further efforts, however, are necessary in order to optimize these microsystems, making them more performing and more rugged at the same time. An approach which seems promising to achieve higher performance is that of employing ferroelectric liquid crystals, which can exhibit much shorter response times than other LC's. As to the ways of increasing ruggedness and compactness of LC devices, a viable route appears to be that of exploiting the characteristics of the polymer-dispersed LC's: these composite materials, besides alleviating the alignment problems, lend themselves to an easy and convenient use in guided-wave structures. Their optical properties, however, still have to be deeply investigated.

ACKNOWLEDGEMENTS. Useful discussions with M.A. Forastiere and S. Pelli, as well as the technical assistance by R. Calzolai (all at IROE CNR) are gratefully acknowledged.

One of us (LS) wishes to thank CNR for his research fellowship.

REFERENCES

1. P.G. de Gennes, The Physics of Liquid Crystals, Clarendon Press, Oxford (1974).
2. S. Chandrasekhar, Liquid Crystal, Cambridge Univ. Press, 2nd Ed. (1992).
3. T.W. Stinson and J.D. Lister, Pretransitional phenomena in the isotropic phase of a nematic liquid crystal, Phys. Rev. Lett. 25: 503 (1970).
4. B. Jérôme, Surface effects and anchoring in liquid crystals, Rep. Prog. Phys. 54: 391 (1991).
5. I.C. Khoo and S.T. Wu, Optics and Nonlinear Optics of Liquid Crystal, Series in Nonlinear Optics, World Scientific, Singapore, (1993).
6. S.T. Wu, A semi-empirical model for liquid crystal refractive index dispersion, J. Appl. Phys. 69: 2080 (1991).
7. N.A. Clark and S.T. Lagerwall, Submicrosecond bistable electro-optic switching in liquid crystals, Appl. Phys. Lett. 36: 899 (1980).
8. G. Andersson, I. Dahl, P. Keller, W. Kucynski, S.T. Lagerwll, K. Skarp and B. Stebler, Submicrosecond electro-optic switching in the liquid crystal smectic A phase: the soft-mode ferroelectric effect, Appl. Phys Lett. 51: 640 (1987).
9. E. Santamato and Y.R. Shen, Liquid crystals for nonlinear optical studies, in: A Guide to Liquid Crystal Research, P. J. Collings and J. S. Patel Eds., Oxford Univ. Press (1996).
10. L. Marrucci and Y.R. Shen, Nonlinear optics of liquid crystals, in: The Optics of Thermotropic Liquid Crystals, edited by R. Sambles e S. Elston, Taylor & Francis, London (1997).
11. I.C. Khoo, Nonlinear optics of liquid crystals, Progress in Optics XXVI: 107 (1988).
12. F. Simoni, Nonlinear optical phenomena in nematics, in Physics of Lliquid Crystals, Gordon & Breach, N.Y., (1989).
13. I.C. Khoo and Y.R. Shen, Liquid crystals: nonlinear optical properties and processes, Opt. Eng., 24: 579 (1985).
14. I.C. Khoo and F. Simoni, Physics of Liquid Crystalline Materials, Gordon and Breach Science Publishers (1988).
15. J.W. Doane, N.A. Vaz, B.G. Wu and S. Zumer, Field controlled light scattering from nematic microdroplets, Appl. Phys. Lett. 48: 269 (1986).
16. S. Zumer, Light scattering from nematic droplets: anomalous-diffraction approach, Phys. Rev. A 37: 4006 (1988).
17. J. Beard, Liquid crystals take the temperature in tests, New Scientist (ISSN:0262-4079) 139: 20 (1993).
18. Liquid-crystal devices sense temperature, Laser Focus (ISSN:0740-2511) 23: 48 (1987).
19. A T Augousti, K.T.V. Grattan and A.W. Palmer, A liquid crystal fibre-optic temperature switch, J. Phys.E: Sci. Instrum. 21: 817 (1988).
20. A.T. Augousti and J. Mason, Temperature sensing using a hybrid optoelectronic multivibrating systm under full computer control, Meas. Sci.& Technol 5: 736 (1994).
21. D.S. Parmar, A novel technique for response function determination of shear sensitive cholesteric liquid crystals for boundary layer investigation, Rev. Sci. Instrum. 62: 1596 (1991).
22. D.S. Parmar, A Novel boundary layer sensor utilizing domain switching in ferroelectric liquid crystals, Rev. Sci. Instrum. 62: 474 (1991).
23. D.S. Parmar and J.J. Singh, Partially exposed polymer dispersed liquid crystals for boundary layer investigations, Appl. Phys. Lett. 61: 2039 (1992).
24. D.S. Parmar and H.K. Holmes, Pressure dependence of the electro-optic response function in partially exposed polymer dispersed ferroelectric liquid cystals, Appl. Phys. Lett. 63: 21 (1993).
25. S. Lacroix, R. Bourbonnais, F. Gonthier and J. Bures, Tapered monomeode optical fibers:understanding large transfer, Appl. Opt. 25: 4421 (1986).
26. S. Lacroix, R.J. Black, C. Veilleux and J. Lapierre, Tapered single-mode fibres:external refractive index dependence, Appl. Opt 25: 2468 (1986).
27. C. Veilleux, R.J. Black and J. Lapierre. Nematic liquid crystal clad tapered optical fiber with temperature sensing properties, J. Appl. Phys. 67: 6648 (1990).
28. J.F. Henninot, D. Louvergneaux, N. Tabyrian and M. Warenghem, Controlled leakage of a tapered optical fiber with liquid crystal cladding, Mol. Cryst. Liq. Cryst. 282:297 (1996).

29. B. Picart, L. Dugoujon, O. Petit, C. Destrade, C. Leon, H.T. Nguyen and J.P. Marcerou, Proc. SPIE 1080: 131 (1989).
30. J.F. Henninot and M. Warenghem, Leakography: visualization of losses in fibers and fiber optics systems using liquid crystals, to be published in Mol. Cryst. Liq Cryst..
31. P.K. Tien, R.J. Martin, R. Wolfe, R.C. LeCrew, and S.L. Blank, Switching and modulation of light in magneto-optical waveguides, Appl. Phys. Lett. 21: 394 (1972).
32. K. Miyano and Y.R. Shen, Domain pattern excited by surface acoustic waves in a nematic film, Appl. Phys. Lett. 28: 472 (1976).
33. K.M. Johnson, D.J. McKnight and I. Underwoood, Smart spatial light modulator using liquid crystals on silicon, IEEE J. Quantum El. 29: 699 (1993).
34. T.G. Giallorenzi and J.P. Sheridan, Light scattering from nematic liquid crystal waveguide, J. Appl. Phys. 46: 1271 (1975).
35. M. Green and S.J. Madden, Low loss nematic liquid crystal cored fiber waveguides, Appl. Opt. 28: 5202 (1989).
36. T.P. Sosnowski, Polarization mode filters for integrated optics, Optics Commun. 4: 408, (1972).
37. D.J. Chanin, Optical waveguide modulation using nematic liquid crystal, Appl. Phys. Lett. 21: 365 (1973).
38. J.P. Sheridan, J. M. Schnur, and T.G. Giallorenzi, Electro-optic switching in low-loss liquid crystal waveguides, Appl. Phys. Lett. 22: 560 (1973).
39. Y. Okamura, K. Kitatani and S. Yamamoto, Electrooptic leaky anisotropic waveguides using nematic liquid crystal overlayers, J. Lightw. Techn. LT-2: 292 (1984).
40. Y. Okamura, K. Kitatani and S. Yamamoto, Low-voltage driving in nematic liquid crystal overlayers waveguide, J. Lightw. Techn. LT-4: 360 (1986).
41. B.X. Chen, X. Tong Li, Y. Limura and S. Kobayashi, High-contrat channel waveguide switch with two nematic sections of a nematic liquid crystal covering, Appl. Opt. 32: 6018 (1993).
42. T. G. Giallorenzi, J. A. Weiss and J. P. Sheridan, Light Scattering from smectic liquid-crystal waveguides, J. Appl. Phys. 47: 1820 (1976).
43. S.K. Lo, L.M. Galarneau, D.J. Rogers and S.R. Flom, Smectic liquid crystal waveguide with cilindrical geometry, Mol. Cryst. Liq. Cryst. 201: 137 (1991).
44. N.A. Clark and M.A. Handshy, Surface-stabilized ferroelectric liquid crystal electro-optic waveguide switch, Appl. Phys. Lett. 57: 1852 (1990).
45. M. Ozaki, Y. Sadohara, T. Hatai and K. Yoshino, Fast optical switching in polymer waveguide using ferroelectric liquid crystal, Jpn. J. Appl. Phys 29: L843 (1990).
46 D.B. Walker, E. N. Glytsis and T.K. Gaylord, Ferroelectric liquid crystal waveguide modulation based on a switchable uniaxial- uniaxial interface, Appl. Opt. 35: 3016 (1996).
47. G. Scalia, D.S. Hermann, G. Abbate, L. Komitov, P. Mormile, G.C. Righini, L. Sirleto, Integrated Electro-optical switch based on a ferroelectric liquid crystal waveguide, submitted to Mol. Cryst. Liq. Cryst.
48. G. Abbate, F. Castaldo, L. De Stefano, E. Santamato, P. Mormile, M.A. Forastiere, G.C. Righini, L. Sirleto, Mode coupling between glassy and liquid crystal planar waveguide, Proc. SPIE 2954: 180 (1996).
49. G.C. Righini and S. Pelli, Sol-gel glasses waveguides. J. Sol-Gel Techn. 8: 991(97).
50. L. Sirleto, M. Forastiere, G.C. Righini, A. Verciani, G. Abbate, E. Santamato, P. Mormile, G. Pierattini, Integrated optical switches using liquid crystal cells, in "PHOTONICS-96", J.P. Raina and P.R. Vaya Eds. (Tata McGraw Hill), 1: 629, (1996).
51. G. Abbate, P. Mormile, L. Petti, G.C. Righini, L. Sirleto, Integrated electro-optical switch based on nematic liquid crystal, Proc. SPIE 3278: 84 (1998).

INDIUM TIN OXIDE FILMS FOR OPTICAL SENSORS

C. Calì, and M. Mosca

Istituto Nazionale di Fisica della Materia and
Dipartimento di Ingegneria Elettrica dell'Università degli Studi di Palermo
Viale delle Scienze
90128 Palermo, Italy

1. INTRODUCTION

Recently we have successfully produced Indium Tin Oxide (ITO) films by pulsed laser deposition.[1] The grown films are transparent to visible and near infrared radiation and have a low resistivity. The dependence of the ITO resistivity on light radiation in some wavelength ranges, as Hamberg and Granqvist reported,[2] induced us to further investigate the properties of our films. The present paper reports on this investigation and on the possibility of designing sensors with suitably realized films.

2. CHARACTERISTICS OF INDIUM TIN OXIDE

ITO is an n-type semiconductor that behaves as a good electric conductor, transparent in the visible and in the infrared spectrum. The principal properties of this material are, in general, similar to those exhibited by other transparent semiconductor oxides; however, ITO compares favorably with the others and, for this reason, is more widely used for all those applications that require high electric conductivity and low absorption of light.

Table 1. Percentage variations of conductivity after 90 seconds for samples deposited on different conditions.

Density of energy [J/cm^2]	6			8			10.5			19.5		
Conditions: T substrate [°C] Ext. field	25 no	250 no	25 yes	25 no	250 no	25 yes	25 no	250 no	25 yes	25 no	250 no	25 yes
Variations of conductivity [%]	1.7	0.0	1.3	0.8	0.0	0.6	0.7	0.3	0.4	0.3	0.3	0.4

Optical Sensors and Microsystems: New Concepts, Materials, Technologies
Edited by Martellucci et al., Kluwer Academic / Plenum Publishers, New York, 2000.

In ITO, as in all semiconductors, the resistivity decreases with doping, that is with increasing density of the conduction electrons. ITO films are transparent in the visible and near infrared regions while they reflect very well in the medium and far infrared. The transmittivity can reach values up to 90% and is approximately constant in the visible region. A strong absorption is instead present in the UV, beginning from 300nm. This agrees with the gap between valence and conduction bands, equal to 3.75 eV in the intrinsic ITO.[2] In addition, doping increases the gap considerably, causing a shift of the absorption curve toward the ultraviolet.

3. DEPOSITION PROCESS

ITO thin films have been deposited using different techniques. We employed a photoablation process that uses harmonically tripled radiation from a pulsed Nd:YAG laser.[1] Depositions have been made on BK7 glass substrates inside a chamber where a high vacuum or a controlled pressure of oxygen introduced through an adjustable valve is maintained. The oxygen pressure is always below atmospheric pressure. The laser beam, entering the bell jar through a quartz window, is focused to a specified density of energy onto the ITO tablet. The chamber contains a halogen lamp able to heat the substrate to 250°C, an electrode placed 1 cm above the target to apply a strong ionizing field and a glow discharge system for the final cleaning of the substrate.

It was observed that the properties of the grown films strongly depend on the conditions of deposition. In particular, it is possible to vary the optical and electrical properties of the films by acting on the pressure of oxygen inside the bell jar, on the temperature, on the intensity of the external field applying a suitable voltage to the electrode and on the energy density of the laser radiation.

4. PHOTOCONDUCTIVE EFFECT

It has been demonstrated that the conductivity of an ITO film can be modified when it is irradiated with UV light. Specifically, the conductivity of the irradiated film increases. This phenomenon is reversible, if some conditions are verified.[3] The same phenomenon was also observed in intrinsic In_2O_3.[4]

In our experiments we have observed that the photoconductive effect depends strongly on the initial conditions of the film, that is on the conditions in which it is deposited. In the films grown with an oxygen pressure optimized for lowest resistivity, the photoconductive effect is not very evident. We deposited samples in different conditions, varying the temperature of the substrate, density of energy of the laser radiation and intensity of the external applied field but keeping the same, optimal pressure of oxygen (10^{-2} mbar), all of equal thickness to around 120 nm. All samples were transparent (n= 1.95 and k< 0.01) and good electric conductors (resistivity 10^{-4} ohm·cm), but showed a weak photoconductive effect, as Table 1. shows. This table reports the variations (in percent) of the conductivity due to the 355 nm radiation when the beam was suitably enlarged and stopped down to have a density of energy of the order of 1 mJ/cm^2.

Table 1. shows percentage variations of conductivity less than 1%, except for two samples, which had been deposited on a room temperature substrate and with a low density of laser energy. The latter two conditions, as reported in a previous work,[1] are those that lead to higher resistivity. We have therefore concluded that the more sensitive films (in terms of photoconductive effect) are those less conductive. In all the films examined the phenomenon is only partially reversible and quite slow.

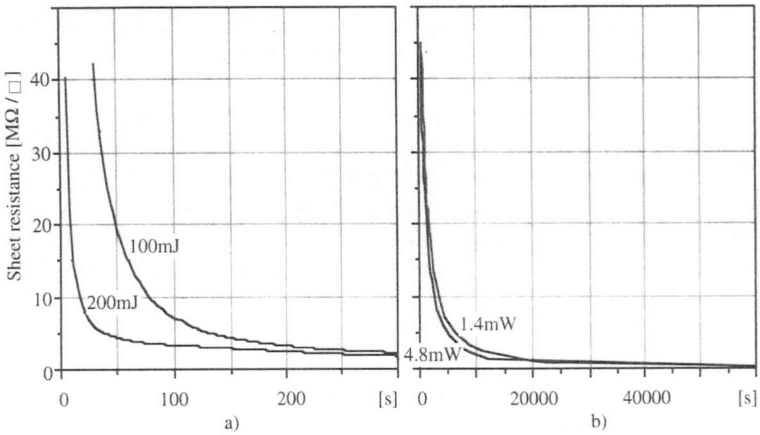

Figure 1. Measured sheet resistence versus time illuminating the sample with a) the harmonically tripled radiation from a pulsed Nd:YAG laser (100 and 200 mJ pulse energy) and b) the UV radiation emitted from a mercury lamp (1.4 and 4.8 mW power).

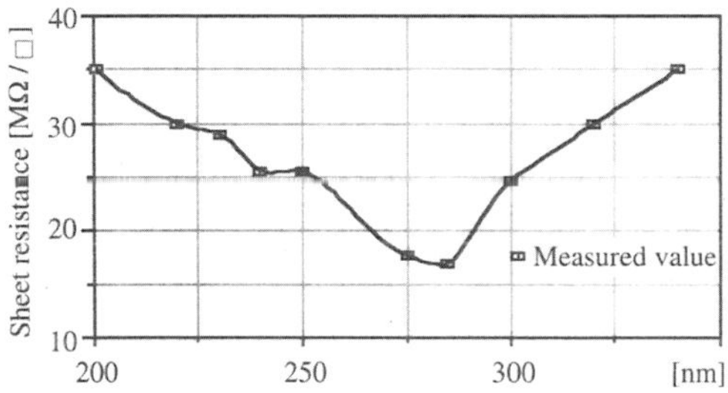

Figure 2. Measured sheet resistence versus wavelength (saturation values).

Since in our previous investigations we found that the resistivity increases with decreasing laser energy density and increasing oxygen pressure during the process of deposition, we deposited a film of ITO using a lower density of energy and a higher pressure of oxygen ($5 \cdot 10^{-2}$ mbar) compared to the optimal one. We obtained a sample, 80 nm thick, very transparent and almost insulating, which exibited an appreciable reduction in resistivity when illuminated with UV radiation, as shown in Figures 1a. and 1b.

The diagrams of Figure 1a. have been obtained by illuminating the sample with the same radiation used to analyze the previous samples while those of Figure 1b. applie when illuminating the sample with the UV radiation emitted from a mercury lamp. The intensity has been varied acting on the lamp-sample distance. With the sample so fabricated the phenomenon has resulted perfectly reversible, with a faster response. It is possible to observe from Figures 1a. and 1b. that the speed of response of the resistivity depends on the intensity of the UV source. The saturation value shows little or no dependence on the intensity of the UV light.

The saturation value itself is sensitive to the wavelength of the UV. Using a Xenon lamp and a spectrometer we have been able to analyze the effects of the wavelength variations on the illuminated film. The results are shown in Figure 2.

The saturation resistance is therefore related to the energy of the single photon, i.e. to the wavelength, while it does not change with the number of photons; therefore, theoretically, few photons are sufficient to reach the asymptotic value but, of course, the whole process becomes much slower. As is shown in Figure 2., the photon conductive process has a peak of sensitivity around 285 nm and decreases for smaller wavelengths, presumably because of the absorption of the UV radiation in air. The high cut-off wavelength is certainly correlated with the band gap and the absorption of the film.

We have also deposited ITO films in an oxygen-free atmosphere. The samples obtained show a metallic color, and are opaque, absorbing strongly in the near infrared. In these films the photoconductive effect was almost absent and they manifested a reversible increase in resistivity when illuminated with radiation in the near infrared (1.06 micron) or when heated. This is the typical behavior of metals. Those effects depend also in this case on the conditions during deposition of the film, as Nath et al. reported.[6]

5. PHYSICAL INTERPRETATION OF THE OBSERVED PHENOMENA

The valence electrons essentially determine the conductivity and optic properties of the ITO, for photon energies close to or greater than the energy gap. The conductivity in a transparent oxide semiconductor depends on oxigen vacancies. In the case of a film of intrinsic or little-doped ITO, oxygen vacancies act like donors and the conductivity of the sample can be modified by controlling the deficiency of oxygen. The exposure of the sample to an UV radiation causes a chemical photoreductive effect that increases the density of holes and therefore the conductivity. For highly conductive ITO films, the mechanism of conduction depends mainly on the doping concentration of and on the activation of single doping atoms, rather than on oxygen vacancies. Therefore the photosensitive behavior appears stronger for higher initial resistivity of the film, simply because in this case the conduction is mainly due to oxygen vacancies.

To obtain a film which is less dependent on the characteristics of the doping it is sufficient to deposit the ITO film in presence of a higher pressure of oxygen; in this way the resistivity increases[5] and the electric characteristics become worse: since the collisions between the particles of indium, tin and oxygen increase, these particles no longer have a uniform speed, as is the case for the optimal pressure of oxygen, and they are deposited with a lower energy, in a such way that the doping impurities are not activated. A further consequence of this loss of energy is that the speed of deposition is small.

6. REVERSIBILITY OF THE VARIATIONS OF THE RESISTIVITY AND FACTORS THAT INFLUENCE THE SPEED OF RETURN TO THE INITIAL CONDITIONS

To return to the original conditions it is necessary to oxidize the sample in way to oppose or annul the process of chemical photoreduction. Obviously this oxidation could happen in air, through the oxygen contained in it, as shown in Figure 3. The oxidation is always present, even during UV irradiation; therefore the saturation value reached by the resistivity represents the point of equilibrium between the oxidizing and reducing processes.

The process of oxidation can be faster if the film is placed inside a plasma of oxygen. We have verified this using the system of glow discharge for the cleaning of the substrate's surface that is present in our apparatus of deposition.

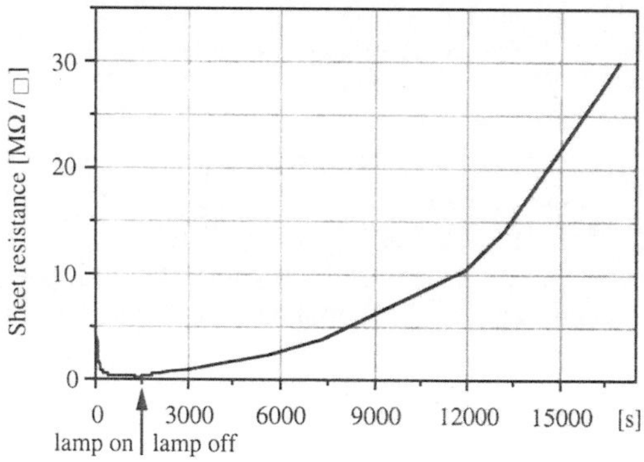

Figure 3. Measured sheet resistence versus on/off irradiation time (lamp power 8mW) as proof of the reversibility of the phenomenon.

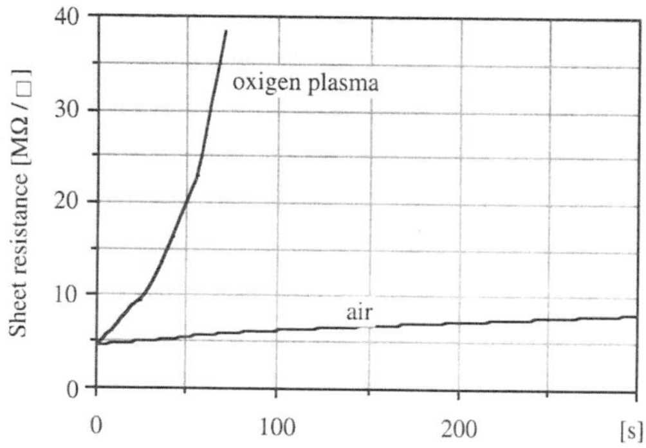

Figure 4. Measured sheet resistence versus time of return to the initial conditions in different environments.

The sample is positioned so that the ITO film is not directly exposed to the plasma, thus avoiding phenomena of ablation or sputtering; we used a weak discharge current (< 50 mA) at a pressure of oxygen of $7 \cdot 10^{-2}$ mbar.

The presence of the oxygen is necessary for return to the initial conditions; the oxygen could be that available in the air or that, ionized and concentrated, produced by the glow discharge process; the difference consists only in the rapidity of the response, as is shown in Figure 4. To confirm this, we have placed an UV irradiated sample in vacuum; in this case the increase in resistance halts until oxygen is introduced inside the vacuum chamber.

The oxidation is faster if the concentration of ionized oxygen is higher. In fact, using two different values for the glow discharge current, it is observed that a higher current leads to a smaller time constant for the resistance recovery curve.

A direct observation and a precise measurement of the phenomena of oxidation should be made holding the sample in vacuum. Irradiation of a sample with the same UV lamp, either in high vacuum conditions or in a room atmosphere, leads to different values for the saturation resistance, as can be seen from Figure 5. It must be considered that in air the UV radiation produces ozone that reacts oxidizing the film; therefore the saturation value is determined by the equilibrium between the two conflicting processes, while in the vacuum the oxidizing action of the oxygen is absent. With reference to the Figures 1a. and 1b., the two saturation values differ, because the first was measured at atmospheric pressure, the second one in vacuum.

It was observed that, when the sample is irradiated in air, the resistivity goes up again after reaching the minimum value: see Figure 6. It is therefore more correct to speak of a minimum value rather than of an asymptotic value. This may be due to the presence of a layer of adsorbed gas on the surface of the film. The adsorbed gases may include oxygen that is slowly activated by the UV radiation. This hypothesis is confirmed by the absence of this phenomenon when the film is irradiated in vacuum. In this last case the saturation value is constant and persists for as long as the sample is irradiated.

7. POSSIBLE APPLICATIONS OF THE PHOTOCONDUCTIVE EFFECT

The photoconductive effect in ITO films could be effectively employed to realize thin film sensors for UV radiation. A sensor of this kind occupies a small space and can have a small cost, once the phases of the process are established and set up. Also fabrication is very simple because, once the film is deposited, it is sufficient to make two thin ohmic contacts above the film to obtain a working device. The reliability of the device is very high, since ITO films deposited with the pulsed laser deposition technique are very hard and adherent (adhesion better than 80 kg/cm^2, as reported in a previous work[1]); the device can therefore bear mechanical shock, sudden rise of temperature, high irradiating powers and different kinds of external agents. Also the sensitivity of the device is very high because it is possible to detect very small UV powers. We have carried out some tests by implementing two ohmic contacts with silver glue on an ITO film sample produced with a high pressure of oxygen. This device is able to detect the UV component of the solar radiation, even in days with covered sun. Better devices could be fabricated by sealing the film in a vacuum case, transparent to the UV radiation.

Conversely, with a UV source of constant intensity it is possible to use the ITO film to detect the presence of oxygen or ozone.

Obviously, by suitably controlling the process of deposition, it is possible to get devices able to distinguish IR radiation from UV radiation, that are transparent and insensitive to the visible light. But such conditions must still be studied and further investigation of the nature of the phenomena described above is currently in progress.

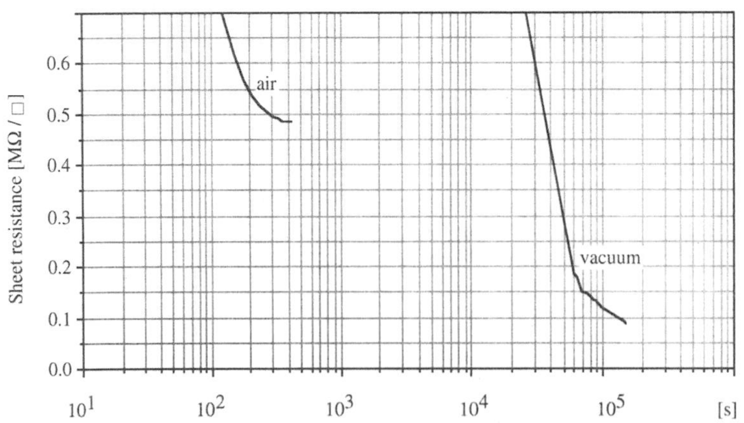

Figure 5. Measured sheet resistence versus irradiation time in different environments.

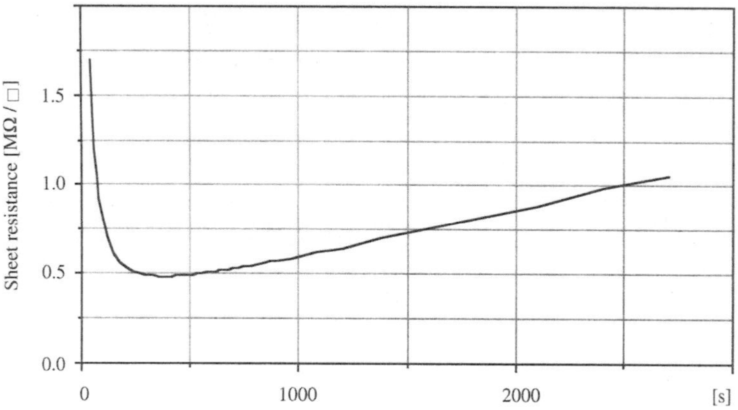

Figure 6. Measured sheet resistence versus irradiation time in air.

8. CONCLUSIONS

The experiments and results presented here show that it is possible to implement sensors, based on ITO thin films, capable of detecting radiation of opportune wavelength. The realization is simple and economic and could yield robust and reliable devices. In this work we have examined the processing conditions, suitable to realize a film of ITO that can work as a sensor. The behavior of the film is connected to the photochemical properties of the sample, that is to oxidation-reduction processes, and in turn to the morphological structure and the stoicheometry of the deposited films, determined by process parameters.

REFERENCES

1. C. Calì, M. Mosca and G. Targia, Deposition of Indium Tin Oxide films by laser ablation: processing and characterization, Accepted for publication in *Solid-State Electronics*.
2. I. Hamberg and C.G. Granqvist, Evaporated Sn-doped In_3O_3 films: basic optical properties and application to energy-efficient windows, *J. Appl. Phys.*, Vol.60, R123 (1986).
3. S. Mailis et al., Holographic recording in Indium-Oxide (In_3O_3) and Indium-Tin Oxide (In_3O_3:Sn) thin films, *Appl. Phys. Lett.*, Vol.69, 2459 (1996).
4. M. Fritzsche et al., Reversible changes of the optical and electrical properties of amorphous InO_x by photoreduction and oxidation, *Solar Energy Materials and Solar Cells*, Vol.32, 383 (1984).
5. J.P. Zheng and H.S. Kwok, Low resistivity Indium Tin Oxide films by Pulsed Laser Deposition, *Appl. Phys. Lett.*, Vol.63, 1 (1993).
6. P. Nath et al., Electrical and optical properties of In_3O_3:Sn films prepared by activated reactive evaporation, *Thin Solid Films*, Vol.72, 463 (1980).

OPTOELECTRONIC NEURAL NETWORKS

A.W. Domański

Warsaw University of Technology
Institute of Physics
Koszykowa 75
00-662 Warszawa, Poland

1. INTRODUCTION

Neural networks are supposed as a future of computing. They have two main advantages in comparison to conventional operational system of computers.

First, they have possibility of parallel signal processing. Thousands of neurons are able to analyze a problem in the same time and thanks to them the network can solve problems millions times faster then actual used computers.

Second, the neural networks do not need to be programmed. It allows to find a solution with incomplete data and for very complicated algorithms. Plenty kind of architecture and learning methods make neural networks very useful tool to solve many different problems.

Since beginning of 80-th investigations in the area of neural computing have been developing. The most popular models of neural processing are realized in digital electronic technology but there are several attempts to utilize an optoelectronic technique.[1,2,3]

2. OPTOELECTRONIC NEURAL PROCESSING

Neuron is a typical cell in our neural system possessing many inputs and only one output as shown in diagram in Figure 1. Optoelectronic implementation of neuron models along with different types of neural networks may be done by use of combination of liquid crystal cells and optical fibers.[4] A neuron model established in such a way is shown in Figure 2.

The model consist of four semiconductor lasers acting as light sources and pigtailed to multimode lead-up fibers. Four directional couplers separate a part of the light from the fibers to detectors in order to control signal levels. Multimode fibers lead light to liquid crystal (LC) cells.

The light intensity transmitted through the LC cell is controlled by a change in voltage applied to the cell. The lead-out fibers collect light after LC cells and transmit it to only one photodetector. Learning of neuron requires changing in voltage applied to LC cells.

Optical Sensors and Microsystems: New Concepts, Materials, Technologies
Edited by Martellucci et al., Kluwer Academic / Plenum Publishers, New York, 2000.

Table 1. Results of calculations of output signals and results of experiments.

Laser				Transmission				Calculated value	Experimental data
I	II	III	IV	K I	K II	K III	K IV		
0	0	0	0	40	20	10	5	0	0
0	0	0	1					5	6
0	0	1	0					10	11
0	0	1	1					15	16
0	1	0	0					20	20
0	1	0	1					25	25
0	1	1	0					30	30
0	1	1	1					35	34
1	0	0	0					40	40
1	0	0	1					45	47
1	0	1	0					50	51
1	0	1	1					55	56
1	1	0	0					60	61
1	1	0	1					65	65
1	1	1	0					70	72
1	1	1	1					75	77

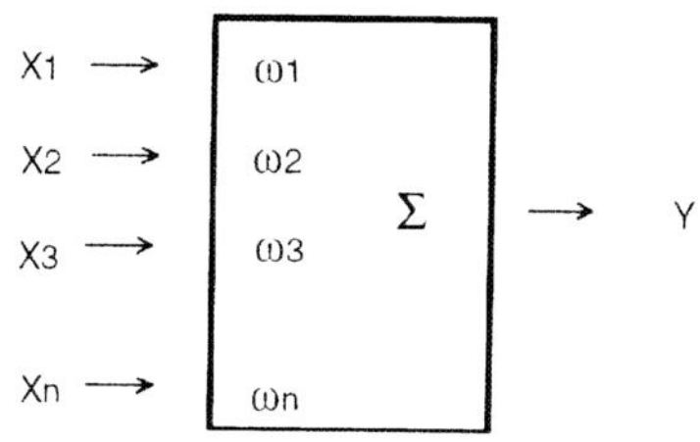

Figure 1. Diagram of neuron.

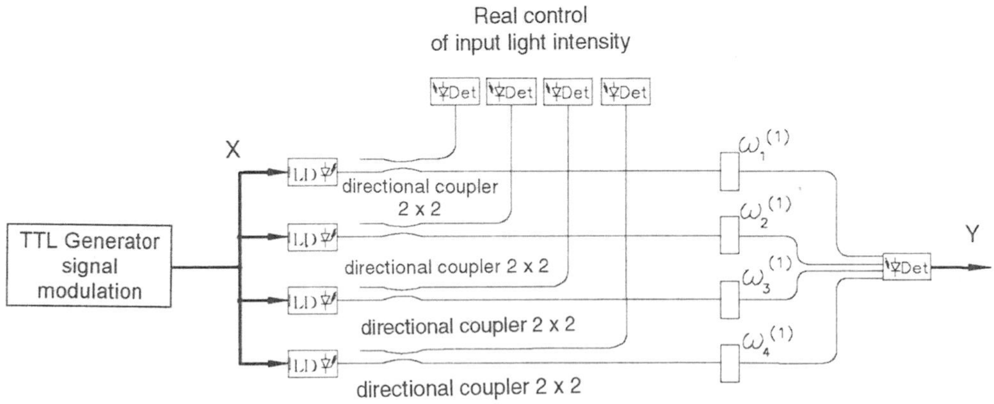

Figure 2. Set-up for experiments with optoelectronic model of neuron.

In our experimental set-up the supervised learning was applied. The scheme of neuron learning can be described as follows:

$$\omega_i^{(j+1)} = \omega_i^{(j)} + \eta(A - Y)x_i \qquad (1)$$

where i - weight number; j - iteration number; A - expected output signal; Y - real output signal; X - input signal.

Neuron operates properly only if input signals are digital (see Table 1.). In the beginning it is necessary to teach neuron to recognize an input vector by changing transmission of liquid crystal cells in order to adapt proper weights. After that an external supervised learning is not necessary.

3. OPTOELECTRONIC NEURAL NETWORKS

Different models of neural networks, based on Hopfield, Kohonen, Grossberg and Counter Propagation (CP) neural system have been proposed in the same way since the neuron is a basic element of all neural networks.[4]

The Grossberg network consists of several neurons (Figure 3.). Output signals of Grossberg layer have to be normalized. Such a layer forced needed answers of the network. Supervised learning of the layer means the modification only these neurons weights that were activated according to the following equation:

$$\omega^{(m)(j+1)} = \omega^{(m)(j)} + \eta\left[A^m - Z^{(m)(j)}\right]\left[Y^{(m)(j)}\right] \qquad (2)$$

where:

η	parameter describing the velocity of learning;
A	desired result;
$Z^{(m)(j)}$	output signal of m neuron in j iteration;
Y	input signal.

Counter propagation network consists of two layers. First layer (Kohonen layer) is submitted unsupervised learning. Unsupervised learning in Kohonen layer bases on competition between neurons. Only winning neuron is activated. This learning process can be described as follows:

$$\omega_i^{(m)(j+1)} = \omega_i^{(m)(j)} + NL\left[X_i + Y_m \cdot NL\left(\frac{1}{\sum_{n\neq m} Y_n}\right)\right] \qquad (3)$$

$$A^{(m)(j+1)} = A^{(m)(j+1)} - NL\left(\sum_{n\neq m} A^{(n)}\right) \qquad (4)$$

where:

i	weight number;
j	iteration number;
m	neuron number;
NL	non-linear function;
$A^{(m)(j)}$	expected output signal of m neuron in i iteration.

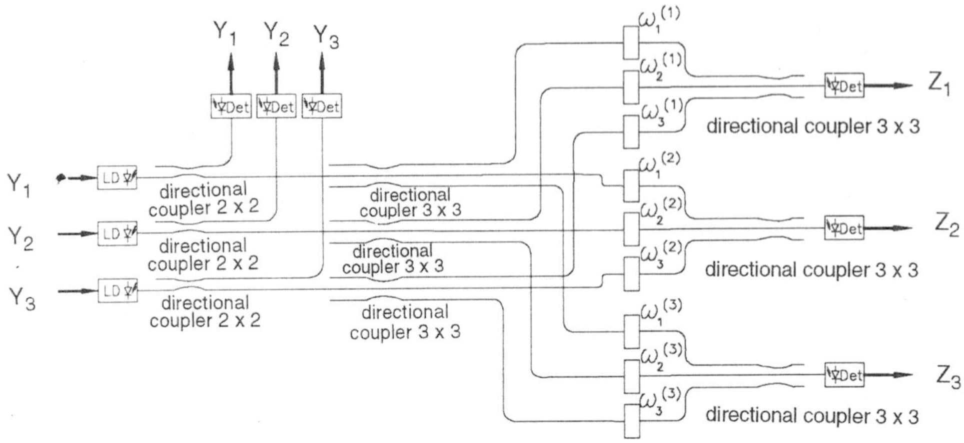

Figure 3. Scheme of linear neuron network with Grossberg layer with liquid crystal cells and fibers for connecting.

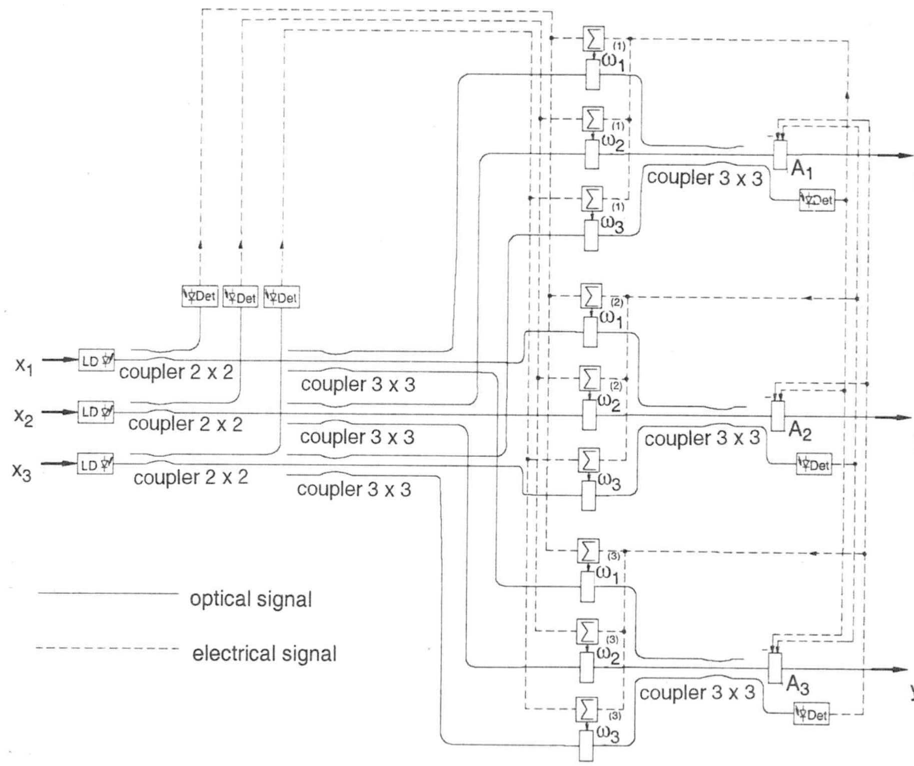

Figure 4. Scheme of neural network with Kohonen layer.

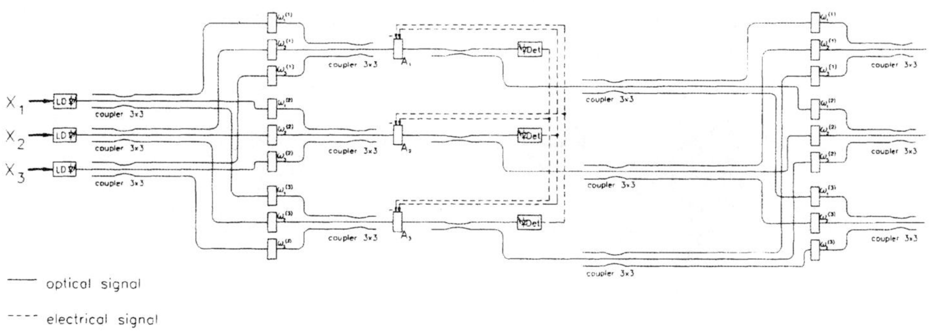

Figure 5. Scheme of neural Counter Propagation network with two layers of neurons.

Second layer, named Grossberg layer, consists of neurons being submitted supervised learning. CP network has possibility of infinite quality representation input X vector in output Z vector. To improve work of networks it is important to normalize input signal and increase neurons quality in Kohonen layer according to length input and output vectors. An example of Kohonen layer which could be realized with optical fibers and LC cells is shown in Figure 4. The scheme of possible CP-network realized in the similar way is shown in Figure 5.

4. SPATIAL LIGHT MODULATORS AND VERTICAL CAVITY SURFACE EMITTING LASERS IN NEURAL SYSTEMS

Application of single liquid crystal cells for parallel neural processing is very inconvenient. Hence two-dimensional spatial light modulators (SLMs) are commonly used. Spatial light modulation may be based on several physical phenomena such as:

- linear electrooptic Pöckel effect;
- quadratic electrooptic Kerr effect;
- acoustooptic Bragg diffraction;
- magnetooptical Faraday effect;
- different types of electro-mechanical effects e.g. micromirrors.

In general SLMs may be optically and electrooptically addressed. The most popular SLMs are based on application of electrically addressed liquid crystals displays working in transmission or reflection configurations. A typical SLM has 256x256 pixels although there are possibility to achieve 1000x1000 individually addressed pixels on area about 50x50 mm. It means that number of parallel channels with independent neural weights may be very high.

Transmission characteristics of each pixel with liquid crystal may be changed by electric signal with not very high speed but applications of ferroelectric liquid crystal can change this parameter from few milliseconds up to several microseconds.

There are other possibilities to build up the neural network with many parallel channels. One of them is to use a matrix of vertical cavity surface emitting laser (VCSELs). In contradiction to edge emitting laser diodes, VCSELs may have a very high density on small surface. Hence there are some attempts to use VCSELs as components of neural network by control of light emitted by each laser-pixel according to requirements of neural system.

VCSELs are also very helpful as light sources for optical fiber sensor system consisting of many sensors. Optical output signals from such systems may be processed by use of optoelectronic neuron or neural network (Figure 6.).

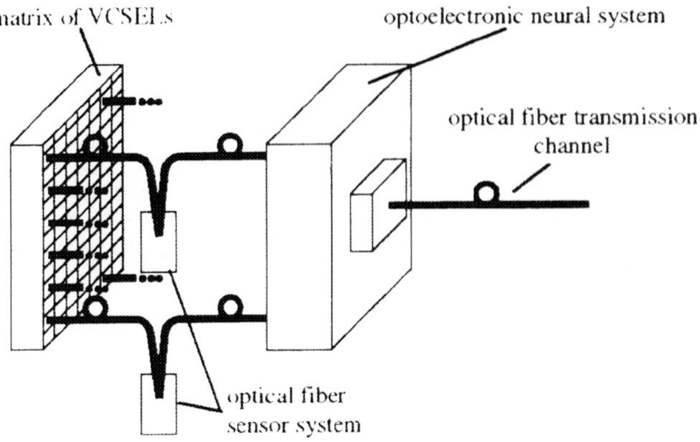

Figure 6. Optical fiber sensor system with VCSELs as light sources and neural processing of output signals.

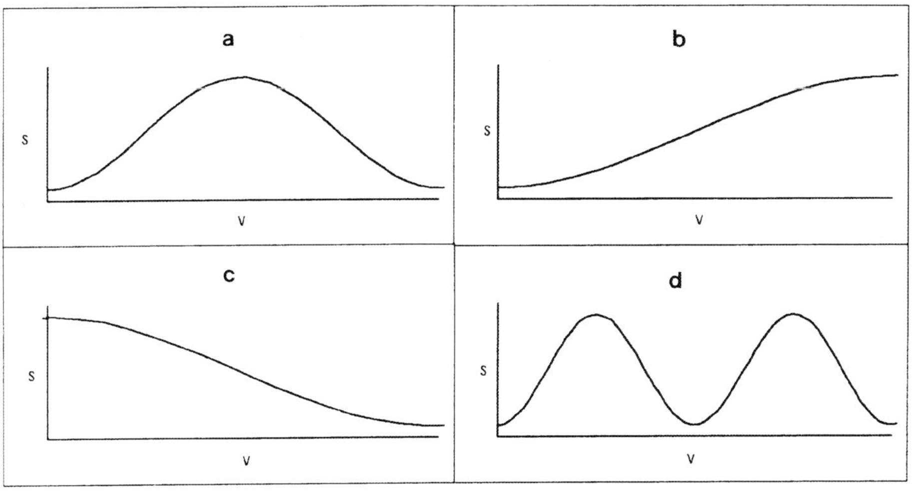

Figure 7. Typical output signals from optical fiber sensors for different variables.

5. OPTOELECTRONIC NEURAL SYSTEM FOR OPTICAL SENSOR SIGNAL PROCESSING

Optical fiber sensor output signals are usually electronically detected and transmitted to the control unit through many communications channels. The signals may be also transformed by microprocessor and transmitted through one channel with high transmission rate. Optical fiber sensor network, for example for environmental monitoring,[5] may have so many sensors that such method of data transmission is not sufficient. Hence an idea of optical fiber sensor signal processing by use of a optoelectronic neural system and then transmission of such prepared data to the control unit with help an ordinary transmission link seems to be very promising.

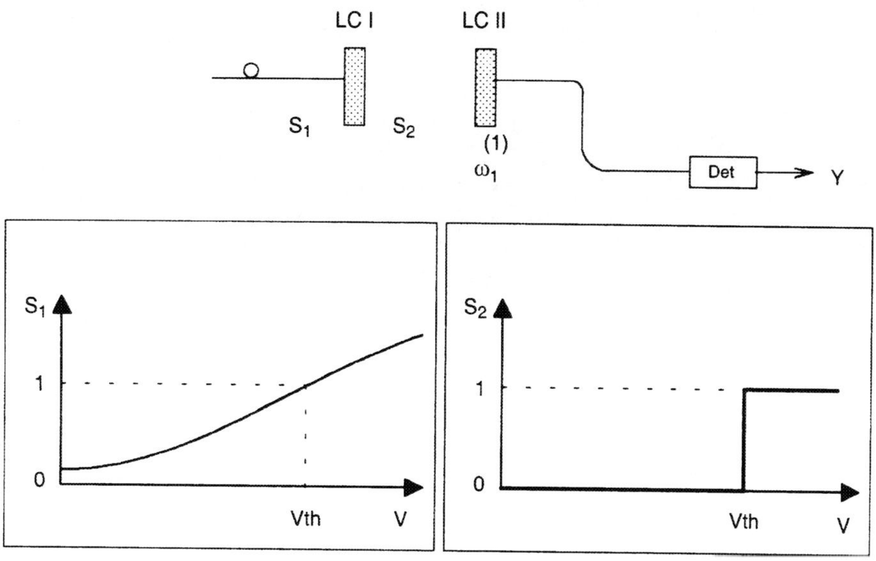

Figure 8. Additional transformation of optical fiber sensor signal for neural processing.

In general, there are several thousands types of optical fiber sensors with different types of output signals. Most of them, belonging to either intensity or phase sensors, have similar output signals (Figure 7.)

The characteristics shown in Figure 7. may be more or less linear and symmetrical with different slopes but they are almost the same for such different variables like pH, temperature proximity, strain, angular movement etc.[6,7] Such output signals cannot be directly applied for neural processing without additional transformation by use additional liquid crystal cells which have threshold transmission characteristics (Figure 8.).

Neural processing of optical fiber sensor signals may be useful for sensor network with threshold system of sensing. It means that slowly changing signals are transferred into step shape ones based on arbitrary chosen threshold levels of variables.

Both system of liquid crystal cells may be replaced by spatial light modulators in which each pixel may play role of active attenuator. Then number of input signals may be limited only by number of pixels.

In our experiments with neuron we used four optical fibers sensors instead of four laser diodes (Figure 9.). The temperature optical fiber sensor employees an idea of proximity sensing. Temperature changes length of rod in terms of length of cylindrical tube in different way due to different dilatation properties of rod and tube materials. We apply also proximity idea for hydrostatic pressure sensing. A membrane changes distance between fiber optic tip in terms of hydrostatic pressure and in the same changes optical power coming to the outgoing fibers.

The fiber optic pH sensors are based on the use of the acid-base indicators which change its absorbency in dependence on pH. The indicators were immobilized inside of cellulose acetate membrane according to the procedure described in.[7] The pH sensors exhibit two different working ranges. In all sensors the bundles of multimode fibers are used and they are coupled to the LEDs working synchronously in all sensors.

Part of light (20%) outgoing from each sensor is detected by photodiode and it is used for control of voltage supplying the liquid crystal cell.

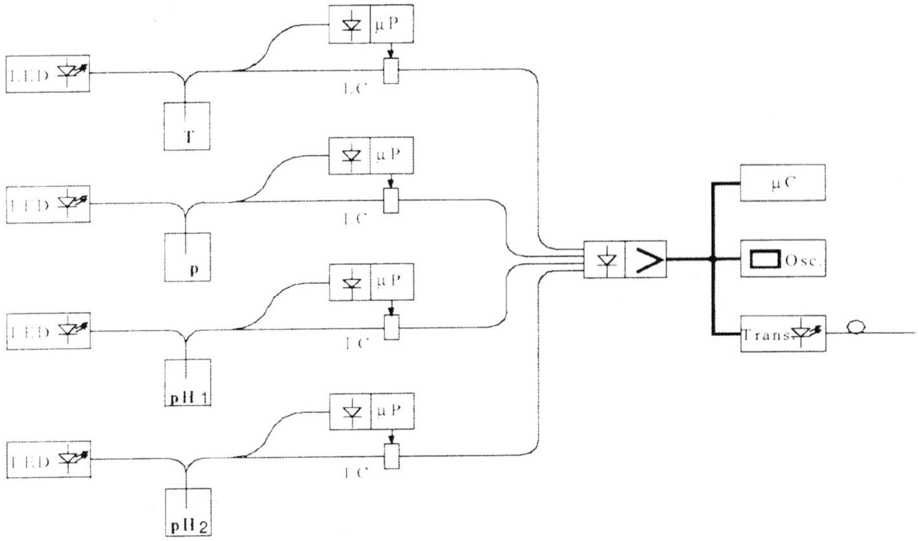

Figure 9. Optoelectronic neuron for output signal processing of optical fiber sensor system; electronic control system (μP) and liquid crystal cell (LC) work as a smart pixel.

The electronic system consists of sensor threshold discriminator which works according to our requirements regarding each sensor and of weight controller necessary for system of operation of the neuron (Figure 10.). Such LC cell with the electronic system is named smart pixel because for many sensors we should apply a spatial light modulator with many smart pixels.

6. RESULTS OF EXPERIMENTS

In experiments with neuron and smart pixels four optical fiber sensors giving output characteristics shown in Figure 11. were used. All characteristics should be transformed into step shape.

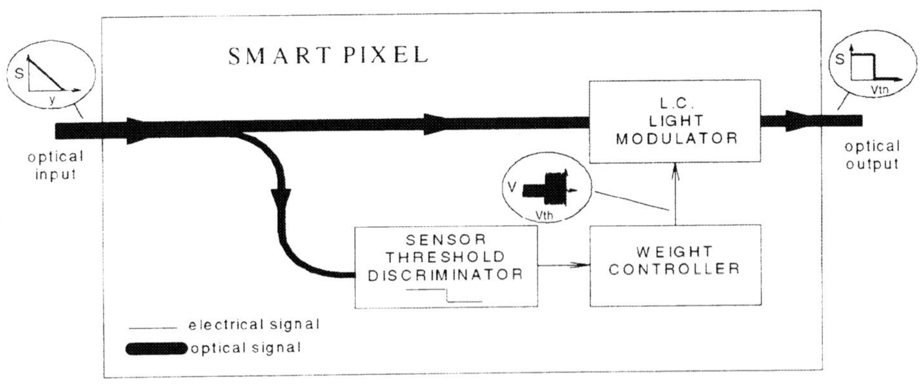

Figure 10. Liquid crystalline light modulator with sensor level discriminator and weight controller as a smart pixel.

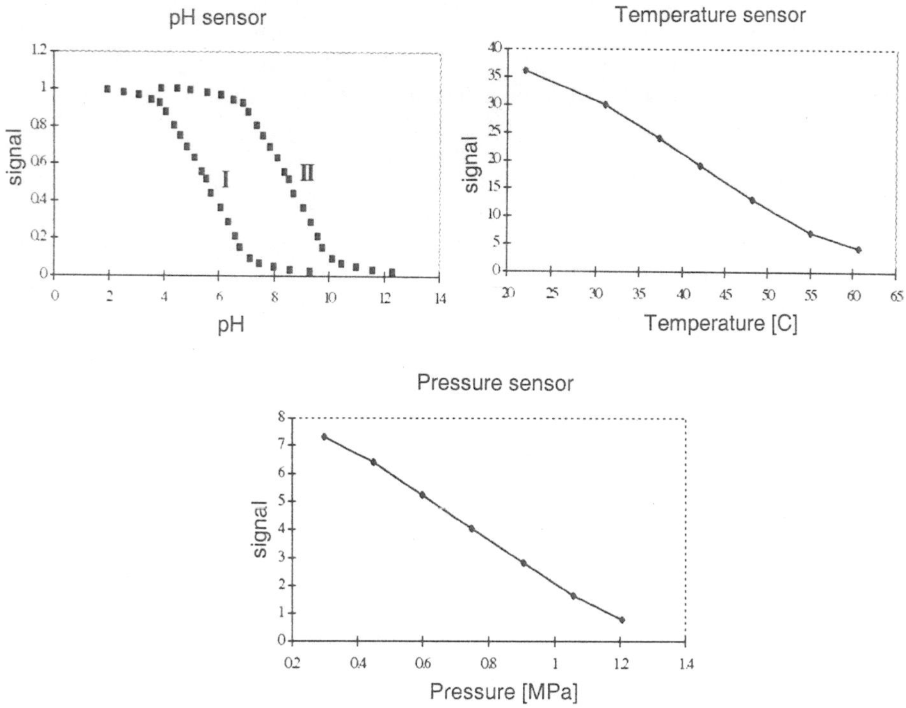

Figure 11. Measuring characteristics of optical fiber sensors used in experiments;two pH-sensors have different range of operation (I and II).

Table 2. Results of experiments with smart pixels working as a part of optoelectronic neuron and helpful in recognition of state of optical fiber sensor system.

States of the sensors				Optical output power				Total optical power		State of the sensor system concluded from measurements
pH_1	pH_2	T	p	pH_1 +smart pixel	pH_2 +smart pixel	T +smart pixel	p +smart pixel	calculated	measured	
pH	pH	°C	MPa					[• W]	[• W]	
3.2	3.2	25	0.1	10	20	40	80	150	148	All sensors below threshold
5.5	5.5	25	0.1	0	20	40	80	140	140	pH_1 over the threshold
9.8	9.8	25	0.1	0	0	40	80	120	122	pH sensors over the thresholds
9.8	9.8	50	0.1	0	0	0	80	80	78	p sensor below threshold. all others over
9.8	9.8	50	1.0	0	0	0	0	0	0	All sensors over threshold level
4.6	4.6	50	1.0	10	20	0	0	30	31	pH sensors below threshold, others over

A threshold value for each sensor was arbitrary chosen: $pH_1 = 5$, $pH_2 = 8$, $T = 40°C$, $p = 0.8$ MPa. Below the threshold value of the measured variable the sensor output optical power is transformed by use of smart pixel into optical power dependent only on requirements of neuron weights. In our case it is equal to 10 μW, 20 μW, 40 μW, 80 μW accordingly for pH_1, pH_2, T and p sensors. Over the threshold value the smart pixel attenuates optical output signal to zero.

7. CONCLUSIONS

Results of experiments show that neuron used for primary processing limits number of output data from the sensor system but such processing limits also applications of the method to the cases in which only threshold is important. The problem may be solved by use more advanced neural processing in which multithreshold system for each sensor could be applied.

System of neuron operation requires different levels of optical output power. The level values of sensing variables differ in geometrical progression. Hence number of sensors working with optoelectronic neuron as a processor are limited and depends of transmission characteristics of LC cell or each pixel of spatial light modulator. The problem may be particularly solved when the cascade of neuron layers creating a neural network are used.

ACKNOWLEDGMENT. The works leading to the paper are supported by grant of Polish State Committee for Scientific Research No: 8 T11B 04813.

REFERENCES

1. A. W. Domański, T. R. Woliński, A. Dybko, T. Sosin, W.J. Bock, Optical Fiber Sensors System for Forest Protection in Canada, *Proc. of SPIE*, Vol. 1716, pp. 87-90, 1992.
2. E. Udd, Editor, Fiber Optic Sensors, John Wiley & Sons, New York, 1991.
3. N.H. Farhat, D. Psaltis, A. Prata, E. Paek, Optical implementation of the Hopfield model, *Applied Optics*, vol. 24, No 10, pp. 1469 - 75, 1985.
4. Y. Nitta, J. Ohta, S. Tai, K. Kyuma, Optical learning neurochip with internal analog memory, *Applied Optics*, vol. 32, No 8, pp. 1264 - 74, 1993.
5. K. Wagner, D. Psaltis, Optical neural networks: an introduction by the feature editors, *Applied Optics*, vol. 32, No 8, pp.1261-74, 1993.
6. A.W. Domański, R. Buczyński, M. Sierakowski, Liquid crystal cells and optical fibers in neural network implementation, *Proc. of SPIE*, vol. 2372, pp. 354 - 359, 1995.
7. W. Wróblewski, E. Rożniecka, A. Dybko and Z. Brzózka „Optimisation of optomembranes based on cellulose acetate" EUROSENSORS XI, Warsaw 1997, Sensors and Actuators B - in press.

COMPLEX *ABCD*-MATRICES: A GENERAL TOOL FOR ANALYZING ARBITRARY OPTICAL SYSTEMS

[*]B. Rose, [†]S.G. Hanson, and [‡]H.T. Yura

[*]IBSEN Micro Structures A/S
Gammelgaardsvej 65
3520 Farum, Denmark

[†]Risø National Laboratory
P.O. Box 49
4000 Roskilde, Denmark

[‡]The Aerospace Corporation
P.O. Box 92957
Los Angeles, CA 90009, USA

1. INTRODUCTION

In this paper, we describe a novel formulation of light beam propagation through any complex optical system that can be described by an *ABCD* ray-transfer matrix.[1,2] Within the framework of complex *ABCD*-optical systems, we then present novel speckle methods for analyzing linear and angular velocities and displacements. All methods rely on the dynamics of speckle patterns, produced by scattering of coherent light off solid surfaces undergoing angular and/or translational displacements.

Coherent light scattered off a rough surface produces a granular diffraction pattern some distance away from the object. This pattern is referred to as speckles,[3] and originates from elementary interference of the waves emanating from many microscopic areas on the surface within the illuminated region. If the illuminated object is in motion, the resulting speckle pattern also evolves with time, i.e., a dynamical speckle pattern results. The target motion or displacement is usually determined by calculating the space- or time-lagged covariance of the detector currents before and after object displacement,[4] and then determine the position where the peak of the covariance attains its maximum.

We first present the basic properties of the *ABCD* ray-matrix method,[1,2] and summarize ray-matrices for simple optical elements,[5] e.g., thin lenses, mirrors, and free-space propagation. Additionally, we present the ray-transfer matrix for a Gaussian shaped aperture.[1,2] Thus, armed with matrix elements describing the most common elements in an optical system, the resulting ray-transfer matrix can be obtained for most systems encountered in practice. The resulting ray-transfer matrix connects the input ray position and slope with the corresponding output parameters. Rather than dealing with rays, a Gaussian Green's function has been developed,[1,2] valid for arbitrary *ABCD* systems, to reveal the transition of the field in the input plane, through the *ABCD*-optical system, to the output plane.

Armed with the Green's function that reveals the field in the output plane of an arbitrary *ABCD*-system, provided the field in the input plane is known, we then present a general equation, valid for arbitrary *ABCD*-optical systems, for the space-time lagged covariance of photocurrent.[4] The position where the Gaussian-shaped covariance attains its maximum reveals information about target velocity or displacement. We discuss, how, in practice, the target velocity/displacement is assessed. The theoretical results alluded to above are then applied to a series of novel optical sensor applications, all described by the *ABCD* kernel.

Two systems for measuring linear velocities, viz. the laser Doppler velocimeter[6] and the laser time-of-flight velocimeter,[7,8] are presented. A compact system for measuring linear surface velocities is presented, where all passive optical components are replaced with a single holographic optical element.[9] Additionally, other novel schemes for further system miniaturization are presented.[10,11]

We then present a new method for measurement of out-of-plane angular displacement in one or two dimensions.[12,13] It is demonstrated that the angular displacement sensor is insensitive to both object shape and target distance, and any transverse or longitudinal movements of the target. It is further shown that the method has a resolution of 0.3 mdeg ($\approx$5 μrad).

A new method for measuring in-plane angular velocities or displacements are then presented.[14] Here, we consider off-axis illumination, and it is shown that, for Fourier transform optical systems, in-plane rotation causes the speckles to translate in a direction perpendicular to the direction of surface motion, whereas for an imaging system, the translation is parallel to the direction of surface motion. Based on this, we discuss a novel method, which is independent of both the optical wavelength and the position of the laser spot on the object, for determining either the angular velocity or the corresponding in-plane displacement of the target object.

The out-of-plane angular displacement sensor can be modified to measure the distribution of static torsion angles of targets undergoing twisting motion.[15] Because the torsion angle sensor is independent of object shape, we measure the distribution of torsion angles in both uniform and non-uniform deformation zones.

Finally, we present a novel method for measuring out-of-plane angular velocities.[16] Besides measuring angular velocities, the sensor can measure, simultaneously, torsional vibrations of the rotating shaft.[17]

2. COMPLEX *ABCD*-MATRICES

In most of the methods, commonly used to calculate the response of optical systems, full expressions are written for the response of the single elements in the optical train. Tracing the field from one element to the next usually requires mathematical approximations in each step. The complex *ABCD*-matrix method attacks the problem differently, because the approximations are performed on the optical system itself avoiding any mathematical approximations in the pursuing calculations. Therefore, the result obtained is valid for an arbitrary *ABCD*-optical system.

We limit our considerations to the case where all illuminating beams are of Gaussian shape (i.e., that obtained from a TEM_{00} laser mode). That is, all intensity and transmission functions are of Gaussian shape. The reason for this choice is, that the Green's functions in the Fresnel regime for free-space propagation as well as the kernel for quadratic phase change introduced by a lens are of complex Gaussian form.

Free space of length L:	$\begin{bmatrix} 1 & L \\ 0 & 1 \end{bmatrix}$
Thin lens with focal length f:	$\begin{bmatrix} 1 & 0 \\ -1/f & 1 \end{bmatrix}$
Flat dielectric interface (refractive indices, n_1 and n_2):	$\begin{bmatrix} 1 & 0 \\ 0 & n_1/n_2 \end{bmatrix}$
Curved mirror (radius of curvature, R):	$\begin{bmatrix} 1 & 0 \\ -2/R & 1 \end{bmatrix}$
Curved dielectric interface of radius (R, n_1, and n_2):	$\begin{bmatrix} 1 & 0 \\ (n_1-n_2)/n_2 R & n_1/n_2 \end{bmatrix}$
Thin lens f followed by free space L:	$\begin{bmatrix} 1-L/f & L \\ -1/f & 1 \end{bmatrix}$

Figure 1. Ray matrices for simple optical elements 5.

Furthermore, correlating, convoluting, multiplying, and integrating Gaussian expressions reveal a new Gaussian expression, and consequently, simplify considerably the underlying mathematical calculations. Both rotationally symmetric systems and systems given in Cartesian coordinates can be analyzed.

The concept of *ABCD* matrices for systems including simple optical elements (i.e., no limiting apertures) have been used extensively for laser cavities.[18] The connection between the input ray position x_{in}, the input ray slope x'_{in} and the corresponding output parameters, x_{out} and x'_{out}, are given by the *ABCD*-matrix following

$$\begin{bmatrix} x_{out} \\ x'_{out} \end{bmatrix} = \begin{bmatrix} A & B \\ C & D \end{bmatrix} \begin{bmatrix} x_{in} \\ x'_{in} \end{bmatrix} \tag{1}$$

where the matrix elements for various quadratic surfaces are given in Figure 1.

A special characteristics for the matrices depicted in Figure 1. is that the determinant of the matrix equals unity (i.e., $AD-BC=1$), if the input- and output media are identical (i.e., $n_{in}=n_{out}$). A system consisting of several elements, each of which are characterized by a matrix, is defined by a matrix given by the product of the individual matrices. The matrix of the first optical element in the train from the input plane to the output plane is the rightmost matrix. Note that all matrix elements are real valued.

We consider the optical setup depicted in Figure 2. Knowing the Green's function for an arbitrary optical system - within the limitations given above - the field in the output plane $U(\mathbf{p})$ is given by[1,2]

$$U(\mathbf{p})=\left(\frac{ik}{2\pi B}\right)\int_{-\infty}^{\infty} d\mathbf{r} U_o(\mathbf{r}) \exp\left[\frac{ik}{2B}\left(A\mathbf{r}^2-2\mathbf{rp}+D\mathbf{p}^2\right)\right] \tag{2}$$

where $U_o(\mathbf{r})$ is the field in the input plane and A, B, and D are the matrix elements for the arbitrary optical system and $\mathbf{r}$ and $\mathbf{p}$ are the two-dimensional vectors corresponding to the input-

and output planes, respectively. In Eq. (2) we have already assumed identical media in the input and output planes, i.e., the C-matrix element is given uniquely by the other three elements.

Although the ray-matrices depicted in Figure 1. cover a broad range of optical elements, most optical systems include limiting apertures, for example, in the optical train itself or in the detector plane (receiver apertures). Limiting apertures, and consequently diffraction effects, can be including in the matrix method by comparing the transmission function for a lens, given by[1,2]

$$T_{lens} = \exp\left[-ik\frac{\mathbf{r}^2}{2f}\right] \tag{3}$$

with the transmission function for a Gaussian-shaped aperture given by

$$T_{LA} = \exp\left[-\frac{\mathbf{r}^2}{\sigma^2}\right] \tag{4}$$

where σ is the $1/e^2$ intensity radius of the aperture. Given the matrix for a lens (see Figure 1.), it is easily seen that the matrix for the limiting aperture is given by

$$M_{LA} = \begin{bmatrix} 1 & 0 \\ -\frac{2i}{k\sigma^2} & 1 \end{bmatrix} \tag{5}$$

where k is the optical wavenumber.

Having introduced the matrix for limiting apertures provides us with a range of optical elements that will cover most systems encountered in practice. Note that the inclusion of limiting apertures reveals, in general that the matrix elements are complex.

The *ABCD* kernel can also be used to include the effect of longitudinal and transversal misalignment of the individual element in the optical train.[19] Additionally, the kernel has been used to analyze both optical propagation through stochastic media[20] (i.e., optical turbulence) and to include the influence of "dirty optics"[21] (e.g., scattering particles or absorbing aerosols on a lens). Finally, the *ABCD* kernel has been used to describe speckle decorrelation in interferometers[22] and to describe first-order speckle statistics.[23]

In the following, we allude the findings given above to a general optical system where light scattered off a solid surface describes the field in the input plane and detectors are used to measure the resulting field in the output plane.

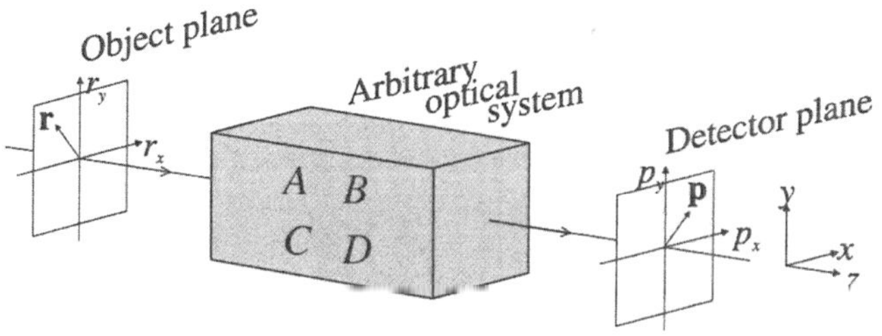

Figure 2. Propagation of reflected light from the input plane through an arbitrary ABCD optical system to the output plane.

3. SPACE-TIME LAGGED COVARIANCE FOR COMPLEX *ABCD* SYSTEMS

We consider light scattered off a rough target in the **r**-plane (i.e., the object plane) through an arbitrary optical system to the **p**-plane (i.e., the detector plane) where the speckle pattern is observed (see Fig. 2). Note, for simplicity, the illuminating light source has been omitted in Figure 2. We seek to determine the space-time lagged covariance of the photocurrent obtained from two spatially separated "point" detectors that are located in an arbitrary observation plane, transverse to the optic axis (here assumed to be the z-axis) of an *ABCD* system.[4] This quantity is given by

$$C_i(\mathbf{p}_1,\mathbf{p}_2;\tau) = \langle i(\mathbf{p}_1,t)i(\mathbf{p}_2,t+\tau)\rangle - \langle i(\mathbf{p}_1,t)\rangle\langle i(\mathbf{p}_2,t+\tau)\rangle \tag{6}$$

where angular brackets denote the ensemble average, i denotes photocurrent, t and τ denote the instantaneous time and time-lag, respectively, and $\mathbf{p}_1$ and $\mathbf{p}_2$ are the two-dimensional vector coordinates of the (point) detectors in a plane transverse to the optics-axis. The instantaneous photocurrent from either detector is given by

$$i_j(\mathbf{p},t) = \alpha\int d\mathbf{p}W_j(\mathbf{p})I(\mathbf{p},t) \qquad j=1,2 \tag{7}$$

where $I(\mathbf{p},t)$ is the corresponding observation plane intensity distribution, $W_j(\mathbf{p})$ is the receiver aperture (intensity) weighting function, where, for simplicity, we assume identical receiver aperture weighting functions for the two detectors. The conversion constant α (power to current) is equal to $q\eta/h\nu$, where q is the electronic charge, η is the detector quantum efficiency, ν is the optical frequency, and h is Planck's constant. For detector-aperture sizes small compared to the speckle size (e.g., individual pixels in a detector array) Eq. (7) becomes

$$i(t) \cong \alpha\, S_d\, I(\mathbf{p},t) \tag{8}$$

where S_d is the area of the detector. The instantaneous intensity can be written as

$$I(\mathbf{p},t) = |U(\mathbf{p},t)|^2 \tag{9}$$

with

$$U(\mathbf{p},t) = \int d\mathbf{r}U_o(\mathbf{r},t)G(\mathbf{r},\mathbf{p}) \tag{10}$$

where $G(\mathbf{r},\mathbf{p})$ is the Green's function for the *ABCD* optical system, given by[1,2]

$$G(\mathbf{r},\mathbf{p}) = -\frac{ik}{2\pi B}\exp\left[-\frac{ik}{2B}\left(A\mathbf{r}^2 - 2\mathbf{r}\cdot\mathbf{p} + D\mathbf{p}^2\right)\right] \tag{11}$$

Note that inserting Eq. (11) into Eq. (10) yields Eq. (2). The quantity $U_o(\mathbf{r},t)$ is the "instantaneous" reflected optical field in the object plane (assumed here to be in the $r_x - r_y$ plane).

We assume that the amplitude of the reflection coefficient of the object surface is constant, while the phase exhibits partial spatial coherence. Specifically, we model the reflected optical field as

$$U_o(r,t) = U_i(\mathbf{r},t)\,\psi(\mathbf{r},t) \tag{12}$$

where $U_i(\mathbf{r},t)$ is the incident field (assumed fixed in space), and $\psi(\mathbf{r},t)$ is the stochastic time varying (due to the motion of the object) complex reflection coefficient. We assume that the magnitude of the reflection coefficient $|\psi(\mathbf{r},t)| = 1$ and that $\langle\psi(\mathbf{r}_1,t)\psi^*(\mathbf{r}_2,t)\rangle = \delta(\mathbf{r}_1 - \mathbf{r}_2)$, where $\delta(\mathbf{r})$ is the Dirac delta function. For an object moving with a constant velocity $\mathbf{v}$, we assume that the time evolution of the reflected phase is given by

$$\psi(\mathbf{r},t+\tau) = \psi(\mathbf{r} - \mathbf{v}\tau, t) \tag{13}$$

That is, the reflected phase at any position in a coordinate system moving along with the diffuser does not change in time (i.e., a Taylor's hypothesis).

On substituting Eqs. (8) and (9) into Eq. (6), we obtain from the first term on the right-hand side of Eq. (6) a very complex term for the intensity covariance of the form

$$\langle U_o(\mathbf{r}_1,t)U_o^*(\mathbf{r}_2',t+\tau)U_o^*(\mathbf{r}_2,t)U_o(\mathbf{r}_1',t+\tau)\rangle, \tag{14}$$

where $\mathbf{r}_1$, $\mathbf{r}_1'$, $\mathbf{r}_2$, and $\mathbf{r}_2'$ are integration variables that result when the substitutions alluded to above are carried out. Primed variables correspond to measurements on a displaced surface. Assuming that the field in the observation plane, to a good approximation, obeys circular complex Gaussian statistics,[3] the average implied by expression (14) can then be written as

$$\begin{aligned}
&\langle U_o(\mathbf{r}_1,t)U_o^*(\mathbf{r}_2',t+\tau)U_o^*(\mathbf{r}_2,t)U_o(\mathbf{r}_1',t+\tau)\rangle = \\
&\langle U_o(\mathbf{r}_1,t)U_o^*(\mathbf{r}_2,t)\rangle\langle U_o^*(\mathbf{r}_2',t+\tau)U_o(\mathbf{r}_1',t+\tau)\rangle \\
&+\langle U_o(\mathbf{r}_1,t)U_o^*(\mathbf{r}_2',t+\tau)\rangle\langle U_o^*(\mathbf{r}_2,t)U_o(\mathbf{r}_1',t+\tau)\rangle.
\end{aligned} \tag{15}$$

From the discussion preceding Eq. (13), it is easily seen that the first term on the right-hand side of Eq. (15) will yield an expression that is identically canceled by the second term on the right-hand side of Eq. (6). As a result, we obtain that

$$C_I(\mathbf{p}_1,\mathbf{p}_2;\tau) = \alpha^2 S_d^2 K(\mathbf{p}_1,\mathbf{p}_2;\tau), \tag{16}$$

where

$$\begin{aligned}
K(\mathbf{p}_1,\mathbf{p}_2;\tau) = &\int d\mathbf{r}_1 \int d\mathbf{r}_2 \int d\mathbf{r}_1' \int d\mathbf{r}_2' \\
&\times G(\mathbf{r}_1,\mathbf{p}_1)G^*(\mathbf{r}_2,\mathbf{p}_2)G(\mathbf{r}_1',\mathbf{p}_2)G^*(\mathbf{r}_2,\mathbf{p}_2) \\
&\times\langle U_o(\mathbf{r}_1,t)U_o^*(\mathbf{r}_2',t+\tau)\rangle \\
&\times\langle U_o^*(\mathbf{r}_2,t)U_o(\mathbf{r}_1',t+\tau)\rangle.
\end{aligned} \tag{17}$$

is the intensity covariance. By employing Eqs. (9), (12) and (13), we obtain that

$$K(\mathbf{p}_1,\mathbf{p}_2;\tau) = \left|a\int d\mathbf{r}_1 U_o(\mathbf{r}_1)G(\mathbf{r}_1,\mathbf{p}_1)\int d\mathbf{r}_{2\tau}U_o^*(\mathbf{r}_{2\tau})G^*(\mathbf{r}_{2\tau},\mathbf{p}_2)\right|^2 \tag{18}$$

where $a = 2\pi/k$ for reflection into half-space and

$$\mathbf{r}_{2\tau} = \mathbf{r}_2 + \mathbf{v}\tau. \tag{19}$$

In order to proceed, we specialize to the case in which the reflected field is of (complex) Gaussian shape (i.e., that obtained from a TEM_{00} laser mode) given by

$$U_i(r) = \sqrt{\frac{2P_i}{\pi r_s^2}} \exp\left[-r^2 \left(\frac{1}{r_s^2} + \frac{ik}{2R_i} \right) \right] \tag{20}$$

where r_s, R_i, and P_i are the $1/e^2$ intensity radius, radius of curvature, and power, respectively of the incident beam (see Figure 3.(a)). Note that a converging beam corresponds to $R_i < 0$ whereas a diverging beam corresponds to $R_i > 0$.

Strictly speaking Eq. (20) is valid for planar objects. For a cylindrical object, of radius R_c, with its axis of symmetry perpendicular to the axis of the illuminating beam (see Figure 3.(b)), the reflected optical field $U_i(r)$ in the initial plane (i.e., $z = 0$) at the coordinate $\{r_x, r_y, 0\}$ contains the multiplicative factor $\exp(2ikbr_x^2)$,[6,16] where $b = 1/2R_c$ as indicated in Figure 3.(b). Thus, for a cylindrical object, of radius R_c, the quantity $1/R_i$ in Eq. (20) is replaced by $1/R_i + 2/R_c$ in the x-direction. Following the derivation of the matrix for the limiting aperture (see the discussion in Chap. 2), a finite aperture radius of the illuminating beam can be modeled, in the matrix terminology, as a lens of focal length $-R_i$ whereas a cylindrical target can be modeled as a first cylindrical lens with focal length $-R_c/2$ in the x-direction.

Substituting Eqs. (11) and (20) into Eq. (18), and performing the resulting integration yields

$$C_i(\mathbf{p}_1, \mathbf{p}_2; \tau) = C_0 \left[1 - \exp(-\sigma_\phi^2)\right]^2 \exp\left[-2 \frac{(p_1^2 + p_2^2)}{\omega_s^2} \right] |\gamma(\Delta\mathbf{p}; \tau)|^2 \tag{21}$$

for the space-time lagged photocurrent covariance, where

$$|\gamma(\Delta\mathbf{p}; \tau)| = \exp\left\{ - \frac{\left(A_i - \frac{2B_r}{kr_s^2} \right)^2 (\mathbf{v}\tau)^2 + \left[\Delta\mathbf{p} - \left(A_r + \frac{2B_i}{kr_s^2} \right) \mathbf{v}\tau \right]^2}{\rho_0^2} \right\} \tag{22}$$

and

$$\rho_0 = \sqrt{\frac{8|B|^2}{k^2 r_s^2} + \frac{4}{k} \mathrm{Im}(BA^*) + r_c^2 \left[|A|^2 + \frac{4|B|^2}{k^2 r_s^4} + \frac{4\,\mathrm{Im}(BA^*)}{kr_s^2} \right]} \tag{23}$$

We will not discuss, in detail, the properties of this result, but instead refer the interested reader to Ref. 4 for a thorough discussion of the results. Here, we seek to determine the position where peak of the space-time lagged covariance of the photocurrents attains its maximum. Examination of Eq. (22) reveals that this quantity is given by[4]

$$p_0 = \left(A_r + \frac{2B_i}{kr_s^2} \right) \mathbf{v}\tau \tag{24}$$

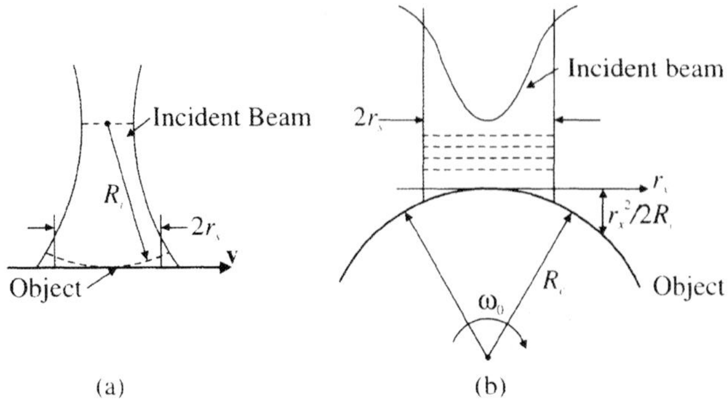

Figure 3. Measurement geometry for (a) a flat object moving with a constant velocity v and (b) for a cylindrical target with radius of curvature R_c moving with constant angular frequency ω_0. The incident beam has a wavefront-curvature radius R_i and a $1/e^2$ spot radius r_s.

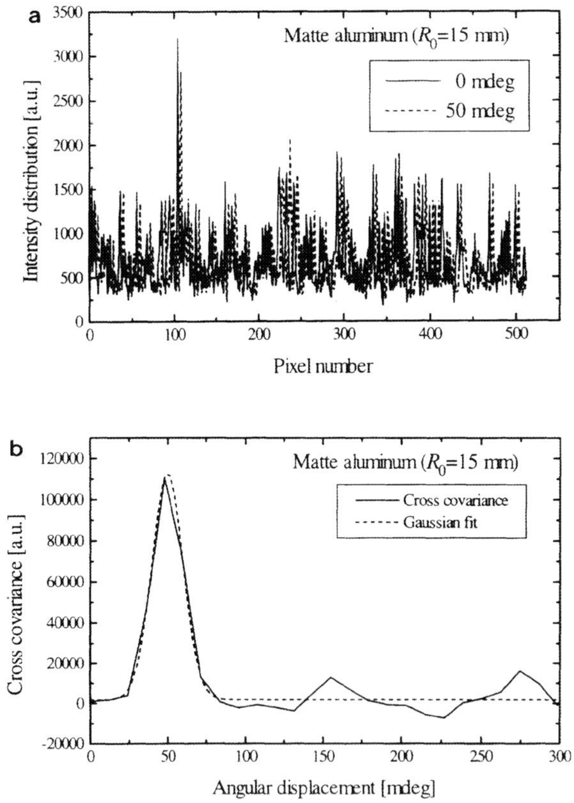

Figure 4. Two speckle patterns with a mutual angular displacement of 50 mdeg (a) and the corresponding cross covariance (b) peaked at p_0 =50 mdeg.

where p_0 is the detector plane displacement, v is the object velocity, τ is a time lag, and $(A,B)_{r,i}$ are the real and imaginary parts of the A- and B-matrix elements of the optical system. Equation (24) reveals that if the target is displaced the distance $v\tau$, then the speckle pattern is displaced the distance $\left(A_r + 2B_i/kr_s^2\right)v\tau$ in the detector plane. We consider two different methods for determining the speckle displacement. Consider first the case where the target is displaced a small distance $\Delta r = v\tau$. The speckle pattern is recorded on a linear image sensor (a CCD-array) before and after displacement of the target. The readout from the image sensor after displacement is then cross correlated with the readout recorded before the target displacement. The position where the covariance attains its maximum then reveals p_0, and for the current optical system (where the A- and B-matrix elements are known), calculating the displacement $v\tau$ is straightforward.

Consider now another case where the target is moving with a constant velocity v. By positioning two detectors with a known mutual distance in the detector plane (i.e., Δp is known) and correlating the two detector currents reveals the time-lag τ_0, when a speckle pattern seen by detector 1 is again seen by detector 2. That is, the time-lag τ_0 is given by $\Delta p/\left(A_r + 2B_i/kr_s^2\right)v$. As the time lag is obtained from the peak of the covariance, calculation of the velocity is straightforward.

Figure 4.(a) illustrates two speckle patterns, recorded in the Fourier plane before and after target displacement, whereas Figure 4.(b) shows the corresponding covariance.

4. LASER DOPPLER AND LASER TIME-OF-FLIGHT VELOCIMETERS

The ABCD kernel has been used to analyze two different methods for measuring linear velocities of solid surfaces, viz. the laser Doppler velocimeter (LDV)6 and the laser time-of-flight velocimeter (LTV).7-11,21 Both systems rely on a clean imaging system[2] as is depicted in Figure 5. (the receiver part) for an LTV. Here, we limit our discussion to the LTV system, and refer the interested reader to Ref. 6 for a thorough analysis of the LDV principle.

The LTV system[7-11,21] illustrated in Figure 5. is based on determining the time it takes for a scattering element on a moving surface to travel from one illuminated region to the next. The spacing between the two regions is known ($2d$) and the velocity v is then given by the ratio of the spacing to the measured time of flight (i.e., $v = 2d/\tau_0$). This result is easily obtained from simple inspection of the optical system depicted in Figure 5. Use of the *ABCD* kernel, where we include the matrix elements for an imaging system (see the receiver part of Figure 5.), reveals the same result.[7]

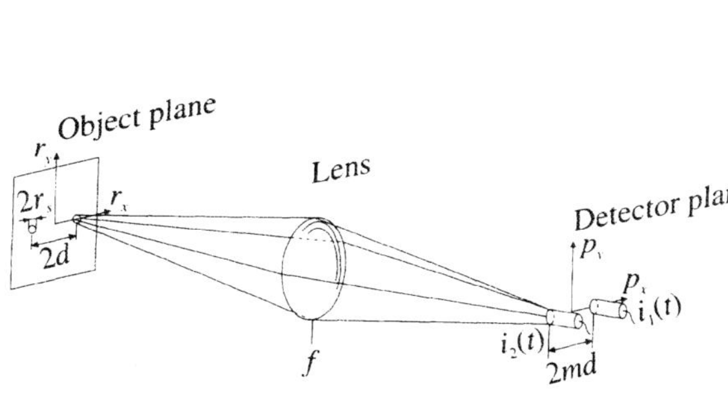

Figure 5. Optical diagram for the LTV-setup. The laser beams from two VCSEL's are focused onto the moving target. The spot size r_s is 10 μm and the mutual separation $2d$ is 330 μm. The focal lengths are f_1 = 60 mm, f_2 = 40 mm, and f_3 = 20 mm. The lasers emitted at 850 nm.

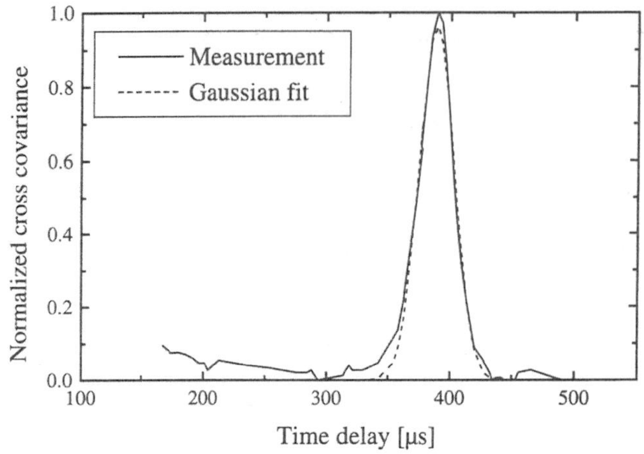

Figure 6. Peak of the covariance for an LTV system centered at τ_0=383 μs, which corresponds to a velocity of 0.86 m/s.

Additionally, because this result is valid for arbitrary *ABCD* systems, it is straightforward to include, for example, defocusing effects, and then perform a thorough trade-off analysis for the LTV system.

Both methods will reveal the same time-of-flight τ_0 that can be obtained by determining the time delay for maximum cross covariance of the detected signals arising from scattered light from the two illuminated regions, which is shown in Figure 6.

The system depicted in Figure 5. can be constructed in a much more compact form.[9-11] An example of a compact velocimeter, where the lenses are integrated into a single holographic optical element are discussed in Ref. 9. In Figure 7., we depict a further miniaturized version of the LTV sensor. This design employs VCSEL's and detectors, both mounted in a 20-pin DIP package, and a combined transmitter/receiver microlens array.

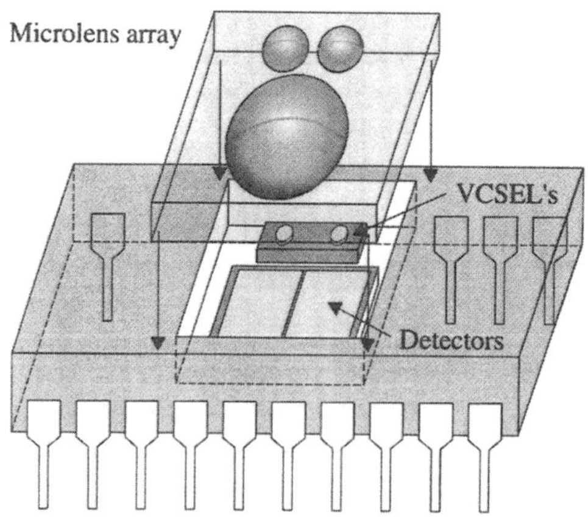

Figure 7. Miniaturized concept for the LTV system employing VCSEL's, a microlens array and a twin-detector. The sensor is housed in a 20-pin lead DIP package, and the total outer dimensions are approximately 8×8×25 mm^3.

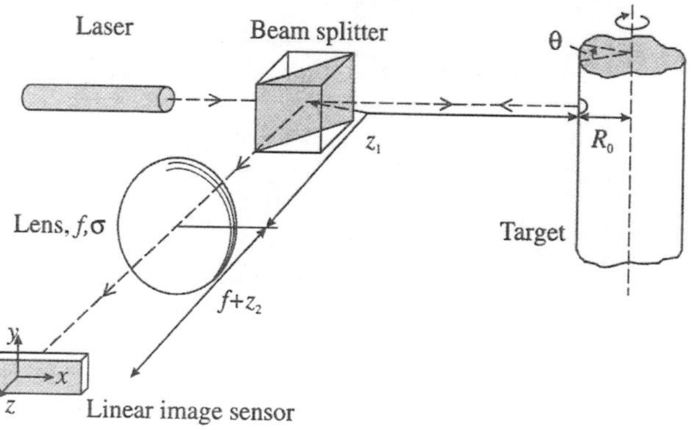

Figure 8. Optical diagram for the angular displacement sensor setup. A collimated HeNe beam illuminates the target with a spot size r_s=0.75 mm. The focal length of the lens is 60 mm, and the image sensor had 512 pixels with a pitch of 25 μm.

5. OUT-OF-PLANE ANGULAR DISPLACEMENT

A schematic of the angular displacement sensor setup that has been used for the measurements is shown in Figure 8.[12,13] A rotating stage with a stated resolution of 1 mdeg provided the angular displacement for the various objects.

The primary parameter of interest for the angular displacement sensor is the relation between applied angular displacement and the measured linear displacement of the speckle pattern on the image sensor. Following the procedure of complex *ABCD*-matrices given above for the optical system depicted in Figure 8., and formally replacing $v\tau$ with $R_0\theta$ in Eq. (24) (which is now the displacement of the target in the x-direction) yields[12,13]

$$p_{x0} = 2\theta\left[f + \frac{z_2}{f}\left(f - \frac{R_0}{2} - z_1\right)\right] \tag{25}$$

where z_1 is the distance from the target to the lens (i.e., the target distance) and z_2 is the distance from the image sensor to the Fourier plane. We note, that for angularly displaced targets, we include the effect of the angular displacement as a first cylindrical lens of focal length $-R_0/2$ in the x-direction, as was discussed following Eq. (20).[12] Note, furthermore that, if the image sensor is placed in the Fourier plane ($z_2 = 0$), we expect the linear displacement in the Fourier plane to be insensitive to the distance from the surface of the target to the axis of rotation (for shafts, the radius of curvature) R_0 both independent of the target distance z_1 and the wavelength.

A set of measurements was performed on a matte aluminum shaft ($R_0 = 15$ mm) in steps of 1 mdeg, which is the smallest possible step of the rotational stage. The results are illustrated in Figure 9.(a), and it is clearly seen that there is a slight negative offset between the measured and applied angular displacements. Other measurements showed positive offsets, which indicates that the offset is caused by play in the rotational stage. This is further emphasized by the measurements as pairs of measurement points seem identical because of play in the stage. After each pair of measurements, the stage catches up with the play. With a precision stage, controlled with interferometry, we performed 25 statistically independent measurements (i.e., on different part of a surface giving rise to fully developed speckle) applying the same angular displacement. These measurements revealed a resolution in angular displacement of 0.28 mdeg (5 μrad).

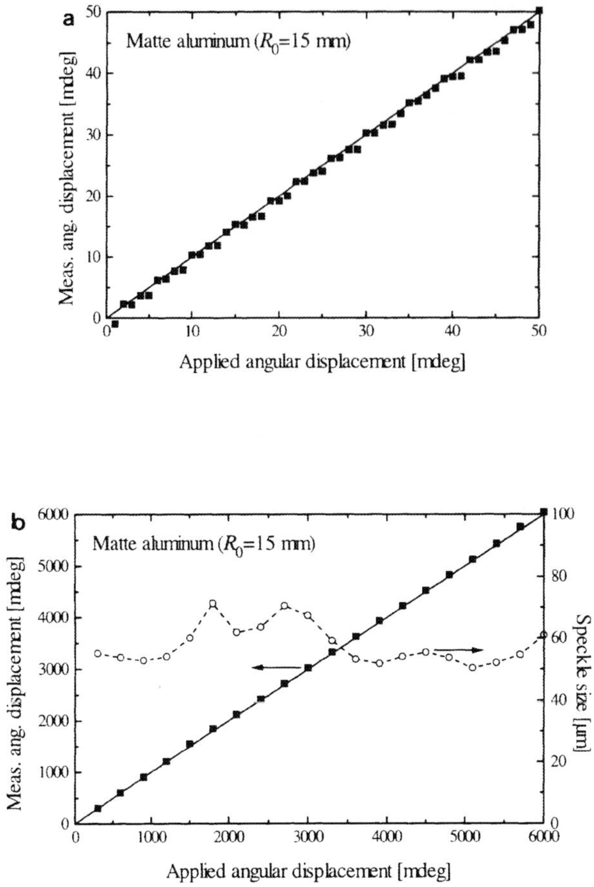

Figure 9. Measurements performed on a matte aluminum shaft (giving rise to fully developed speckle) (a) in steps of 1 mdeg (left) and (b) total angular displacement of 6000 mdeg measured in steps of 300 mdeg (right).

If measurements of larger angular displacements are required, it is necessary to measure in small steps and use each consecutive measurement as the reference for the next iteration and so on. Figure 9.(b) shows the measurements for a total angular displacement of 6000 mdeg performed in 20 steps of 300 mdeg. Note that the measured angular displacements (squares) fit the theory (solid straight curve) without biasing effects.

We rotated the shaft an angle θ and then displaced the shaft linearly Δr_x in the x-direction, perpendicular to the optical axis and the shaft axis.[12] This introduces decorrelation, because the overlap spot areas are not identical, but more importantly, the measurements reveal that the measured angular displacement is independent on linear shaft displacement and can be determined even for a large linear displacement.

For the 2D measurements,[13] we used the same setup as the one depicted in Figure 8., but replaced the linear image sensor with a 2D one. The image sensor had 512×512 pixels with a pitch of 11.7 μm in the x-direction and 8.8 μm in the y-direction, respectively. In addition, in the equations given above, we replace θ with $6 = \sqrt{\theta_x^2 + \theta_y^2}$, where $6_{x,y}$ are the angular displacements about the x- and y-axes, respectively.

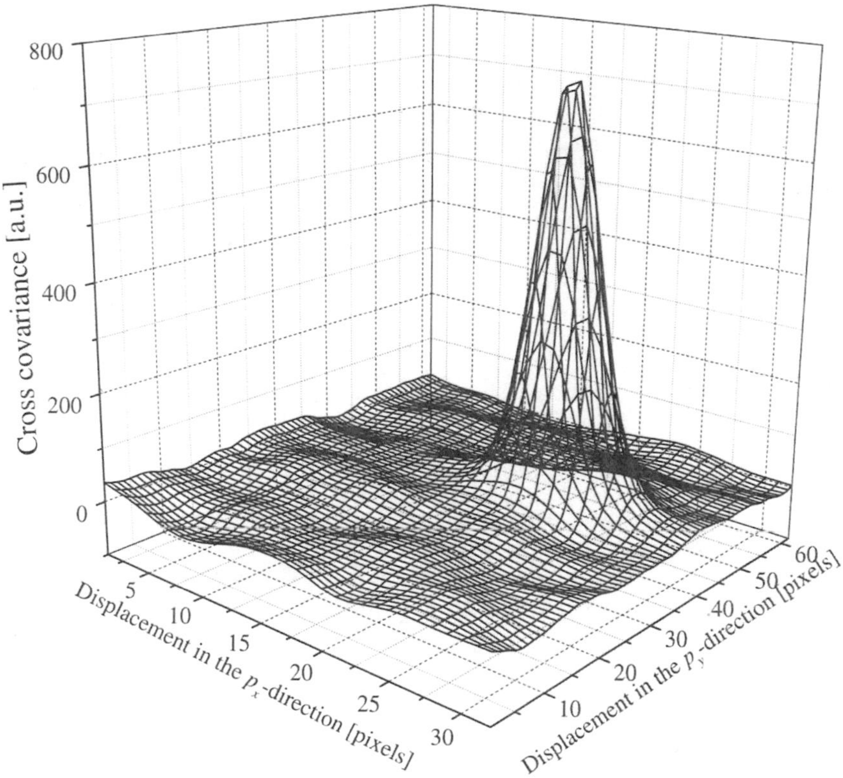

Figure 10. Resulting 2D covariance for an angular displacement of 200 mdeg. The measurement was performed on a matte aluminum shaft with a radius of 15 mm.

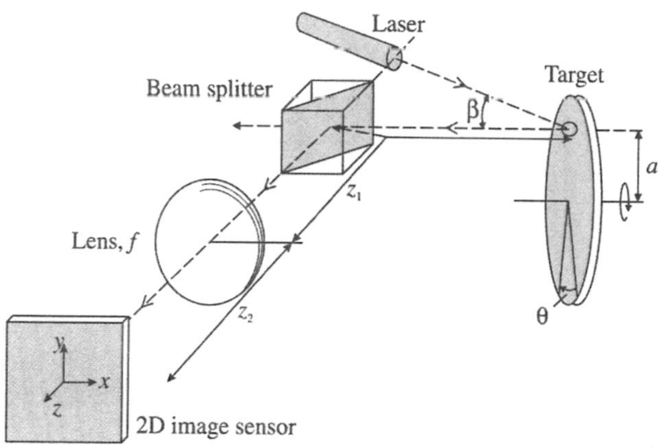

Figure 11. Optical diagram for the in-plane measurement setup. The target is placed such that $z_1 = 2f$, and the spot size, angle of incidence, position of incidence and angular displacement are $r_s = 2.5\ mm$, $\beta = 45°$, $a = 2\ cm$, and $\theta = -100 mdeg$, respectively, whereas the position of the image sensor is varied along the z-axis.

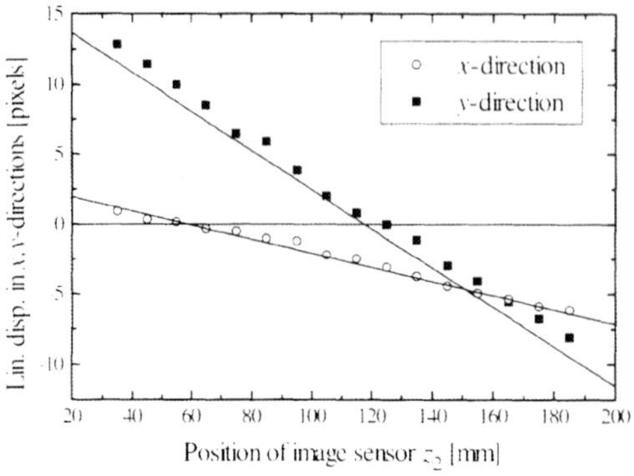

Figure 12. Measured linear displacement in the x-direction (open symbols) and y-direction (solid symbols) versus position of the image sensor ($z_1 = 2f$, $a = 2$ cm, and $\theta = -100$ mdeg).

The resulting 2D covariance for an angular displacement θ of 200 mdeg is illustrated in Figure 10. Besides revealing the magnitude of the angular displacement (200 mdeg), the figure also yields the direction of the angular displacement. For the measurements illustrated in Figure 10, the target is angularly displaced in a direction inclined 52 deg with respect to the x-axis.

6. IN-PLANE ANGULAR DISPLACEMENTS

In contrast to the systems discussed above, we now specialize to the case where the systems are given in Cartesian coordinates.[14] The application can be used to measure in-plane angular velocities and displacements, and is depicted in Figure 11. The peak of the covariance is located at

$$p_{x0} = -a\left(1 - \frac{z_2}{f}\right)\theta \tag{26}$$

and

$$p_{y0} = -\sin\beta\left(z_1 + z_2 - \frac{z_1 z_2}{f}\right)\theta \tag{27}$$

in the x and y-direction, respectively. Note that the wavelength of the illuminating light source does not enter Eqs. (26) and (27), thus, any shift in wavelength does not affect the peak location of the covariance.

The resulting measurements for the linear displacement (caused by in-plane angular displacement of the target) in the x and y-directions are shown in symbols in Figure 12, whereas the theoretical values are illustrated with solid lines. Examination of the figure reveals that for a Fourier transform system ($z_2 = f = 60$ mm) there is no translation in the x-direction, whereas in an imaging configuration ($z_2 = 2f = 120$ mm) there is no translation in the y-direction as is expected physically [see Eqs. (26) and (27)]. Following this discussion, the speckles will translate along a line whose angle with respect to the p_x-axis is given by $\tan^{-1}(p_{y0}/p_{x0})$. This means that the speckle pattern rotates in free space after passing through the lens. The spiraling of the speckle pattern can be thought about in two ways. Either a simple in-plane displacement will translate the speckles as a frozen pattern. An alternate mechanism could arise if we realize that the speckles are in fact three-dimensional objects ("speckle grains") with their major axes tilted slightly with respect to the z-axis. An apparent displacement of the intersection of the three-dimensional speckles with the observation plane will now show up if the speckle grains are displaced in the z-axis.

The two planes, the Fourier- and the image plane, probe different aspects of the object. The first reveals the angular velocity/displacement, whereas the second gives information regarding the surface translation (i.e., the angular velocity/displacement multiplied by the rotation radius a). Recording these two parameters simultaneously provides a means for deducing the distance from the center of the illuminated spot to the axis of rotation.

7. TORSION ANGLE SENSOR

A schematic of the torsion angle sensor setup[15] that has been used for the measurements is shown in Figure 13. Imaging is performed in the y-direction and the Fourier transformation is performed in the x-direction. The plates were fixed at the bottom and clamped at the top to a rotating stage with a stated resolution of 1 mdeg. The rotating stage provided the twist for the flat target.

The primary parameters of interest for the torsion angle sensor are firstly the relation between applied twist and the measured linear displacement on the image sensor in the x-direction. Twisting the target after acquiring the reference measurement introduces a linear displacement of the speckle pattern on the image sensor. Cross correlating the displaced signal with the reference signal gives rise to a correlation peak centered at $p_{x0;b}$, given by

$$p_{x0;b} = A_x K_0 \theta_{y;b} = 2\theta_{y;b} f_2 \qquad (28)$$

for the corresponding surface position b along the y-axis (see Figure 13.).

We performed measurements with the illuminating spot centered 10 mm above the bottom clamping zone ($y = 10$ mm), i.e., in a region where we did not expect a uniform deformation (see Figure 14.). Figure 14. clearly illustrates a non-uniform distribution of torsion angles close to the clamping zone. Especially it can be seen that the plate does not twist in the region closest to the clamping zone (for y-positions below 6 mm) where the measured torsion angles approach zero. Physically it makes sense that a flat plate does not deform uniformly throughout the whole length of the plate. The measurements have shown that not only can the setup be used to measure very small twist angles in uniform deformation zones, but the system can also be used to measure the distribution of torsion angles locally in zones where the deformation is non-uniform. This means that the torsion angle sensor can replace both conventional strain gauges, and very advanced and expensive holographic interferometric methods. The measurements are independent of the target distance.

8. ROTATIONAL SPEED AND TORSIONAL VIBRATIONS OF A ROTATING SHAFT

Finally, we present a system for measuring angular velocities[16] and torsional vibrations.[17] The system, shown in Figure 15., comprises two displaced detectors in the Fourier plane each tracking the speckle pattern as it is translated from one detector to the next caused by the steady rotation of the shaft. Compared with the previous systems, we have replaced the linear detector array with two detectors having a size comparable with the speckle size.

The temporal cross covariance of the two detector signals reveals a peak at the time delay corresponding to the transit time for the speckle pattern going from one detector to the next. Inserting the *ABCD*-matrix elements for the Fourier-transforming system depicted in Figure 15. into Eq. (24) yields[16]

$$\tau_0 = \frac{\Delta p}{\omega_0\left[f + \frac{z_2}{f}\left(f - \frac{R_0}{2} - z_1\right)\right]} \tag{29}$$

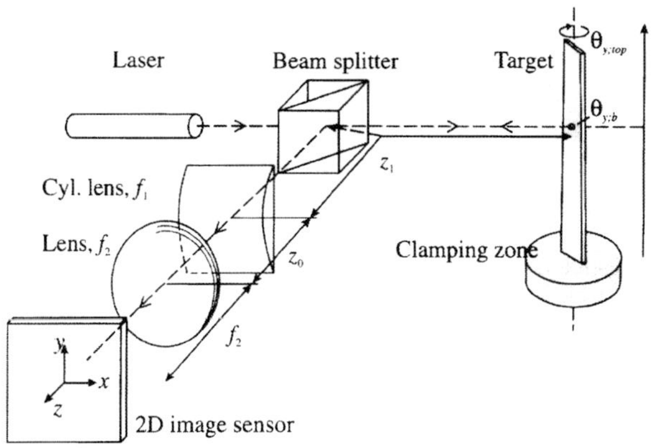

Figure 13. Optical diagram of the torsion angle sensor setup, with an imaging configuration in the y-direction and a Fourier transform system in the x-direction. The spot radius r_s at the target was 6 mm. The cylindrical lens f_1 had a focal length of 200 mm in the y-direction and $f_1 \to \infty$ in the x-direction. The achromatic lens f_2 had a focal length of 60 mm. The 1 mm thick target plates (48×200 mm²) were clamped in both ends. The clamping zones were 48×10 mm², and thus, the size of the plates that were subjected to twist was 48×180 mm².

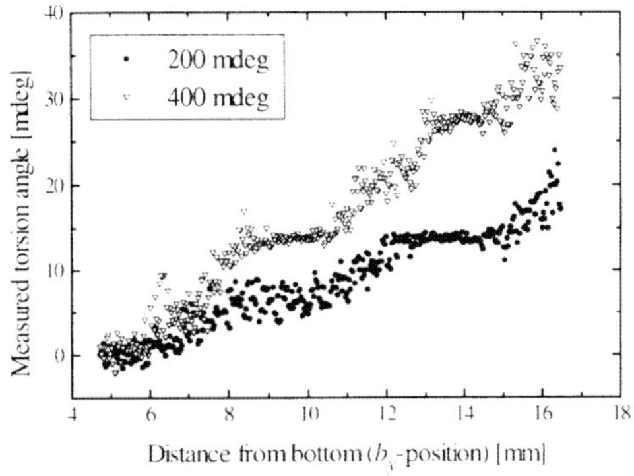

Figure 14. Measured torsion angles versus lateral position for various values of applied twist. The measurements were performed on a matte steel plate (fully developed speckle field) where non-uniform deformation was assumed. At the top of the plate we applied an angular twist of 200 mdeg (solid) and 400 mdeg (open), respectively.

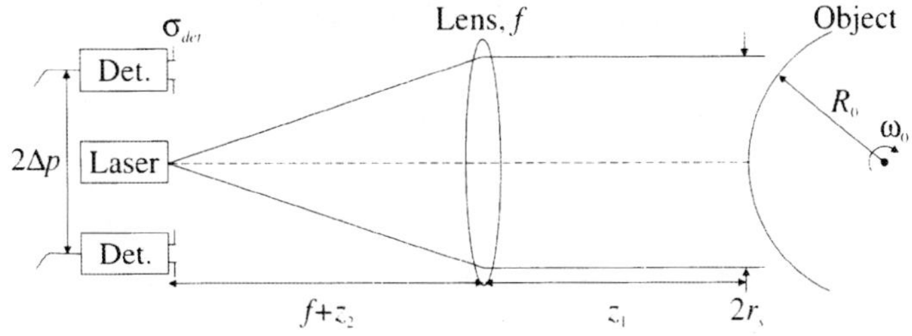

Figure 15. System for dynamic measurement of the rotational speed and torsional vibrations.

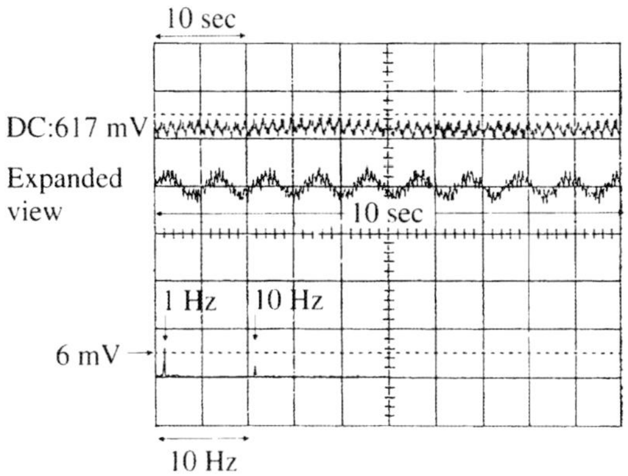

Figure 16. Variations in rotational speed. Upper trace shows the estimate of the rotational speed with DC (617 mV) subtracted. The centered curve is an expanded view of 10 sec of the upper trace. The lower trace shows the spectrum of the variations in rotational speed. The imposed modulation of 1% at 1 Hz is visible as well as the first harmonic of the fundamental rotation, 10.6 Hz.

where ω_0 is the rotational speed of the shaft and $2\Delta p$ is the separation of the detectors. In case the two detectors are placed in the Fourier plane (i.e., $z_2 = 0$), the time delay is independent on the radius of curvature of the rotating shaft and the target distance. Neither is any influence encountered from linear translations of the shaft (see, also, Chap. 5).

Figure 16. shows an oscilloscope trace of the output signal from an electronic unit tracking τ_0.[16] The shaft was rotating at 10.6 Hz and a slow perturbation (the induced torsional vibration) of 1 Hz was introduced via a sinusoidal modulation of the rotational speed. The figure shows a good agreement between the applied velocities and the measurements. This was tested on different surfaces and within a broad range of rotational parameters.

9. CONCLUSION

In conclusion it has been demonstrated that complex *ABCD*-matrix methods together with the Huygens-Fresnel principle have been shown to be a general wave optics tool for analyzing arbitrary optical systems. Several applications in statistical optics of practical interest have been analyzed and discussed. In particular, in all cases, closed form analytic solutions are obtained that, firstly, have direct physical interpretations and, secondly, are useful for parametric systems (trade-off) studies.

The work reported here was supported by the Danish Technical Science Foundation under the *LIS* framework program. The research of H. T. Yura was performed while he was a guest scientist at Risø National Laboratory, Roskilde, Denmark.

REFERENCES

1. A.E. Siegman. *Lasers,* University Science Books, Mill Valley, Calif. (1986), Chap. 20.
2. H.T. Yura and S.G. Hanson, Optical beam wave propagation through complex optical systems, *J. Opt. Soc. Am. A* 4:1931 (1987).
3. see, for example, *Laser Speckle and Related Phenomena,* J. C. Dainty, ed., Springer-Verlag, Berlin, (1984).
4. H.T. Yura, B. Rose, and S.G. Hanson, Dynamic laser speckle in complex *ABCD* optical systems, to appear in *J. Opt. Soc. Am. A* (1998).
5. H. Kogelnik and T. Li, Laser beams and resonators, *Appl. Opt.* 5:1550 (1966).
6. H.T. Yura, S.G. Hanson, and L. Lading, Laser Doppler velocimetry: analytical solution to the optical system including the effects of partial coherence of the target, *J. Opt. Soc. Am. A* 12:2040 (1995).
7. H.T. Yura and S.G. Hanson, Laser-time-of flight velocimetry: analytical solution to the optical system based on *ABCD* matrices, *J. Opt. Soc. Am. A* 10:1918 (1993).
8. B. Rose, H. Imam, S.G. Hanson, and H.T. Yura, Effects of target structure on the performance of laser time-of-flight velocimeter systems, *Appl. Opt.* 36:518 (1997).
9. S.G. Hanson, L.R. Lindvold, and L. Lading, A surface velocimeter based on a holographic optical element and semiconductor components, *Meas. Sci. Technol.* 7:69 (1996).
10. H. Imam, B. Rose, L.R. Lindvold, S.G. Hanson, and L. Lading, Miniaturising and ruggedising laser anemometers," in: *Eighth International Symposium on Applications of Laser Techniques to Fluid Mechanics* 2:40.2.1 (1996).
11. H. Imam, B. Rose, S.G. Hanson, and L. Lading, Laser time-of-flight velocimetry: proposals for miniaturization, in: *Diffractive Optics and Optical Microsystems,* S. Martellucci and A. N. Chester, eds., Plenum Publishing Corporation, New York (1997) 399-409.
12. B. Rose, H. Imam, S.G. Hanson, H.T. Yura, and R.S. Hansen, Laser speckle angular displacement sensor: a theoretical and experimental study, to appear in *Appl. Opt.* (1998).
13. B. Rose, H. Imam, and S.G. Hanson, Non-contact laser speckle sensor for measuring one and two-dimensional angular displacement, to appear in *J. Opt.* (1998).
14. H.T. Yura, B. Rose, and S.G. Hanson, Speckle dynamics from rotating diffuse objects in Complex *ABCD* optical systems, to appear in *J. Opt. Soc. Am. A* (1998).
15. B. Rose, H. Imam, S.G. Hanson, and H.T. Yura, A laser speckle sensor to measure the distribution of static torsion angles of twisted targets, *Meas. Sci. Technol.*, 9:42 (1998).
16. S.G. Hanson and B.H. Hansen, Laser-based measurement scheme for rotational measurement of specularly reflective shafts, *Proc. SPIE* 2292:143 (1994).
17. S.G. Hanson, System for noncontact measurement of torsional vibrations of shafts, *Proc. SPIE* 2729:153 (1995).
18. A.E. Siegman. *An Introduction to Lasers and Masers,* McGraw Hill, New York (1971), Chap. 8.
19. W.G. McKinley, H.T. Yura, and S.G. Hanson, Optical system defect propagation in *ABCD* systems, *Opt. Lett.* 13:333 (1988).
20. H.T. Yura and S.G. Hanson, Second-order statistics for wave propagation through complex optical systems, *J. Opt. Soc. Am. A* 6:564 (1989).
21. H.T. Yura and S.G. Hanson, Effects of receiver optics contamination on the performance on laser velocimeter systems, *J. Opt. Soc. Am. A*, 13:1891 (1996).
22. H.T. Yura, S.G. Hanson, and T.P. Grum, Speckle: statistics and interferometric decorrelation effect in complex *ABCD* optical systems, *J. Opt. Soc. Am. A* 10:316 (1993).
23. R.S. Hansen, H.T. Yura, and S.G. Hanson, First-order statistics: an analytical analysis using *ABCD* matrices, *J. Opt. Soc. Am. A* 14:3093 (1997).

MICROSYSTEMS AND RELATED TECHNOLOGIES

[*]V. Foglietti, [†]A. D' Amico, [†]C. Di Natale, [†]S. Petrocco, and [‡]S. Mengali

[*]Istituto di Elettronica dello Stato Solido del CNR
Via Cineto Romano, 42
00156 Roma, Italy

[†]Università degli Studi di Roma "Tor Vergata"
Dipartimento di Ingegneria Elettronica
Via di Tor Vergata
00133 Roma, Italy
and Progetto Sensori e Microsistemi – CNR
Area di Ricerca del CNR di Tor Vergata
Via Enrico Fermi, 70
00044 Frascati (RM), Italy

[‡]Centro Tecnologie del Vuoto
Via Turanense, 26
67061 Carsoli, Italy

1. INTRODUCTION

After three decades since the introduction of the first integrated circuit we are now witnessing a new step in the continuous technological revolution which has changed many aspects of our lives. This step can be sintesized by the term "Microsystem" or by the acronym "MEMS" , MicroElectroMechanical Systems.

Many definitions are to day available for the word Microsystem, but the most widely accepted one seems to derive from an accurate description of its features: a microsystem is a small apparatus having dimensions less then 1 millimeter, which combine sensing, actuating and processing functions, through the use of electrical, mechanical, optical,chemical and biological principles.

The crystalline silicon shows very remarkable mechanical properties, a Young modulus very close to that of steel a yield strenght two times higher. It does not deform plastically, therefore does not show mechanical hysteresis. All this characteristics combined with the high stress sensitive electrical properties make silicon an ideal material for fabricating micromechanical devices. The microsystem devices store and process information electronically incorporating a

Optical Sensors and Microsystems: New Concepts, Materials, Technologies
Edited by Martellucci et al., Kluwer Academic / Plenum Publishers, New York, 2000.

wide and new set of functionality and enabling a direct interface with the application fields. The new chips can perform physical and chemical sensing, actuation, light processing, fluid motion. The term "chips" is used rather intentionally to stress the main characteristics which these new systems have in common with microelectronics: the size. Complete machines can be fabricated with typical dimension of micrometers. This aspect, other than capturing the imagination of people and let speculations go wild on popular science magazines, has an important consequence: the microsystems can be produced massively using mainly existing integrated circuits manifacturing plants, thus reducing costs of manifacturing and beeing in this way a very competitive product for the market. The area of microsystems is interdisciplinary, for the application horizon and for the knowledges required in the fabrication process. The fields involved ranges from material science to elecronic, fluid and mechanical engineering, from physics and chemistry to medicine. Many successfull products covering these various fields have been fabricated recently: field emitter and projection displays, inertial sensors, effusion nozzels for ink-jet printers, instrumentation for neural stimulation and microsurgery, biological and chemical analysis systems, prosthetic devices and tactile systems.

The common background for a solid activity on microsystems is the technology. As mentioned above the microelectronic technology is the base for every processing flow chart involving MEMS. The MEMS rely on the various steps of the silicon technology, from the design and simulation, through interdisciplinary CAD tools, up to the fabrication of microsystems. The possibility of including in smaller and smaller volumes a variety of functions allows a more flexible design, reliability improvement and a substantial cost reduction. The fabrication of MEMS has obviously required further development in the silicon processing for microelectronics; this new technology is commonly identified as "micromachining". Three basic and well consolidated techniques are identified as: "bulk", "surface" and "mold" micromachining. The following paragraph will give a short overview of these main techniques, highlightening future research and development of related technologies. It must be stressed that the following overview does not pretend to be exaustive of all the emerging technologies that have been recently explored with very promising results for microsystem fabrication; among them the high aspect ratio processing using high plasma density source and the plating techniques. The third paragraph will describe a few example of micromachined device developed at IESS-CNR and at the Department of Electronic Engineering of the University of Tor Vergata.

2. OVERVIEW OF TECHNOLOGIES SUITABLE FOR MICROSYSTEMS.

2.1 Bulk Micromachining

The bulk micromachining has been the earliest technology developed to fabricate mechanical structures by removing material from a silicon wafer. The most important tool is the silicon deep wet or dry etch. In wet etching the etchant is in liquid phase while in dry etch the etch is usually a gas, where an external excitation create chemically active radicals. We can furtherly distinguish between isotropic and anisotropic etching, both able to remove unwanted part of the wafer, leaving the desired mechanical features on the silicon wafer.The wet anisotropic etching has been extensively used in the fabrication of microstructures. The main feature is the orientation dependent etch-rate. The (100) and (110) silicon planes are etched two orders of magnitude faster than the planes (111). Various basic aqueous solutions like hidracine, etilendiamine, KOH and NaOH, etch anisotropically the single crystal silicon. The most commonly used is based on potassium hydroxide because of the high quality achievable in terms of surface smoothing, the high etching rate, reproducibility and safety.Recently an acqueous solution of TMAH (Tetra-Methil Ammonia Hidroxide) have shown interesting qualities, among them the complete compatibility with the integrated circuit CMOS technology. Figure 1. shows an example of anisotropic wet etching performed with KOH a based solution. In the example shown in Figure 1, silicon dioxide has been used as masking layer, although this material is not ideal for masking purposes in a KOH based etch. Very well shown are the planes (111) which constitutes the walls of the square holes, with an inclination of 54.7°. In this example the wafer has been completely etched through is whole thickness. The geometry of three dimensional structures etched through the openings of the masking layer depends on the shape of the opening itself. When the mask has only concave corners, like in Figure 1., then there is not undercut provided that the opening is properly oriented. When the geometry of the openings includes convex corners, irrespective of alignment conditions, these corners will lead to underetching. This phenomenon leads to a deformation of the structure and special compensation calculations must be performed in order to adjust the final shape to the wanted structure.

Several programs are commercially available to perform the corner compensation. The undercutting is very well visible in Figure 2., where the circular geometry of the openings cannot be mantained during the etching. As the etch proceeds, the walls of the holes intersect various family of cristallographic planes which are etched at a different rate. Rough walls appear which may smooth out when the (111) planes are encountered by the etching front, as shown in Figure 2. The final shape of a circular opening is always a square hole delimited by the (111) planes touching the circular shape. All the geometric dimension should be precisely controlled in order that the fabricated device shows the desired specifications. For example, the sensitivity of piezoresistive pressure sensors is inversely proportional to the square of the diaphragm thickess. The dimension in the wafer plane can be defined by photolithography, but the dimension in the perpendicular dimension are more difficult to control accurately. In order to provide methods for this control several techniques have been proposed for automatic etch stop: high concentration boron doping,[1] buried layers of mask materials,[2] electrochemical etch stop.[3] The physical and chemical mechanism governing the anisotropy of the etch rate has been matter of various debates. Seidel has proposed a model[4] which can explain quite well also the etch stop mechanism of boron doped silicon.[5] In this model the OH(-) groups play a key role in the chemical reaction. Other models take into account the kinetic of crystal growth, in this model is the particularly flat face of the (111) planes which determines the main difference in the etch rate of various planes.[6]

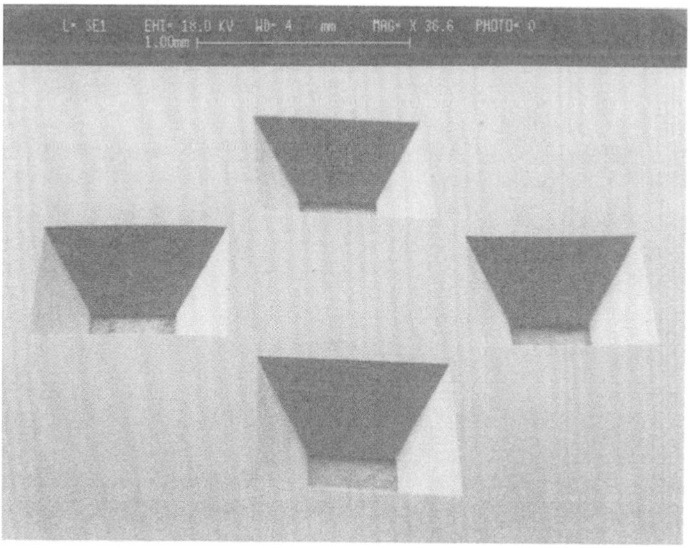

Figure 1. Four square holes etched on a (100) oriented silicon wafer.

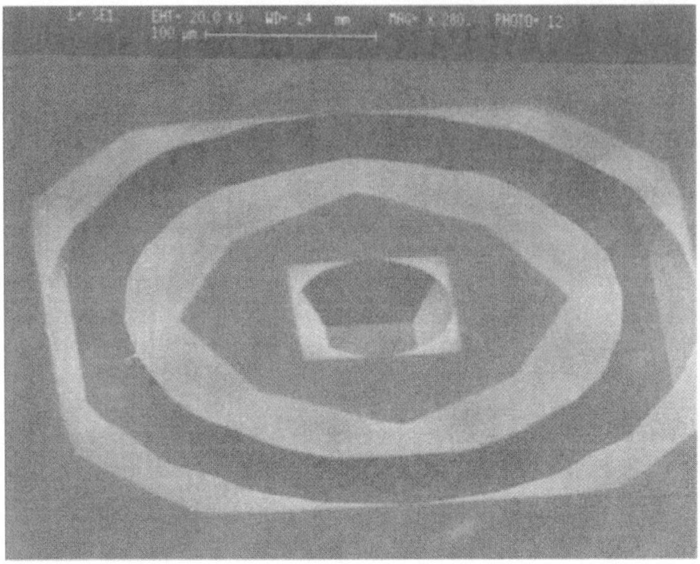

Figure 2. Anisotropic wet etch performed on circular openings. The light images are the self standing silicon oxide masking layer. The underetching, the smoothness and roughness of walls is clearly visible.

It must be stressed that these models do not explain all the experimental evidence and furtherly, there is a group of foundamental question still unanswered concerning the wet etching mechanism. For example the (111) etch rate is not known with good precision and the measure determines the maximum anisotropy obtainable.

Wafer to wafer bonding is another key step in bulk micromachining. It adds new possibilities for device fabrication and enables the fabrication of complex structures, beyond the capabilities of single wafer processing. For example, in the fabrication of an accelerometer, it permits to bulk micromachine the seismic mass in one wafer and the damping cavities in a two separate wafers. Afterwards the three wafers are bonded. Silicon Direct Bonding[7] attempts the bonding of silicon wafers without the use of any adhesive material at temperatures low enough to prevent damage on metallized wafers. The technique uses wet cleaning treatments to induce an hydrophilic character to the wafer's surface.

Most of the accelerometers on the automotive market are today fabricated using the bulk micromachining.[8] Also many chemical gas sensing and pressure sensor devices that rely on thin membranes of silicon or silicon nitride are fabricated using this technology.[9] It is a solid and reliable technology which has today overcome the major drawback consisting in the lack of full compatibility with the strict requirements of CMOS technology.

2.2 Surface Micromachining

The surface micromachining enables the fabrication of micromechanical elements using thin films ranging from 1 to 10 μm, as structural materials.The main step is the selective etching of a sacrificial layer underneath the structural layer.The sacrificial material is etched or dissolved in a chemical etchant resulting in freely movable structures. The process requires a compatible set of structural materials and sacrificial layers. Mechanical structures can be built by surface micromachining using one sided processing.[10,11] The residual stress and stress gradient of the structural layer are of primary importance. Since the first demonstration of the sacrificial layer technique by Nathason,[12] surface micromachining technique have been developed and greatly improved, resulting in a large variety of structural materials and sacrificial layer combinations. Figure 3. shows an example of a suspended structure of chromium metal film deposited on top of silicon oxide. The structure is downward bended due to a stress gradient in the film.

Polysilicon surface micromaching is the best documented technique. It meets the requirements of compatibility with sacrificial layer and etchants and can be easily integrated with IC processes. The schematic process is shown in Figure 4. The first step (a) is the deposition and patterning of a silicon dioxide sacrificial layer. This is followed by deposition and patterning of polysilicon (b). The sacrificial layer is then selectively etched in hydrofluoridric acid solution. This is followed by a rinsing and a final drying procedures that leaves the polysilicon released (c).

A critical factor in surface micromaching is the control of mechanical properties of the structural material. Residual stress can lead to warpage, buckling and unwanted deflections. Cracking can occurr from excessive internal stress, blistering and peeling can result from poor adhesion to the substrate. The residual stress in polysilicon can be controlled varying the deposition parameters, annealing procedures and by doping. It is generated from a mismatch in thermal expansion between the film and the substrate and from effects of grain growth and deposition techniques. Polysilicon is usually deposited by Low Pressure Chemical Vapour Deposition from pyrolysis of silane at temperature around 600° and pressure of several hundred mTorr. The residual stress varies significantly with processing conditions, especially with deposition temperature and silane pressure. Hydrofluoridric acid is used to etch the sacrificial layer. This takes advantage from of the isotropic nature and high selectivity of wet etching to release overlying structural films. A reaction model for the etching of silicon dioxide has been presented by Monk.[13] The polysilicon can be patterned with a very high precision using reactive ion etching or improved dry etching technology, like Electron Cyclotron Resonance (ECR) or Inductively Coupled Plasma (ICP). This dry etching techniques are also becoming very important for microsensors fabrication in special application like field emitter display and high aspect ratio tips for atomic force microscopy.

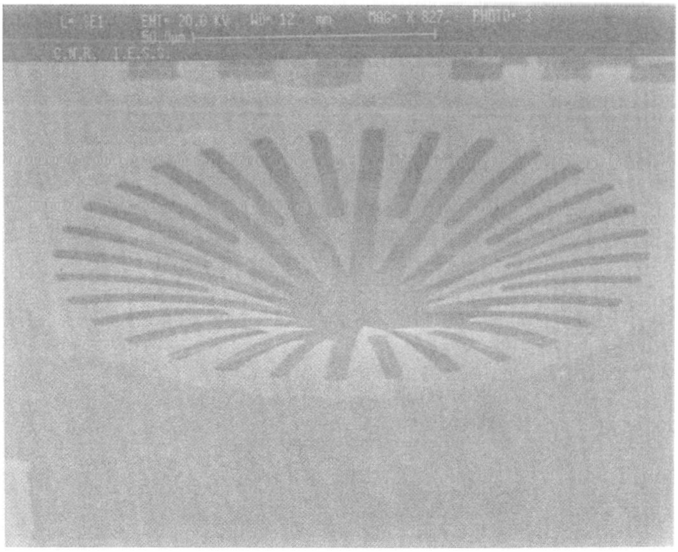

Figure 3. Suspended structure of chromium thin film deposited on top of silicon dioxide. The structure has been released by plasma etching.

A crucial point of the surface micromaching is that large surfaces tend to be sticked to the underneath surface in the cleaning step after the etching. Standard drying procedure such as spin drying produce capillary forces during drying that cause "pull-down" of the thin film structure. Once physical contact has been established, standard forces come into play resulting in permanent attachment. A number of technological recipes have been proposed to mitigate the "stiction" problem.[14,15]

The precision in patterning the structural layer leads to improved performance in microsensors, field emitters and optical devices. As a result many successfull devices have been fabricated using surface micromaching, among them: polysilicon micromotors for data storage,[16] wobble motors,[17] optical gratings.[18] The full compatibily with IC CMOS technology makes this technique a challenging technology provided that the stiction problem be reliably solved.

2.3 Mold Micromachining

Another reliable technology to fabricate MEMS is mold micromachining. In this technology the mechanical structures are created by first creating a mold. The mold is filled with a structural material to create the desired mechanical device. The most famous mold technology is the LIGA. LIGA is an acronym for the German words: lithography, electroforming and molding. It was initially developed at FZK in Karlsruhe, Germany.[19] There are three basic steps in the LIGA process. In the first one the pattern on an x-ray mask is transferred into a thick resist layer by shadow printing with synchrotron radiation. The radiation is emitted into a narrow band with a vertical emission angle of typically 0.5 mrad. The thick resist layer is applied either on a massive metal plate or on an insulating plate covered with a plating layer. In the second step a microstructure pattern is electroplated using the conductive layer of the substrate as a pattern. Metals like Cu, Ni or Au are used to create the complementary microstructure pattern. The electroplating may reach up to a few millimiters above the height of the primary structure, producing a robust mold insert. In the third step the final device is fabricated by standard molding techniques, such as reaction injection molding,[20] hot embossing[21] into polymers and so on. The process can create very high aspect ratio structures as a consequence of the high depth of focus of the x-ray lithography. The final device can be made using various materials that fit more closely the specifications, introducing a very important degree of freedom in the design of microstructures. The technology is limited in that it requires a high-energy synchrotron radiation source and such a source is not part of the standard integrated circuit infrastructure.

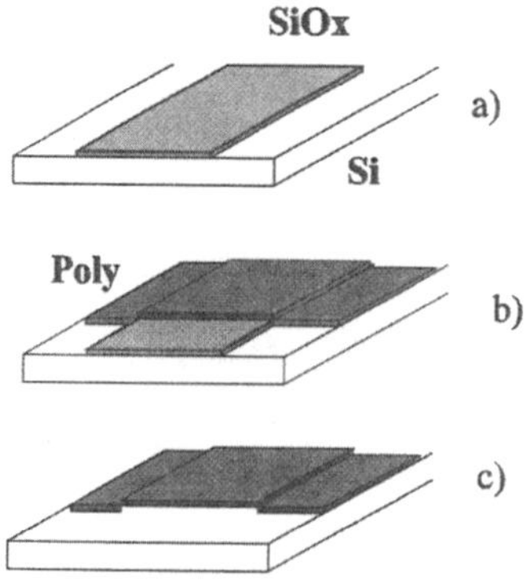

Figure 4. Schematic of surface micromachining polysilicon steps.

Where the special LIGA properties are not indispensable to the characteristics of the product, other processes are preferred for mold insert fabrication.Recently, other mold technologies have been developed that are much more compatible with IC fabrication.[22] Molding micromaching is currently succesfully used to develop complex microactuators like micromotors, microturbines[23] and so on.

2.4 Porous silicon

Porous silicon[24] can be considered an interesting and useful material in the frame of microsystem technologies due to its capacity of inducing a greater flexibility in the process design of integrated microstructures.

It can be produced by electrochemical dissolution of silicon in HF. Once silicon is immersed in HF, in presence of an electrical current, an anodization process takes place and silicon atoms are dissolved. Under control of certain parameters, a development of pores can be observed, as schematically shown in Figure 5. In particular the surface of the silicon cristal presents a morphology of a spunge.This kind of silicon structure has captured the attention of many researchers who are trying to better understand its behaviour in terms of transport properties and luminescence effects.[25] Porous silicon on the other hand has some characteristic which make it an interesting material for sacrificial layer technology.

In fact it can be etched to form thick layers for the whole thickness of the wafer, it can sustain other microelectronic processes and removed by a ultra weak KOH solution and finally it is compatible with the CMOS process.

The main interest to day around the porous silicon in the optoelectronic and electronic area including scientific aspects related to the quantum size effects of the surface structure. It can also be employed as: antireflective coating for silicon solar cells, in lasers technology and to improve the characteristics of infrared silicon detectors.

In addition there are also other interesting features related to this material. In fact as its sponge-like structure is in the nm dimension range, it can be considered in both the visible range and infrared as an homogeneous material characterized by a well defined refractive index n whose value is located between that of bulk silicon and that of the air (unity).

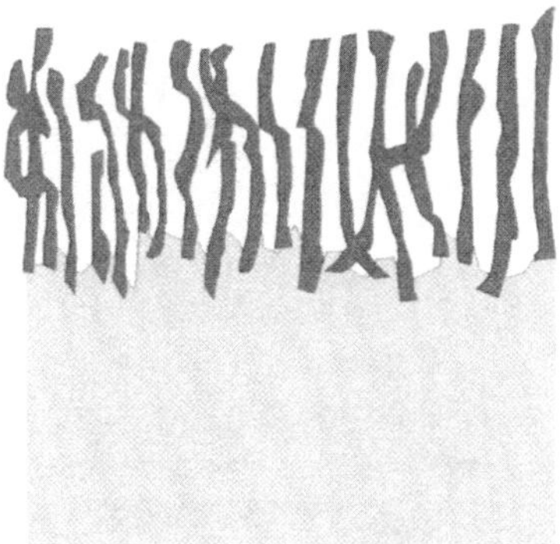

Figure 5. Schematic structure of porous silicon layer.

Also the refractive index n is proved to be porosity dependent; for instance, with a porosity density changing from 30% up to 70% ,n can span between 2.3 and 1.1, in 0.02 ohm-cm silicon.

The main features of this porous layer can be controlled by changing the current value or its time dependence, and also by varying the dissolution process time.

In order to locate the importance of porous silicon in the micromachining context, we can see the Figure 6., where three ways used to obtain free standing structure are represented.

2.5 Gallium arsenide: an interesting material for micromechanics

This material and related III-V compounds have been utilized in the past mainly for high frequency and optoelectronic applications, but since a few years attention is payed towards its use in micromechanical devices context. This is due to both the interesting etching properties of this material and to the presence of piezo and optical characteristics which are seen of relevant importance also for actuator applications and for the possibility to operate at higher temperature if compared to silicon.

The following Table 1. illustrates some features of four important materials employed for the development of micromechanics.

Table 1. Some features of relevant micromachining materials.

	GaAs	AlAs	Si	SiO_2
Density (g/cm^3)	5.31	3.72	2.33	2.65
Lattice constant (Å)	5.65	5.66	5.43	4.93
Melting point (°C)	1238	1740	1413	1710
Thermal expansion (ppm/K)	6.1	4.2	2.6	13.6
Thermal resistivity (K·cm/W)	2.28	1.12	0.64	7.7-16

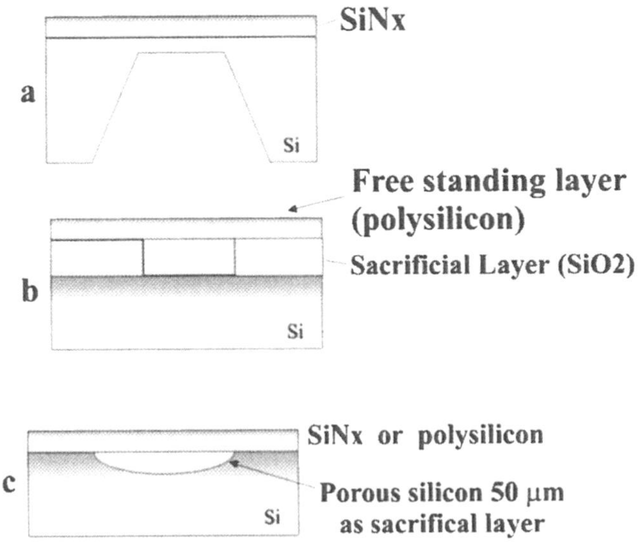

Figure 6. a) Bulk anisotropic etching in KOH, b) Surface micromaching, c) Porous silicon technology.

There is a high number of etchants that have been suggested by the literature which can be succesfully employed for the GaAs micromachining. Some of them are indicated in the following together with the references.

1% Br_2: CH_3OH (for mirrorlike etched surfaces; slowly attacking quartz);[26]
NH_3OH:H_2O_2:H_2O (widely employed selective etch);[27]
NaOH:H_2O (Anodic spray etch with very high selectivity).[28]

A short comparison between GaAs, silicon and quartz would show that GaAs incorporates the main features of both silicon and quartz. GaAs is a semiconductor, like silicon, and it can be worked by advanced micromachining techniques. Like the quartz, it shows piezoelectric behaviour and, in addition, it permits the integration of lasers and photodiodes on the same chip; and finally it can be used in high temperature applications. As micromechanical material it is foreseen to represent a promising candidate of large unexploited potentials.

2.6 Thin film technology

Among the fabrication process segments of micromechanical devices, some are dealing with thin film deposition of insulators (doped oxides) metals and semiconductors mainly in polycristalline forms.The insulator layers can be utilized, for instance, as interlayer dielectrics or passivation layers. There are many reasons related to the use of doped oxides, in comparison to the undoped ones, for instance: 1) the possibility to get low stress layers, 2)easy planarization of the surfaces, 3) getters for mobile ion concentration and 4) the possibility to use this kind of layer as so called "sacrificial layers" in the context of surface micromachining.

The widely employed technique by which these films are deposite is generally called CVD (chemical vapor deposition), because all the constituents of the layer are introduced in the reactor in gas phase.The energy needed for the film formation comes from the heated substrate and/or from the plasma that is electrically generated into the chamber.

Examples of CVD films are: polysilicon, silicon dioxide, borosilicate glass (BSG), phosphosilicate glass (PSG), borophosphosilicate glass (BPSG), other semiconductor such as the III-V compounds.

The CVD process can be activated at atmospheric pressure (APCVD), or at low pressure (LPCVD). Furthermore there exist the following methods for enhancing the deposition rate: Plasma Enhanced CVD (PECVD), Photo Enhanced CVD (PHECVD), and the Laser assisted CVD (LACVD) of which some advantages (**ad.**) and disadvantages (**dis.**) are also indicated.

APCVD (**ad.**: higher gas density, high deposition rate, high throughput; **dis.**: difficulty to get good uniformity of the chemical physical characteristics of the layer).

LPCVD (**ad.**: high uniformity rather independent on the reagents distribution; **dis.**: possible shadowing in small geometries).

PECVD (**ad.**:low deposition temperature ;**dis.**: residual stress, difficulties to characterize the reactants).

3. AN EXAMPLE OF MICROSTRUCTURES : A MICROHEATER ARRAY FOR SELECTIVE GAS SENSING

3.1 Design and simulation

The development and fabrication of a gas sensor need an array structure design to obtain competitive results: this kind of sensor should be able to evaluate the concentration of one or more gases in an unknown ambient. Thus, every chemical species could cause changes in the sensor's output that depends on both the concentration and the kind of material that is sensed: this causes the need of making either qualitative or quantitative analyses. Mathematical optimization methods[29] force us to use a number of sensors greater than that of the chemical species present in the ambient because of the use of sensing materials with low selectivity to obtain cheap devices. A sensor like this has as response: $y = \sum S_i C_i$ where S_i is the sensitivity of the sensing material toward the ith chemical species and C_i is the concentration of the ith material. To invert this equation we need a number of equations greater than that of the unknown concentrations of the chemical species to be sensed.

The structure we use to realize the sensor is a gold interdigitated fingers device covered with the sensing materials (in this case metallo-porphyrines) that give us the chance to make either capacitive or resistive measurements and a large sensing zone. To exploit the characteristics of the sensing materials used to sense gases, we need a micro-hotplate that allows us to find the better fit to obtain the higher sensibility and to study the behaviour of the chemically interactive materials chosen toward the temperature: we have then realized a platinum serpentine of about 300Ω that could be used also as temperature sensor simple to use because of the linear behaviour of the resistivity of this metal toward the temperature.

In this work we deal with the structure used to realize the array and we show the ANSYS simulations[30] that we have done to choose the better solution to fabricate a micro-hotplates that should heat uniformly the sensing zone to obtain valid gas measurements. The basic structure that is the unitary cell of the array that we have fabricated is shown in Figure 7.

The array is formed by eight interdigitated fingers and eight hotplates in a row (array 1x8) that are insulated from each other thermically anchoring the silicon substrate to room temperature: in this way we can use the different chemically interactive materials in the best conditions. Every sensor and hotplate covers a square area of 2mm of side and the resulting array is about 4.5 mm wide and 4.8 cm long. This geometry has been preferred to a 2x4 one because the latter forbids the deposition with the Langmuir-Blodgett method with which could be obtained uniform and controlled layers of sensing materials and then reproducible results. ANSYS is a FEA program that allows to make thermal, electrical, magnetic and many other kind of analyses. We have put in input the geometries and the features of the structures (thermal exchange coefficient, thermal conductivity) and we have provided the right thermal power on the hotplate in order to obtain a temperature of 100°C in the sensor's centre neglecting the presence of the gold fingers and the sensing material on the upper surface and forcing the device's sides to 20°C. We have obtained the desired temperature with only 10.9mW (equivalent to 1.8V) in the case of silicon nitride as insulator layer, the distribution of temperature is similar when using between the sensor and the hotplate, as shown in Figure 8. Moreover the structure with 2000 Å SiO_2 need a power of 14.7mW equivalent to 2.1V, to reach the same maximum temperature.

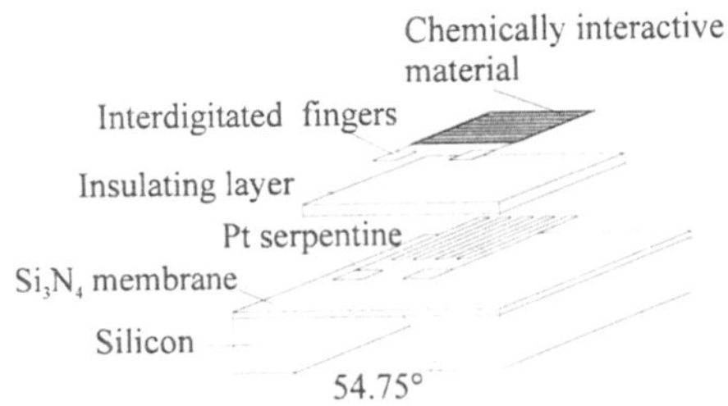

Figure 7. Unit cell structures of the array fabricated.

Finally we have tested the case of 2000 Å diamond like layer: now we need to provide the serpentine with more power (44mW that are equivalent to 3.63V), but we have obtained a very uniform temperature distribution in the middle of the device on which there will be the chemical sensor, as shown in Figure 9. The reason of these differences is the high thermal conductivity of the diamond like material that is 1000 W/m•K instead of 1.5W/m•K in SiO_2. That makes the temperature uniform on the entire layer, but the price to pay is greater thermal power to reach the same temperature. We have then chosen this last geometry because we consider it more suitable to the development of a gas sensor.

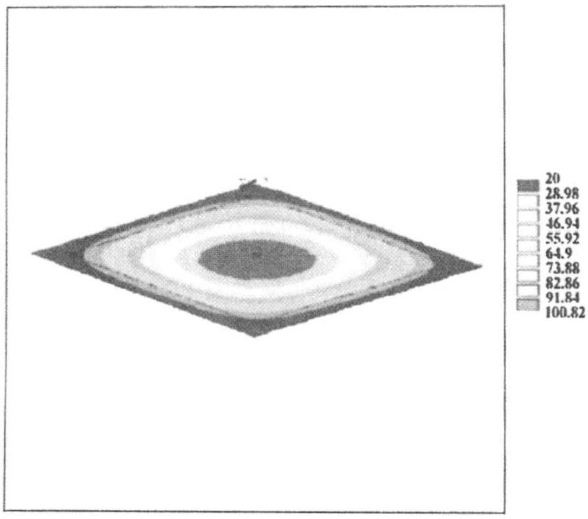

Figure 8. ANSYS simulation of the device with SiO_2 as insulating material.

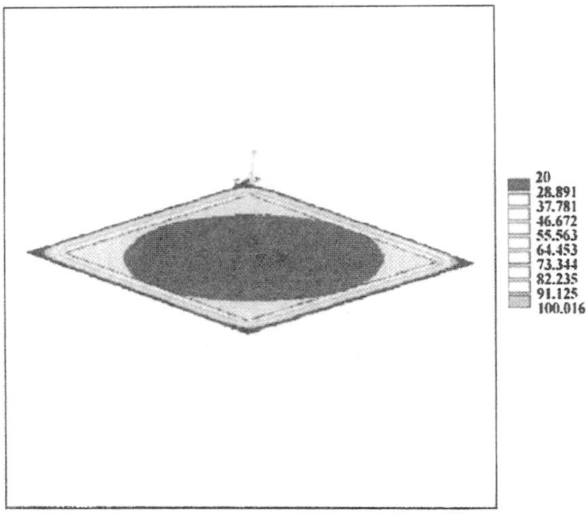

Figure 9. ANSYS simulation of the device with diamond like as insulating material.

Moreover, using metallo-porphyrines as chemically interactive materials, interdicts us from overcoming 180-200°C: anyway, with this structure we can easily reach very high temperature only by providing the hotplate with voltage of few volts and then we would be able to utilize chemically interactive materials different from those now used. In the further analyses we have neglected the influence of the metallic contacts to provide the hotplate and to make the measurements on the interdigital fingers device.

3.2 Device fabrication and performance

The device is fabricated using a 380 μm silicon wafer covered with a 1 μm Si_3N_4 layer. The silicon nitride membrane is obtained etching the silicon with KOH. The insulation between the heater and the interdigitated fingers is realized by using a 0.2 μm diamond like insulation layer, Alternatively, the silicon nitride layer could be used as insulator between the two devices: this solution is difficult to realize because of membrane's fragility. We decided to use a high thermal conductivity diamond like film for reason that will be clarified below. The film has been deposited at CETEV by plasma ehnanced chemical vapour deposition, using methane and hydrogen as precursors. In order to fabricate platinum resistor and gold interdigital fingers devices, lift-off technique has been used. AZ5214E positive resist from Hoechst has been used with the image reversal technique which gives mushroom shaped profiles that are ideal for lift-off technique. The interdigital fingers structure is made of 115 fingers 20m wide and 1200Å thick (200Å Cr and 1000Å Au) will be covered with metallo-porphyrins or a proper sensitive polymer to create a chemical sensor. The platinum resistor is made of a film 2500Å thick (200Å Cr and 2300Å Pt) and 100 m wide. The resistivity of platinum film is a factor 1.5 the bulk platinum resistivity $1.1 \cdot 10^{-7}\Omega m$, because of film's granularity that increase the resistivity.

Figure 10. shows an optical microscope image of a part of the single cell microheater. The silicon nitride and the diamond like films are optically transparent, this enable us to see through the whole device. The photograph has been taken from the backside of the chip. The platinum serpentine heater partially cover the interdigitated gold fingers that cross normally the heater.

Figure 10. Photograph of the single cell microheater.

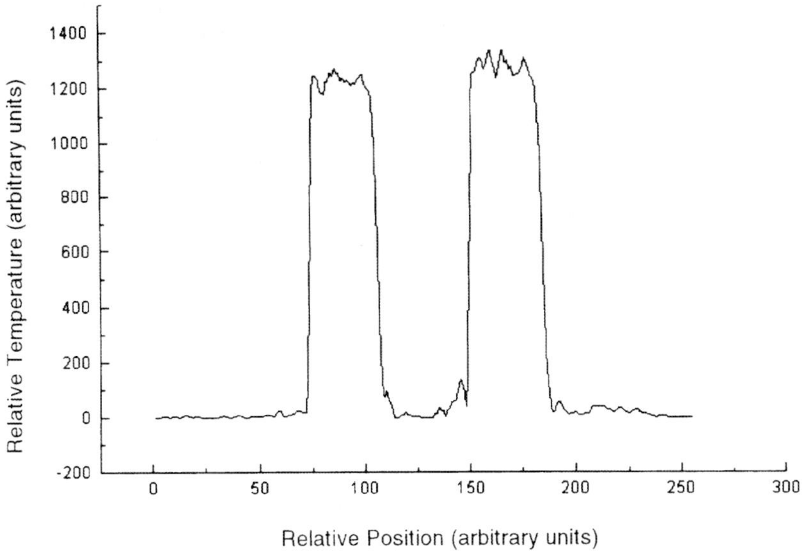

Figure 11. Thermal scan on a line crossing two contiguous powered heaters.

The thermal measurement with a thermal scanner on two contiguous powered heaters show the complete thermal isolation of the heaters (Figure 11.) with respect to the silicon that frame the array, the temperature in the membrane is ≅150°. The thermal distribution over the membrane is reasonably good.

4. CONCLUSIONS

This paper gives a brief overview of the techiques and materials employed in the context of micromachining technology. Surface, bulk and bold micromachining methodologies have been discussed and illustrated with the intent of showing the huge amount of possibilities available to day to design and fabricate microsystems. A microheater is also illustrated as an example of microstructure, together with its main thermal distribution characteristic, to indicate a way of producing devices with the capability of hosting temperature dependent sensing features.

ACKNOWLEDGEMENTS. Authors want to give a particular thank to Prof. F. Scudieri of the Departement of Ingegneria Meccanica for the thermal measurements related to Figure 11.

REFERENCES

1. N.F. Raley, Y. Sugiyama, T. Van Duzer, "(100) silicon etch rate dependence on boron concentration in EDP-water solutions", J. Electrochem. Soc., **131**, 162, (1984).
2. I.G. Stoev, R.A. Yankov, G. Jeynes, "Formation of etch stop structures utilizing ion-beam syntesized buried oxide and nitride layers in silicon", Sensor and actuators, **19**, 183, (1989).
3. E.D. Palik, O.J. Glembocki, R.E. Stahlbush, "Fabrication and characterizations of silicon membranes", J. Electrochem. Soc., **135**, 3126, (1988).

4. H. Seidel, L. Csepregi, A. Heuberger, H. Baumgartel, "Anisotropic etching of crystalline silicon in alkaline solutions I , orientation dependance and behavior of passivation layers", J. Electrochem. Soc., **137**, 3612, (1990).
5. H. Seidel, L. Csepregi, A. Heuberger, H. Baumgartel, "Anisotropic etching of crystalline silicon in alkaline solutions II, influence of dopants", J. Electrochem. Soc., **137**, 3626, (1990).
6. M. Elwenspoeck, "The form of etch rate minima in wet chemical anisotropic etching of silicon", J. Microm. Microeng., **6**, 405, (1996).
7. J.M. Shimbo, K. Furukawa, K. Fukuda, K. Tanzawa, "Silicon to silicon direct bonding method", J. Appl. Phys., **60**, 2987, (1986).
8. See for exampleAnalog Device and SensoNor products.
9. See for example Motorola pressure sensor products.
10. L.S. Fan, Y.C. Tai, R.S. Muller, "Integrated movable micromechanical structures for sensors and actuators", IEEE Trans. on El. Dev., **35**, 724, (1988).
11. M. Mehregany, K.J. Gabriel, W.S. N. Trimmer, "Integrated fabrication of polysilicon mechanisms", IEEE Trans. on El. Dev, **35**, 719, (1988).
12. H.C. Nathason, W.E. Newell, R.A. Wickstrom, J.R. Davis, "The resonant gate transistor", IEEE Trans on El. Dev. **ED 14**, 117, (1967).
13. D.J. Monk, D.S. Soane, R.T. Howe, "Sacrificial layer SiO_2 wet etching for micromachining applications", Proc. 8th Int. cnf. Solid State Sensor and Actuators (Transducers 91), San Francisco, USA, p. 647, (1991).
14. F. Kozlowsky, N. Lindmair, Th. Sheiter, C. Hierold, W. Lang, "A novel method to avoid sticking of surface micromachined structures", Proc. 8th Int. cnf. Solid State Sensor and Actuators (Transducers 95), Stockholm, Sweden, Vol 1, 220, (1995).
15. T. Abe, W.C. Messner, M.L. Reed, "Effective methods to prevent stiction during post release etch processing", Proc. 8th IEEE Micro Electro Mechanical Systems, Amsterdams, The Netherlands, p. 94, (1995).
16. N.J. Mourlas, K.C. Stark, M. Mehregany, S.M. Phillips, "Exploring polysilicon micromotors for data storage micro disks", Proc.9th IEEE Micro Electro Mechanical Systems, San Diego California, p. 198, (1996).
17. R. Legtemberg, E. Barenshot, J. van Baar, Th. Lammerink, M. Elwenspoek,"An electrostatic lower stator axial gap wobble motor: design and fabrication", Proc. 8th Int. cnf. Solid State Sensor and Actuators (Transducers 95), Stockholm, Sweden, p. 404, (1995).
18. D.E. Sene, J.W. Grantham, V.M. Bright, J.H. Comtois, "Development and characterization of micromechanical gratings for optical modulation", Proc. 9th IEEE Micro Electro Mechanical Systems, San Diego California, p. 222, (1996).
19. P. Bley, J. Mohr, "The LIGA Process, a microfabrication technology" FED Journal, 5, Suppl.1, 34, (1994).
20. L. Weber, W. Ehrfeld, H. Freimuth, M. Lacher, H. Lehr, B. Pech, "Micro molding, powerfull tool for the large scale producion of precise microstructures", Proc. Symp. Micromachining and microfabrication, 156, (1996).
21. M. Eckele, W. Bacher,, "Modular equipment for quasi monolithic integration of micromechanical and microelectronic components", Mechatronics, 501, (1996).
22. G. Keller and T. Howe, "Nickel filled hexsil thermally actuated tweezers", 8th International Conference on Solid State sensors and actuators, Digest of technical papers, p. 376, (1995).
23. U. Wallrabe, J. Mohr, I. Tesari, K. Wulff, "Power characteristics of 3d operated microturbines for minimally invasive terapy", Proc. 9th IEEE Micro Electro Mechanical Systems, San Diego, California, p. 462, (1996).
24. G. Bomchil, A. Halimaoui, and R.Herino, "Porous Silicon: The material and its applications to SOI Technologies", Microelectronic Eng., Vol.8, pp. 293-310, (1988).
25. N. Koshida, H. Koyama, "Visible Electroluminescence from porous silicon", Appl. Phys. Letters, vol. 60, pp. 347-349, (1992).
26. J. Gannon J. Neuse, "A chemical etchant for the selective removal of GaAs through SiO_2 mask", I. Electroch. Soc., vol. 21, p. 1215-1219, (1974).
27. J.J. Le Porc, "An improvement technique for selective etching of GaAs and AlGaAs", J. Appl. Phys., vol. 51, p. 6441-6442, (1980).
28. C.J. Nuese J.J. Cannon, "Electrolitic removal of p-type GaAs substrates from thin n-type semiconductor layers", J. Electroch. Soc., vol 117, p. 1094-1097, (1970).
29. C. Di Natale, F.A.M. Davide, "Multicomponent analysis in chemical sensing", Proceedings of the First European School on Sensors (ESS '94) "Sensors for domestic application", Editors A. D'Amico, G. Sberveglieri, World Scientific.
30. ANSYS is a FEA program, registered trademark of ANSYS Inc.

THE STRETCH-AND-WRITE TECHNIQUE FOR FABRICATION OF FIBER BRAGG-GRATING ARRAYS

R. Falciai, R. Fontana, and A. Schena

Electromagnetic Research Institute "Nello Carrara" (IROE CNR)
Via Panciatichi, 64
50127 Firenze, Italy

1. INTRODUCTION

Fiber Bragg grating (FBG) technology is currently generating a great deal of interest for applications both in telecommunications[1, 2] and sensing systems,[3, 4] because of its capability of producing devices such as fiber-based dispersive elements, wavelength division multiplexing devices, wavelength selective filters and couplers and sensor networks for measuring different parameters such as temperature, strain, pressure, vibration.

The FBG sensors have some advantages with respect to the traditional fiber sensors: the wavelength-encoded response, that makes the sensor independent of fluctuating light levels; the possibility of realizing arrays of sensor elements, that can be addressed in the time or spectral domain, or a combination of both; easy processing with respect to interferometric sensors.

Two methods are commonly used to write FBGs: the holographic method[5], which makes use of a two-arms interferometer, and the photolithographic method[6], which employs phase masks.

The photolithographic method has significant advantages compared to the holographic one,[7] and thereby has rapidly become the most used for reproducible grating fabrication. Its main disadvantage is the need of different phase masks for each Bragg wavelength required. However, this drawback can be almost in part overcome by stretching the fiber during the writing process, thus providing a means of changing the effective Bragg-grating wavelength.[8]

In the present chapter a description of the stretch-and-write technique is reported, which allows to fabricate spatially and spectrally well separated, as well as, adjacent and spectrally concatenated grating arrays, together with examples of fabricated devices which can find applications both in sensor and telecommunication systems.

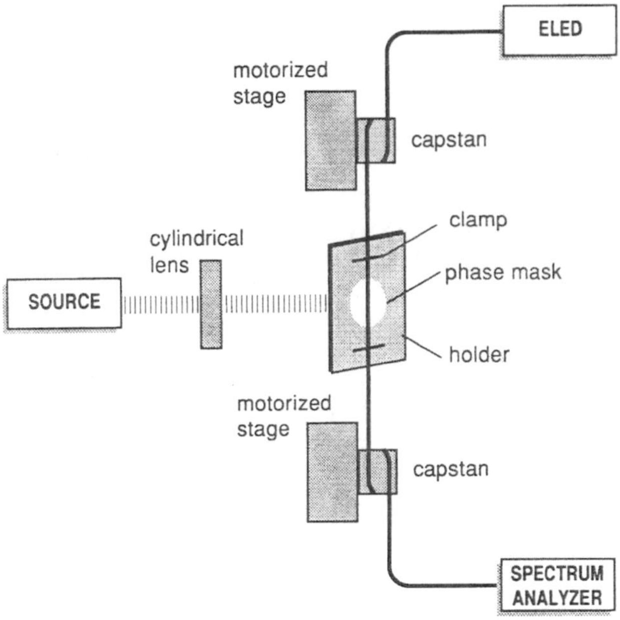

Figure 1. Sketch of the stretch-and-write apparatus.

2. FBG-ARRAY FABRICATION AND CHARACTERIZATION

The photolithographic method with the improvement of the stretch-and-write technique was utilized to fabricate FBG arrays. For this purpose, a properly designed apparatus, as sketched in Figure 1. was built.

An UV excimer KrF laser source ($\lambda = 248$ nm) was used, along with two different phase masks with grating periods of 1071 nm and 1078 nm, respectively. The bare fiber length to be written was placed in close contact behind the phase mask, with the fiber axis perpendicular to the mask grooves. Gratings were written either on a 3M fiber properly designed for FBG technology (2r = 7.5 nm, NA = 0.160) or on a special Pirelli Coupler fiber (2r = 4.5 nm, NA = 0.165). The pitches Λ_{pm} of the two phase masks corresponded to writing wavelengths λ_B near 1547 nm and 1557 nm for the 3M fiber, and near 1550 nm and 1560 nm for the Pirelli fiber, respectively, according to the formula[9, 7]

$$\lambda_B = n_{eff} \Lambda_{pm}, \qquad (1)$$

where n_{eff} is the effective index of the fiber.

The fibers were exposed, over regions of 5÷10 mm lengths, to a fluence level ranging from 300 to 450 mJ/cm^2 per pulse, with a repetition rate of 100 Hz, during different exposure times. The peak growth was monitored during the writing process by using an Optical Spectrum Analyzer and illuminating the fiber with an ELED source. All the gratings were spectrally characterized, in order to determine the whole spectral features of the gratings (peak shape, height, width and Bragg wavelength). FBGs with peak reflectivity up to 99% and bandwidth between 0.12 and 0.35 nm were written.

In order to fabricate gratings, distributed along a fiber length and operating at different Bragg wavelengths, the fiber is rolled up and fixed by two capstans on two spools mounted

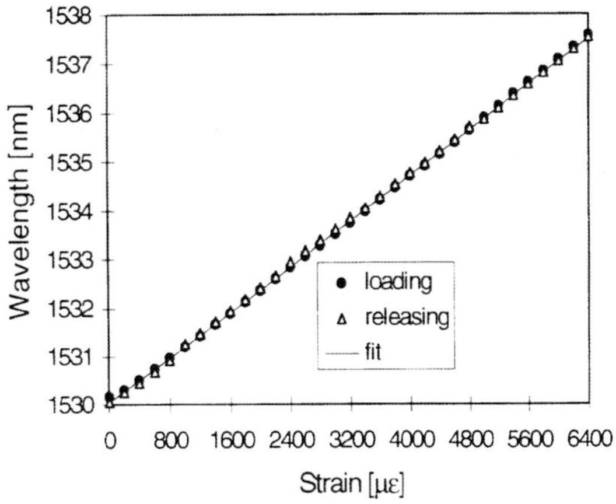

Figure 2. Calibration curve of the strain sensor on a strain cycle from zero up to 6400 με and back, in steps of 200 με.

on two motorized translation stages, so to change the exposure region in a simple and safe way. By suitably setting the stage displacement, it is possible to stretch the fiber to a given strain value. In this condition, the written grating has the same peak wavelength, but when the fiber is relaxed, the grating compresses and the peak moves towards shorter wavelengths. To the purpose to obtain a given shift, it is necessary to know the strain value to be applied to the fiber. The fractional change $\Delta\lambda_B/\lambda_B$ induced in the Bragg wavelength λ_B in response to a strain variation $\Delta\varepsilon$, is given by[10]

$$\frac{\Delta\lambda_B}{\lambda_B} = (1 - P_e)\Delta\varepsilon, \qquad (2)$$

where P_e is an effective photoelastic constant. A strain calibration curve, necessary to pre-set the stretch apparatus, was obtained by utilizing the same experimental set-up used for the on-line FBG characterization during the writing process. The spectral features of a number of gratings were examined during a strain cycle, in order to verify the absence of any hysteresis. One strain cycle consisted of several static strains in steps of 20 με from zero up to the maximum strain and back. Some samples were strained over wider intervals, and in some cases until breakage (approximately 10.000 με). The $\Delta\lambda_B/\lambda_B$ behavior resulted linear over all the investigated strain intervals showing a sensitivity of $1.2\text{x}10^{-3}$ nm/με for both 3M and Pirelli fibers. Figure 2. shows as an example the Bragg wavelength shift as a function of the applied strain for a 3M fiber on a cycle from 0 to 6400 με.

A further check of the separation between two consecutive Bragg gratings was carried out on-line, in order to consider possible failures in the stretching apparatus, due, for example, to a fiber sliding on the capstans. The resultant error in the operative grating wavelength was about 0.02 nm.

Arrays of up to five gratings, with different peak position and distributed over about 1 m length of fiber, were fabricated. Figure 3. shows the transmission spectrum of an array of five FBGs, with a peak separation of about 0.88 nm: the difference in the adjacent peak spacing is due to an experimental set-up error of ± 10 με. In principle, it is possible to write FBGs with peaks over a 12 nm span, which corresponds to the maximum strain that the fiber can undergo before breakage.

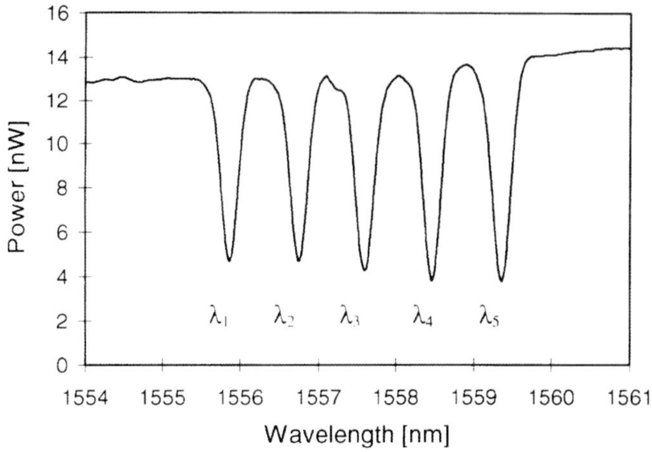

Figure 3. Transmission spectrum of five FBG arrays. $\Delta\lambda_{21} = 0.88$ nm; $\Delta\lambda_{32} = 0.86$ nm; $\Delta\lambda_{43} = 0.86$ nm; $\Delta\lambda_{54} = 0.90$ nm.

3. FBG SENSING APPLICATIONS

The main application of gratings as sensing elements is the strain measurement on wide materials and structures. Grating arrays distributed along a single fiber allow to obtain strain mapping for testing and survey purposes.

To the aim of setting up a sensor system for monitoring the critical regions of a vehicle structure, as a first step a FBG sensor was applied on a steel sample, comparing its response with that of a traditional strain gauge, applied to the same sample.

FBGs were applied to a number of steel specimens consisting of standard 1 mm thick FEP04 plates, employed for car body fabrication (supplied by the Fiat Research Center, Italy). The FBG sensor and the electrical strain gauge were centrally fixed, aligned along the longitudinal axis of the specimen on its two opposite surfaces, in order to compare their performance. Loads were applied in the 0÷1700 N range corresponding to 0÷600 $\mu\varepsilon$: this upper limit was set by steel enervation).

The steel samples were strained through a standard traction machine placed in a temperature-stabilized room, while the output signals from the two sensors were simultaneously recorded. Figure 4. reports the grating sensor response plotted versus the strain gauge response, showing a very good agreement over the whole investigated range between the two sensors and comparable resolution.

Further steps will be strain mapping in dynamic conditions through grating arrays applied on specimens or on vehicle parts.

4. FBG-ARRAYS FOR TELECOMMUNICATION SYSTEMS

FBGs find application also for the realization of passive components for telecommunication systems. An important device for such systems is the pass-band filter which, for instance, can be constituted by two identical gratings at the two output arms of a 3 dB coupler, which must be carefully balanced.[11] Arrays of adjacent and spectrally concatenated gratings, fabricated with the stretch-and-write technique above described, have been utilized for constructing a transmission pass-band filter.

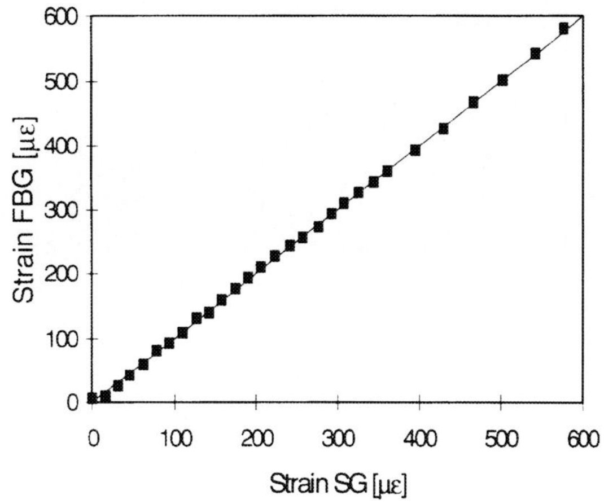

Figure 4. Strain values measured with FBG sensor versus the corresponding value of the strain gauge.

Several adjacent FBGs were written in such a way that their transmission spectra overlap, so to have uniform stop-bands with rejection as good as possible. The pass-band filter was obtained by means of two stop-bands suitably frequency spaced. As an application, the constructed pass-band filter was aimed to select the principal emission line of a commercial laser diode, that has its central line at λ = 1547.3 nm. The two lateral stop-bands of the transmission filter were properly chosen for the suppression of the secondary lines immediately adjacent to the main peak. The transmission spectrum of the filter is shown in Figure 5.

The stop-band in correspondence with the first right hand side lobe of the laser diode was obtained by writing 8 gratings. The first grating (5 mm long) was written at a peak wavelength of 1548.5 nm, with bandwidth of about 0.35 nm and transmittance lower than 5%. After a 5 mm shift, the fiber was stretched by 100 microstrains, and a second grating with the same characteristics of the former was written: the strain level applied corresponded to a peak shift, when the fiber was released, of 0.12 nm.

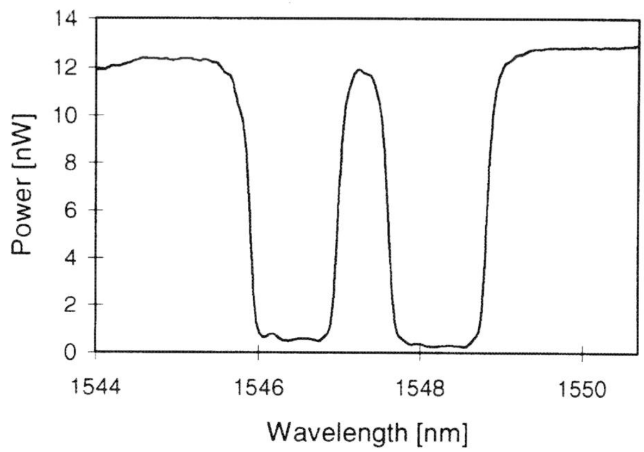

Figure 5. Transmission spectrum of the pass-band filter.

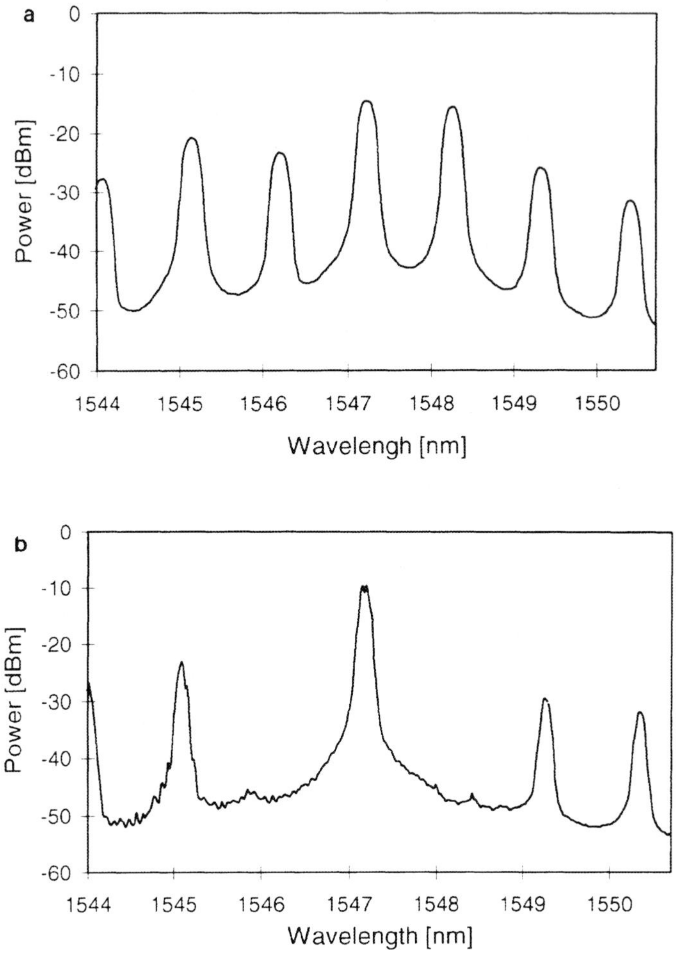

Figure 6. Reflection spectra of the laser diode *a*) before and *b*) after application of the pass-band filter.

By repeating this process up to writing 8 gratings, a ~ 1 nm wide stop-band was obtained capable of suppressing the wavelength band between 1547.7 and 1548.7 nm. An analogous procedure was followed for suppressing the wavelength band between 1546 and 1547 nm, after having applied to the fiber 780 microstrains that corresponded to a pass-band width of 0.6 nm.

The good performance of the filter is shown in Figure 6., where the transmission spectra of the laser diode are plotted before (curve *a*) and after (curve *b*) the application of the pass-band filter.

5. CONCLUSIONS

The photolithographic method with the improvement of the stretch-and-write technique was applied for the fabrication of fiber-Bragg grating arrays. This procedure allows to change the grating peak wavelength utilizing a single phase mask. A properly designed apparatus was described to write spatially and spectrally well separated grating arrays, distributed on a length of fiber, as well as arrays of adjacent gratings spectrally concatenated.

Two different types of FBG arrays were constructed to be used in two different fields of application, that is in sensing and telecommunication systems. In particular an interesting field of application was considered for strain sensing in the automotive industry, to monitor various critical components of vehicles in operating conditions. The replacement of electrical strain gauges with FBG sensors presents certain advantages in terms of linearity, precision, repeatability and the possibility of using only one detecting system due to easy multiplexing with grating arrays.

Another example of application presented is the construction of a transmission pass-band filter, applied to suppress the first side lobes in a laser diode spectrum. The used technique could allow, in principle, to realize superstructure fiber gratings that exhibit a comb-like response in transmission, while in the literature only comb-like reflection response devices are reported.[12]

ACKNOWLEDGMENTS. The authors would like to acknowledge RTM S.p.A. - Italy, who kindly supplied the excimer laser source, and the FIAT Research Center, Italy, for collaborating in the strain test performance on car steel specimens. The authors would also like to thank Prof. A.M. Scheggi for fruitful discussions in the work planning.

REFERENCES

1. R. Kashyap, Optical fiber Bragg gratings for applications in telecommunications, *Proc. 21th Eur. Conf. on Opt. Comm. (ECOC '95 - Brussels)*, 23-26 (1995)
2. P.A. Morton, V. Mizrahi, S.G. Kosinski, L.F. Mollenauer, T. Tanbun-Ek, R.A. Logan, D.L. Coblentz, A.M. Sergent and K.W. Wecht, Hybrid soliton pulse source with fiber external cavity and Bragg reflector, *El. Lett.*, Vol. 28, No. 6, 561-562 (1992)
3. R.M. Measures, T. Alavie, R. Maaskant, S. Huang and M. LeBlanc, Bragg grating fiber optic sensing for bridges and other structures, *2nd European Conf. on Smart Structures and Materials, Glasgow*, 162-167 (1994)
4. M.J. Chawki, L. Berthou, E. Delevaque, E. Gay and I. Le Gac, Evaluation of an optical boosted add/drop multiplexer OBADM including circulators and fiber grating filters, *Proc. 21th Eur. Conf. on Opt. Comm. (ECOC '95 - Brussels)*, 47-50 (1995)
5. G. Meltz, W.W. Morey and W.H. Glenn, Formation of Bragg gratings in optical fibers by a transverse holographic method, *Optics Lett.*, Vol. 14, No. 15, 823-825 (1989)
6. K.O. Hill, B. Malo, F. Bilodeau, D.C. Johnson and J. Albert, Bragg gratings fabricated in monomode photosensitive optical fiber by UV exposure through a phase mask, *Appl. Phys. Lett.*, Vol. 62, No. 10, 1035-1037 (1993)
7. I. Bennion, J.A.R. Williams, L. Zhang, K. Sugden and N.J. Doran, UV-written in-fiber Bragg gratings, *Optical and Quantum Electronics*, 28, 93-135 (1996)
8. K.C. Byron and H.N. Rourke, Fabrication of chirped fiber gratings by novel stretch and write technique, *Electron. Lett.*, 31, 60-61 (1995)
9. R.J. Campbell and R. Kashyap, The properties and applications of photosensitive germanosilicate fiber, *Int. J. of Optoelectronics*, Vol. 9, No. 1, 33-57 (1994)
10. W.W. Morey, G. Meltz and W.H. Glenn, Fiber optic Bragg grating sensors, *Proc. SPIE Fiber Optic and Laser Sensors VII*, Vol. 1169, 98-107 (1989)
11. F. Bilodeau, K.O. Hill, B. Malo, D.C. Johnson and J. Albert, High-return-loss narrow-band all-fiber band-pass Bragg transmission filter, *IEEE Phot. Tech. Lett.*, Vol. 6, No. 1, 80-82 (1994)
12. A. Othonos, X. Lee and R.M. Measures, Superimposed multiple Bragg gratings, *El. Lett.*, Vol. 30, No. 23, 1972-1974 (1994)

APPLICATIONS

FLUORESCENCE LIFETIME-BASED SENSING FOR BIOPROCESS AND BIOMEDICAL APPLICATIONS

G. Rao, L. Eichhorn, and Q. Chang

Department of Molecular Biology
Chemical and Biochemical Engineering
University of Maryland
252 TRC
Baltimore MD, 21250 USA

1. INTRODUCTION

Optical sensing of analytes is an active area of research.[1,2] Optical sensors have the advantage of being robust and generally reliable in comparison with electrochemical sensors. Within optical sensing, several principles such as absorbance, reflectance, light scatter, fluorescence etc. have been employed to construct sensors. We focus here on the issue of fluorescence and in particular, fluorescence lifetimes for sensing several analytes.

Fluorescence lifetime-based sensing can circumvent the limitations of conventional as well as fluorescence intensity-based sensors. Phase-shift lifetime measuremens can be accomplished in seconds and are very robust with respect to amplitude changes, electronic noise and other signal loss problems. Steady-state intensity measurements have not found wide acceptance due to lack of stability and the need for frequent recalibration. Furthermore, intensity measurements perform reliably only in optically clean systems. Wavelength-ratiometric measurements are desirable for chemical sensing, but there are few available intensity-ratio probes. As a rare example, the Ca^{2+} probe Fura-2 displays a fluorescence spectral shift upon binding calcium, which allows wavelength ratiometric measurements independent of probe concentration to be made.

Decay time sensing is advantageous because such measurements are independent of probe concentration and/or intensity, are mostly independent of absorbance and scattering of the samples, and do not require dual emission from the probe. Among the two possible methods, time-domain or phase-modulation, we prefer the latter because one can construct low-cost instruments to measure even subnanosecond lifetimes. In phase fluorometry, sinusoidally modulated excitation light interacts with a chemical probe and excites it. The probe fluoresces in response but is affected by the local analyte concentration. The emission therefore displays a phase angle shift in relation to the excitation, which is dependent on the analyte concentration around the probe. For further details, the reader is referred to Lakowicz.[3] Finally, multi-frequency

Optical Sensors and Microsystems: New Concepts, Materials, Technologies
Edited by Martellucci et al., Kluwer Academic / Plenum Publishers, New York, 2000.

phase and modulation instruments which operate at preset frequencies for a specific application are easily designed.

We describe below the fabrication and use of a pCO_2 sensor that is steam sterilizable and suitable for use in an aqueous environment.

2. pCO_2 RESONANCE ENERGY TRANSFER SENSOR

At the present time, all the known pCO_2 sensors work in an indirect fashion. One relies on making a pCO_2 dependent pH measurement by having a pH sensitive dye immobilized in a plastic membrane such that CO_2 diffuses in or out and changes the local pH. The requirements demanded of a single fluorescent probe molecule (say, for pH sensing) can be reduced through the phenomenon of fluorescence resonance energy transfer (FRET). Two molecular species, a donor and an acceptor are required. Donors can be selected for excitability with inexpensive sources. Acceptors with a pH-dependent absorbance that spectrally overlaps the donor fluorescence are added to the sensor. FRET from the donor to the acceptor will quench the fluorescence and alter both the fluorescence intensity and lifetime.[4]

By using the phenomena of FRET, the donor need not be sensitive to pH, and the pH-sensitive acceptor does not need to be fluorescent. This energy transfer takes place without the appearance of a photon and the transfer rate depends primarily on the extent of overlap of the emission spectrum of the donor with that of the absorption spectrum of the acceptor, and the distance between the two.

For FRET to occur, the distance between the donor and acceptor must be less than the Förster distance, typically 40 Å, and require acceptor concentrations of 1-10 mM. Covalently linking the donor/acceptor pairs could certainly achieve this distance, but would preclude facile substitutions of either donor or acceptor, since a new synthetic methodology would be required for each discrete donor/acceptor pair. We developed a FRET pH sensor based on the donor/acceptor pair M-Cresol Purple (MCP) and Sulforhodamine 101 (SR 101). As can be seen in Figure 1., the MCP acceptor has a pH sensitive absorption spectrum that overlaps the emission spectrum of the SR 101 donor.

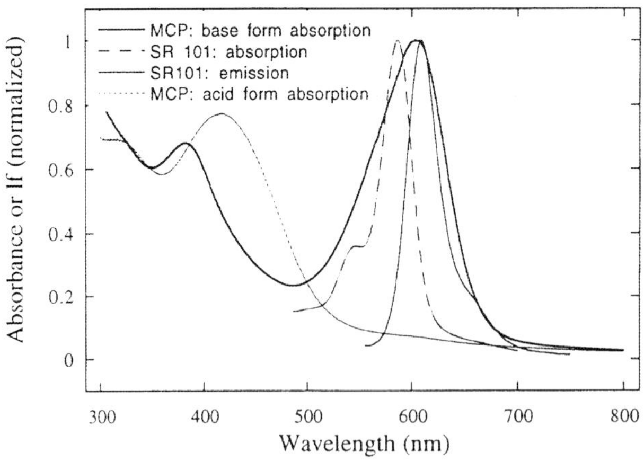

Figure 1. Excitation and emission spectra for Sulforhodamine 101 (SR 101) and absorption spectra for M-Cresol Purple (MCP).

3. pCO_2 SENSOR FABRICATION

For use in an aqueous system, a hydrophobic matrix is required to prevent cross-sensitivity to protons in the sample. Typically, the preparation of the CO_2 sensing film involved three steps: preparing the silicone film, preparing the dye solution, and doping the film with the dye solution. To prepare the silicone film, 0.9 g of polydimethylsiloxane, vinyldimethyl terminated (PS443) was mixed with 66 mg of hydrogen hexachloroplatinate-saturated PS443 and 43 mg of (30-35%) methylhydro-(65-70%) dimethylsiloxane copolymer (PS123). After shaking for 10 min, the viscous solution was smeared on a glass slide, and kept under nitrogen for a week. The curing of the polymer typically started in about one hour, and total finished in a week. The resulting film was very clear and non tacky when touched. Before use, the film was razored off the glass slide, and shaped into the desired size. To prepare the dye solution, a 0.5 M solution of cetyltrimethylammonium hydroxide (CTMAOH) in methanol was first prepared. Then, the acceptor solution prepared by dissolving 2.2 mg of MCP in 1.5 ml of chloroform, with addition of 20 μl of 0.5 M CTMAOH solution. Similarly, the donor solution was made by dissolving 3.5 mg of SR101 in 1.5 ml of chloroform, with addition of 20 μl of 0.5 M CTMAOH solution. The final dye solution consisted of 0.5 ml of the acceptor solution, 0.2 ml of the donor solution, 0.1 ml of the 0.5 M CTMAOH solution, and 0.3 ml of chloroform.

To dope the silicone film with the dye solution, the silicone film was first soaked in chloroform, and then moved into the dye solution. After shaking for one minute, the expanded, dyed film was moved out of the solution, and allowed to dry in ambient atmosphere. The dried film shrank to its original size, and was washed with methanol until the washout was colorless. After drying again, a blue transparent film was formed, which displayed fluorescence intensity and lifetime change when exposed to CO_2. Our CO_2 sensor is based on a previously reported colorimetric plastic film sensor.[5] Entrapping the donor and acceptor dyes in a plastic film, rather than covalently linking them to a substrate, allows facile substitution of different excitation sources. Figure 2. shows phase angle pCO_2 sensing with probe molecules immobilized in a plastic film.

The CO_2 sensor was further tested against an electrochemical analyzer in order to validate its performance. As shown in Figure 3., by passing air or CO_2 through a buffer solution, one can see the response of the optical sensor is similar to the off-line CO_2 analyzer with the advantage of offering continuous real time measurement without requiring discrete sampling. Such a sensor would have applications in fermentation/cell-culture and in blood gas measurements.

In summary, using fluorescence lifetime-based instruments offers the possibility of making reliable, rapid and convenient measurements at the bedside for biomedical applications and in a number of bioprocesses. The technology is generic and a number of analytes may eventually be measured using the same technology. Calibration will be minimal and instrument drift will be eliminated. This technology is also economical due to recent advances in optics and electronics and can potentially have a significant effect on health care cost, by lower cost testing, by shorter hospital stays for critical patients, and immediate decision-making by the physician.

ACKNOWLEDGEMENTS. Funding from the NSF and NIH is gratefully acknowledged.

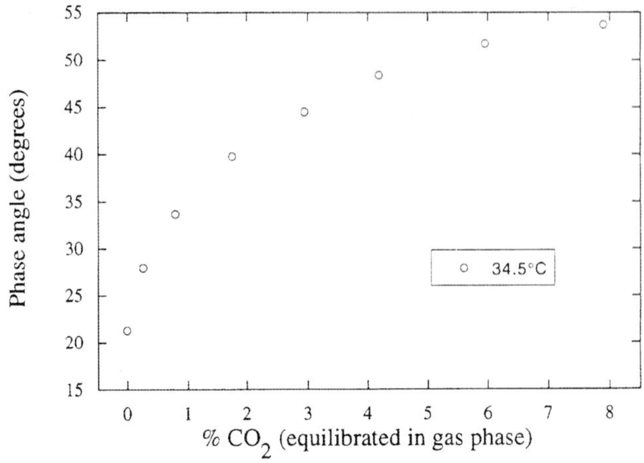

Figure 2. Typical calibration curve for the pCO_2 sensing film.

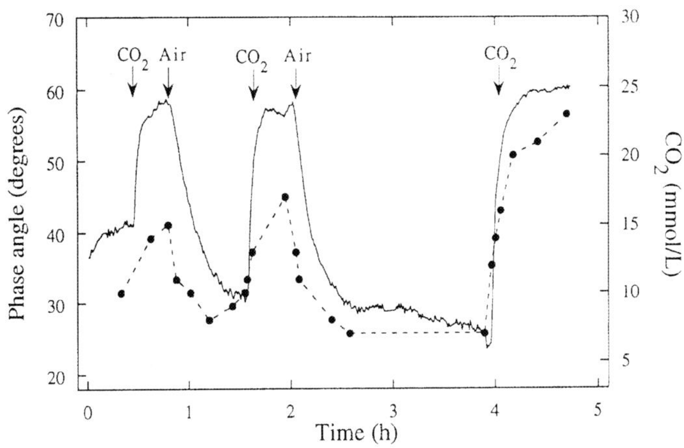

Figure 3. Response of the optical sensor to cycling between CO_2 and air (solid lines). The filled circles are off line readings taken from discrete samples and analyzed on a wet chemistry analyzer.

REFERENCES

1. O. Wolfbeis. Proceedings of the 1st European Conference on Optical Chemical Sensors and Biosensors. Europt(R)ode I. Sensors and Actuators B (1992).
2. F. Baldini. Proceedings of the 1st European Conference on Optical Chemical Sensors and Biosensors. Europt(R)ode II. Sensors and Actuators B (1994).
3. J. R. Lakowicz. Principles of Fluorescence Spectroscopy. Plenum Press, New York (1982).
4. C. C. Herbert. Resonance energy transfer in Topics in Fluorescence Spectroscopy, Volume 2: Principles. J. R. Lakowicz Ed., Plenum Press, New York (1991).
5. A. Mills, Q. Chang and N. McMurray. Equilibrium studies on colorimetric plastic film sensors for carbon dioxide. Anal. Chem. 13:1383 (1992).

A PIEZOELECTRIC BIOSENSOR AS A DIRECT AFFINITY SENSOR

M. Minunni, and M. Mascini

Università degli Studi di Firenze
Via Gino Capponi, 9
50121 Firenze, Italy

1. INTRODUCTION

It is well-known that the resonant frequency of an oscillating piezoelectric crystal can be affected by a change in mass at the crystal surface.[1] This method can be used as sensor for protein adsorption studies and for direct immunosensing. Up to now these studies have been performed by measuring the frequency in dry state, i.e. with the "dip and dry" technique which is rather cumbersome and time consuming.

Recently we obtained some results with piezoelectric crystals used directly in liquid solutions.

We will discuss a real-time monitoring of: (i) adsorption of the human immunoglobulin Ig G (h-Ig G); (ii) the affinity reaction between covalently immobilized antigen (the pesticide 2,4-D) and specific monoclonal antibodies (Mab anti-2,4-D) from two different clones (clone F6C10 and clone E2G2); and, (iii) immunoreactions between immobilized antigen and antibodies performing a competitive assay.

All experiments show how the reaction under study is linked to a mass increase which can be monitored continuously in real time.

Direct antigen-antibody interaction can, thus, be studied without any kind of label.

A microprocessor controlled piezoelectric detector as sensor was employed to monitor in real time protein adsorption and immunoreactions using piezoelectric quartz crystals (AT-cut) with basic resonant frequency of 10 MHz. The adsorbed protein was an immunoglobulin (h-Ig G); in the immunosensing a covalent immobilized molecule (the pesticide 2,4-D) formed the receptor for the immobilized ligand sample (Mab anti-2,4-D) in a competitive assay.

2. SENSOR PRINCIPLE

The crystal most commonly used are 5-9-10 MHz quartz in the form of 10-16 mm disks. The quartz wafer are sandwiched between two electrodes to the wafer surface. These electrodes are used to induce an oscillating electric field perpendicular to the surface of the wafer. The electric field then produces a mechanical oscillation, a standing wave, in the bulk

Optical Sensors and Microsystems: New Concepts, Materials, Technologies
Edited by Martellucci et al., Kluwer Academic / Plenum Publishers, New York, 2000.

of the quartz wafer. Mechanical oscillation of a crystal is maximum where the electrode pads overlaps and diminished rapidly in areas where the oscillating electrodes do not overlap. For most applications, the gold electrode is used because of its inertness.[1]

2.1. Measurement of the Resonant Frequency

When placed in an electronic circuit, the portion of the quartz wafer located between the electrodes vibrates with its precise fundamental frequency.

The frequency output from the oscillator, which is identical to the resonant frequency of the crystal, can be measured by a frequency counter

The change Δf in the resonant frequency, f_0, of the crystal may be directly related to the deposited mass Δm on the surface area A by means of the Sauerbrey Eq. (2, 3):

$$\Delta f = -2.26 \times 10^{-6} f_0^2 \Delta m/A.$$

The rate of this thickness-shear vibration is a function of the natural resonant frequency of the quartz and it also depends on the mass changes resulting from deposition of substances on the surface of the electrodes. Piezoelectric quartz crystals, therefore, can be used for mass measurements. The first analytical application of a PZ-crystal was reported by King [4] and during the next two decades, intensive research was directed to develop organic and inorganic coating for the detection of and determination of various toxic agents in the environment.[5]

Since biologically active materials such antibodies, enzymes and antigens are highly specific, they have been used as coatings, leading to a new class of PZ biosensor. Such crystals have been used for detecting water in gases,[6] organic pollutants,[7] toluene,[8] organophosphorus compounds in air and traces of iodide and silver and metals in solutions[6,9] as well for clinical applications. König and Grätzel reports a piezoimmunosensor for human Granulocytes detection in whole blood and detection of viruses (Rota-and Adenovirus) and bacteria (Salmonella, Shigella, Camylobacter, Escherichia coli) in stool specimen of infected babies.[10,11] In most of the applications the piezoelectric sensor is applied in the vapour phase. The utilization of the piezosensors based on reactions in solution and measuring the frequency shift after drying is time consuming and sensitive to errors due to hydratation and humidity. It was reported that the piezoelectric crystal in a liquid oscillates, the frequency being dependent of viscosity, density and conductivity of the solution.[12,13] Oscillation is even possible when only one side of the crystal contacting the solution and a piezoelectric crystal was proposed as detector in the liquid chromatography.[14] A more exaustive treatment about the theoretical background on the performance of the piezoesensor in liquids could be found in the literature.[15-21]

2.2. Measurement Procedure

The resonant frequency of piezoelectric crystals was determined using the detector PZ-105 (Universal Sensor, New Orleans LO). The frequency data were transferred to a notebook Toshiba T200SX using the RS232C interface and own software was used for grafic presentation and data manipulation. The cristals (AT-cut) with fundamental frequency of 10 MHz were purchased from Universal Sensor (New Orleans, LO).

The crystal was fixed inside an external flow-through thin layer cell (internal volume 30 µl) using two rubber O-rings, only one electrode was in contact with the flowing liquid (Figure1.). The peristatic pump Minipulse 3 (Gilson, France) was positioned before the cell, a silicon tube (0.03 mm diameter) was used for all connections. Flow rate 70 µl/min was used. The electrodes of the crystal were connected to the detector using wires and their

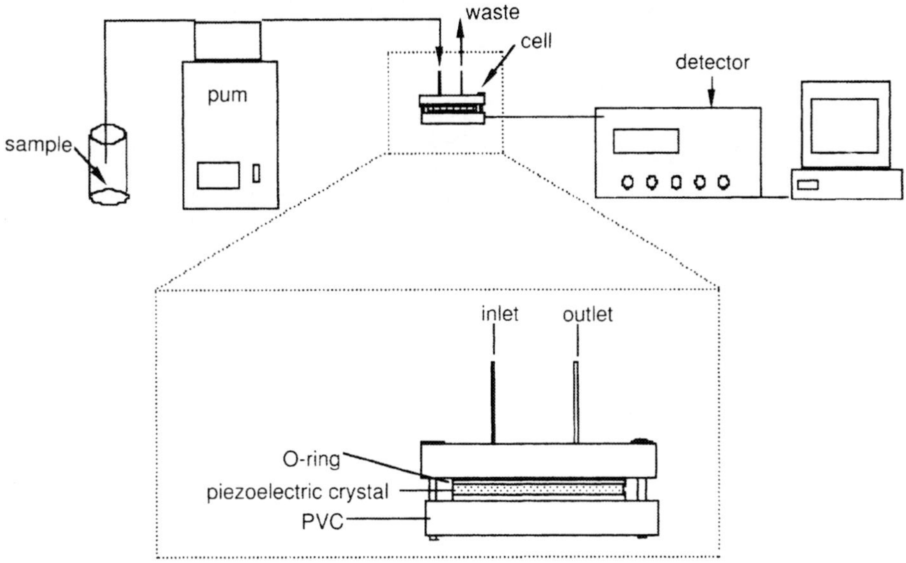

Figure 1. Apparatus for flow mode measurement.

shielding cover connected to metal frame of the detector. All experiment were performed at room temperature (25°C).

2.3. Chemicals

G-aminopropyltriethoxysilane (APTES) was from Sigma (St. Louis, MO); glutaraldheyde (GA), dioxane from Merck (Darmstat, Germany), h-IgG from Sigma (St. Louis MO); 2,4-D, tributylamine and isobutylcloroformiate were from Serva (Heidelberg, Germany), anti-2,4-D monclonal antibodies (MAbs) clone E2B5, F6C10, E2G2 were kindly supplied by Dr. Franek (Biochemistry Dep. University of Brno, Cechia)

3. ADSORPTION EXPERIMENTS

We studied the adsorption of h-IgG on the gold electrode of the piezoelectric crystal. We found that the protein is adsorbed irreversibly and that preserve their functionality reacting with the corresponding ligand (anti h-Ig) (data not shown).

Figure 2. shows the h-IgG adsorption assay. The measurements were made in continuous-flow mode with a flow rate (70 µl/min). The gold electrodes were treated sequentially with the following solutions: 1,2 N NaOH (20 min), washed with water, 1,2 N HCl (5 min), concentrated HCl (2 min), washed and air dried (30 min). Then crystal was mounted in the cell. The buffer solution flowed over the surface recording a fast drop in the frequency due to the liquid mass loading, successively stabilised. The protein concentration was increased from c=1 µg/ml in steps by a factor 10 to c=10 mg/ml. The resonant frequency of the crystal with the adsorbed protein decreases with the increase of the protein concentration.

In Figure 3. the frequency shift versus the protein concentration is reported. The shift is calculated as the difference between the initial frequency of the crystal (no protein adsorbed) subtracted by the resonant frequency obtained after the adsorption: $\Delta f = f_o - f_c$.

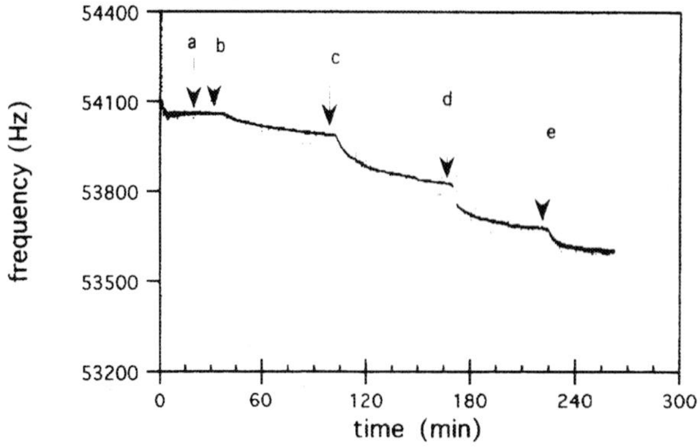

Figure 2. Adsorption of h-Ig G on the gold surface of the electrode; a) 1µg/ml, b) 10 µg/ml, c) 100 µg/ml, d) 1 mg/ml, e) 10 mg/ml in 50 mM phosphate buffer pH 7.0. We used 10 MHz crystal but the first three numbers (100) are omitted in the plot for graphic needs.

The adsorption of the protein is a slow processs. The analysis time in our experiment was of about 5 hours.

4. AFFINITY SENSING EXPERIMENT

For an affinity sensor receptor molecules have to be immobilized on the surface and, in particular, on the gold electrodes of the quartz disk which presents the maximum mechanical oscillation. The corresponding binding molecule, if present in the sample solution, will bind specifically to the receptor of the surface; the sensor directly responds to the formation of the receptor-ligand complex. When the ligand and the receptor are an antigen and the relative antibody their interaction will lead to an immunocomplex.

In the indirect assay the antigen is immobilized on the surface and the analyte presents in solution compete for the binding site of the antibody with the antigen immobilized. In the absence of analyte in solution all the antibody bind the surface; increasing the analyte concentration in the sample the amount of antibodies free for the surface binding is decreased then a minor amount of immunocomplex is formed. The signal recorded in this case is inversely related to the analyte concentration in the sample.

We analysed the binding between a covalently immobilized small molecule and its

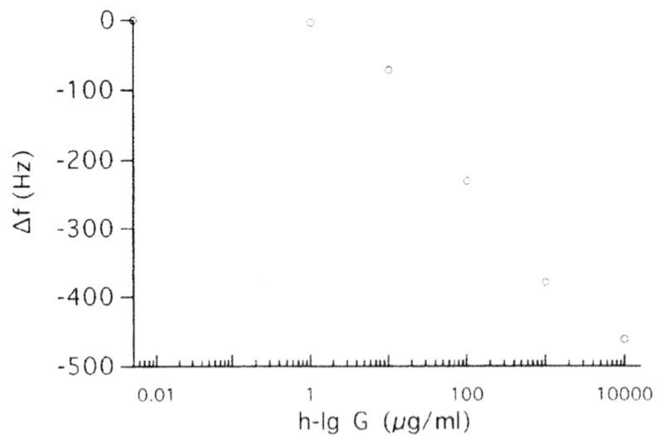

Figure 3. Frequency shit due to the adsorption of h-Ig G on the gold surface of the electrode.

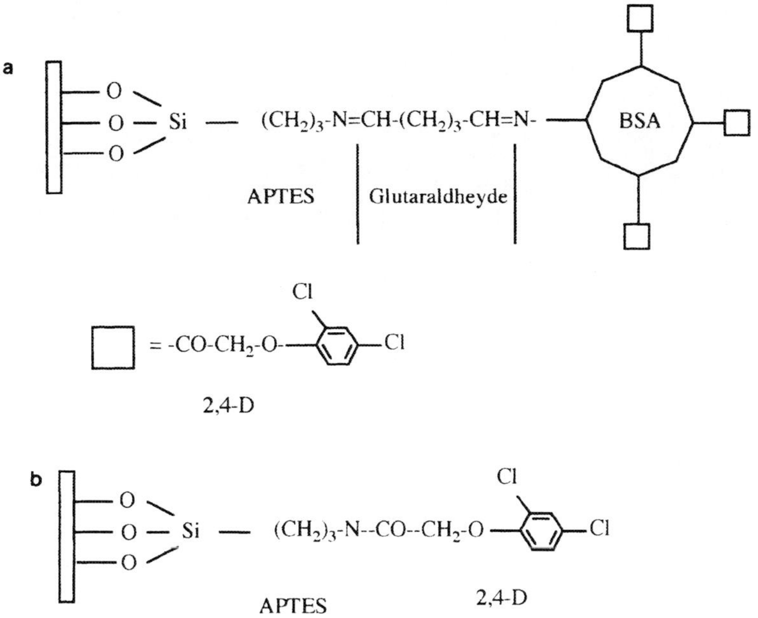

Figure 4. Immobilization chemistry for the 2,4-D; a) the pesticide is immobilized via Ga and BSA coupling; b) the pesticide is immobilized directly to the silanized surface.

relative antibodies. We chose the pesticide 2,4-D (2,4-dichlorophenoxyacetic acid) as model. The formation of the binding between the immobilized antigen and the antibody present in the solution was studied with two different clones of monoclonal antibodies (clone F6C10 and clone E2G2).

First of all the 2,4-D was immobilized on the surface.

4.1. Covalent immobilization of the 2,4-D:

The surface was first treated with 5% g-APTES acetone solution (2 hours), dried at 100°C for 1 hour, immersed in a 2.5 % GA 100 mM Phopsphate solution pH 7 (1 h) and air dried. Then 50 mg/ml BSA (bovin serum albumin) in 100 mM phosphate buffer pH 7 were applied on the crystal and incubated over night. The crystals were washed with water and immersed in a solution of previously activated 2,4-D solution and incubated over night, washed and stored at 4°C. The structure of the obtained product is shown in Figure 4.

4.2. Modification of the 2,4-D for coupling

300 mg of 2,4-D were dissolved in 7 ml of dioxane. 600 μl of tributylamine and the solution is cooled in ice bath at 10°C. While stirring 150 μl isobutylcloroformiate were added slowly. The solution was stirred for 30 min. Then 25 ml cold dioxane, 35 ml water and 3 ml 1.2 N NaOH were added (resulting in a pH 10 to 13). This solution was applied to the crystal.

The binding of anti-2,4-D antibodies to the surface was investigated. The formation of the immunocomplex was described by changes in the frequency. The surface capacity was tested with 1 mg/ml clone E2B5. A frequency shift of 113 Hz was found. When the antibody solution was replaced by the buffer the frequency shift changed very little indicating

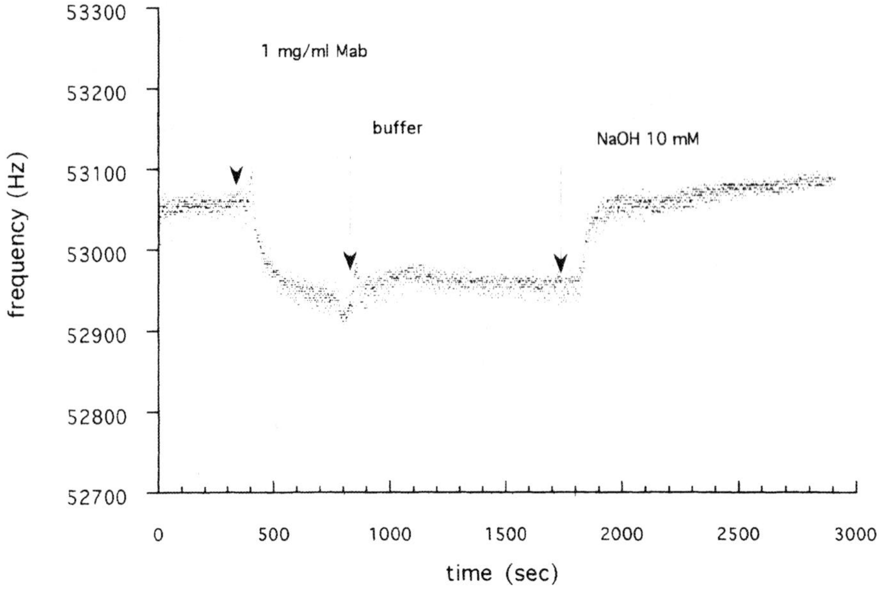

Figure 5. Affinity reaction between anti 2,4-D Mab (clone E2B5) 1 mg/ml and the 2,4-D immobilized on the surface. After the binding a solution of 50 mM phosphate buffer pH 7.0 flows over the surface. Then the surface is regenerated with a solution of NaoH 10 mM. We used 10 MHz crystal but the first three numbers (100) of the frequency are omitted in the plot for graphic needs.

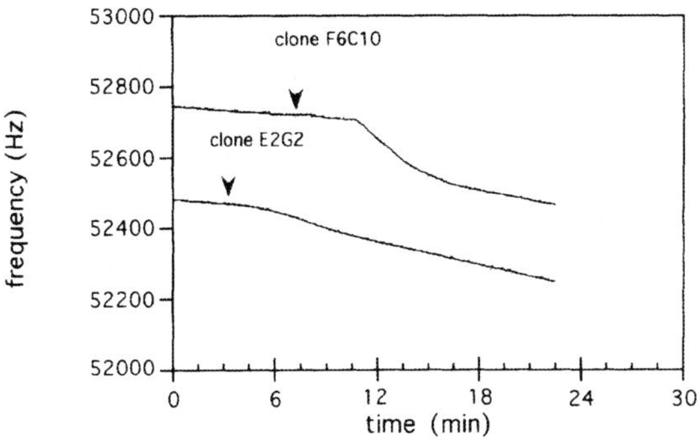

Figure 6. Reaction kinetic for two different clones (F6C10 and E2B2) to the surface with immobilized 2,4-D. We used 10 MHz crystal but the first three numbers (100) of the frequency are omitted in the plot for graphic needs.

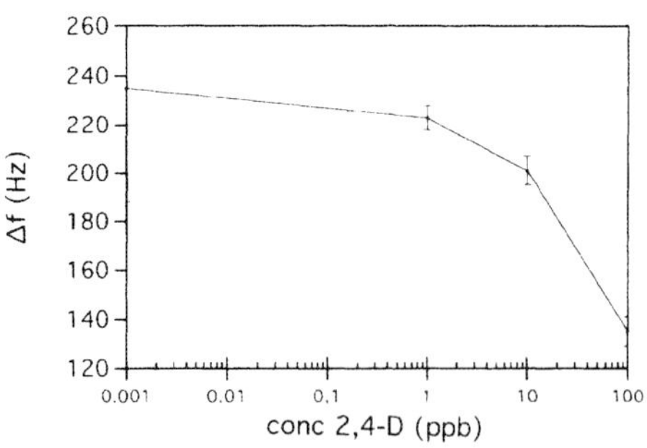

Figure 7. Results from the competitive assay. Different of concentrations of 2,4-D and a fixed amount (10 mg/ml) Mab anti-2,4-D (clone E2B5).

that the binding between the antigen immobilized and the antibody was stable and only a regenerating agent like 10 mM NaOH dissociated this binding as observed in Figure 5. The affinity reaction is relatively fast: in about 10 min. (500 sec) the interaction occurs. A control for unspecific binding with h-IgG was performed (data not shown).

Two different clones (F6C10 and E2G2) at a concentration of 1 mg/ml were used in order to study their ability to bind the surface. The kinetic of the binding is reported in Figure 6.

Clone F6C10 seems to binds more rapidly the crystal than E2G2 giving a frequency shift of 269 Hz and 313 Hz respectively, indicating an higher affinity for the surface of the first clone.

An indirect assay for the 2,4-D in tap water was performed. The surface, in this case, was modified binding the 2,4-D directly to the silanized surface (Figure 4b.). The crystals were incubated in the presence of anti-2,4-D antibodies (10 µg/ml) and different concentrations of 2,4-D (sample). The competition between free and bound 2,4-D for limited amount of IgG binding sites occurs and the resulting frequency decrease is indirectly proportional to the concentration of free pesticide. The results are shown in Figure 7. No matrix effect has been observed with the tap water. The analysis was performed in flow mode. The correlation between the frequency shift and the amount of analyte is evident. The analysis time was of 30 min for each measurement corresponding to the incubation time of the antibody and the surface.

5. CONCLUSIONS

We have demonstrated the potential of the piezoelectric detector for real-time monitoring of adsorption process and affinity-immunoreactions in liquid phase. Results for h-Ig G adsorption indicates the relation between the surface mass and the frequency shift. The adsorption process does not bind covalently the molecules to the surface. In the case of a stable binding for the receptor to the surface is needed, a covalent immobilization is recommended. A pesticide, the 2,4-D, has been bound to the surface and an affinity reaction between the bound 2,4-D and antibodies anti-2,4-D was performed. The stability of the interaction was evident. A qualitative comparison between two different clones is provided suggesting a possible application of this device to studies of affinity constant in the analysis where a ligand and a receptor are involved. A quantitative evaluation is given for the 2,4-D analysis in tap water. The pesticide could be detected at ppb levels. What characterise the adsorption and the affinity experiments are the analysis time. When the protein adsorption occurs the process requests hours, on the contrary, when the interaction between the antigen and the antibody takes place the decrease in frequency is rapidly evident.

REFERENCES

1. G.G. Guilbault and J.H. Luong "Piezoelectric immunosensors and their applications in food analysis" in *Food Biosensor Analysis*, G. Wagner and G.G. Guilbault (eds.), Marcel Dekker, New York (1993).
2. G.Z. Sauerbrey, "Verwendung von schwingquarzen zur wagung dunner schichten und zur mikrowagung" *Z. Physik*, 155, 206 (1959).
3. G.Z. Sauerbrey, *Z. Physik*, 178, 457 (1964).
4. W.H. King, "Piezoelectric sorption detector" *Anal. Chem.* 36, 1735 (1964).
5. C.W. Lee, Y.S. Fung and K. WL. Fung, "A piezoelectric crystal detector for water in gases" *Anal. Chim. Acta* 135, 277 (1982).
6. T.E. Edmonds and T.S. West, *Anal. Chim. Acta*, 117, 147 (1982).
7. M.H. Ho, G.G. Guilbault and B. Rietz, "Continuos detection of toluene in ambient air with a coated piezoelectric crystal" *Anal Chem.*, 52, 1489 (1980).
8. C.W. Lee, Y.S. Fung and K.WL. Fung, *Anal. Chim. Acta*, 135, 277 (1982).

9. J.H. Luong and G.G. Guilbault, "Analytical application of piezoelectric crystal biosensors" in *Biosensor Principles and Applications*, L.J. Blum and P.R. Coulet (eds.), M. Dekker, New York, 107-136.
10. B. König and M. Grätzel, "Human granulocytes detected with a piezoimmunosensor" *Anal. Letters*, 26(11), 2313 (1993).
11. B. König and M. Grätzel, "Detection of viruses and bacteria with piezoelectric immunosensors" *Anal. Letters*, 26(8), 1567 (1993).
12. C.S. Lu, "Mass determination with piezoelectric quartz crystal resonators" *J. Vac. Sci. Technol.*, 12, 578 (1975).
13. T. Nomura and A. Minemura, "Behaviour of a piezoelectric quartz crystal in aqueous solution and application to determination of minute amounts of cyanide" *Nippon Kagaku Kaishi*, 1261 (1980).
14. P. Konash and G.J. Baastians, "Piezoelectric crystals as detectors in liquid chromatography" *Anal. Chem.*, 52, 1929 (1980),
15. T. Nomura and Okuhara, "Frequency shifts of piezoelectric quartz crystals immersed in organic liquids" *Anal Chim. Acta.*, 141, 201 (1982).
16. S.Z. Yao, S.L. Dan and L.H. Nie, "Selective determination of silver in solution by adsorption on a piezoelectric quartz crystal" *Anal. Chim. Acta*, 209, 213 (1988).
17. S. Kurosawa, E. Tawara, Kamo and Y. Kobatake, "Oscillating frequency of piezoelectric quartz crystal in solutions" *Anal. Chim. Acta*, 230, 41 (1990).
18. T. Nomura and K. Tsuge, "Determination of silver in solution with a piezoelectric detector after electrodeposition" *Anal. Chim. Acta*, 169, 257 (1985).
19. T. Nomura and M. Fujisawa, "Electrolytic determination of mercury(II) in water with a piezoelectric quartz crystal" *Anal. Chim. Acta*, 182, 267 (1986).
20. H.E. Hager, "Fluid property evaluation by piezoelectric-crystals operating in the thickness shear mode" *Chem. Eng. Comm.*, 43, 25 (1986).
21. S.J. Martin, V.E. Granstaff and G.C. Frye, "Characterization of a quartz crystal microbalance with simultaneous mass and liquid loading" *Anal. Chem.*, 63, 2272 (1991).

THE COMPLEX PHASE TRACING BASED SHAPE EVALUATION SYSTEM FOR ORTHOPAEDIC APPLICATION

J. Kozlowski, and G. Serra

Associazione Istituzione Libera Università Nuorese
Via della Resistenza,39
I-08100 Nuoro, Italy

1. INTRODUCTION

In orthopaedics, the systems for shape measurement of the human back are important as the automation tool for youth screening and for objective monitoring of the medical and physiotherapeutic treatments. For these purposes a lot of different instruments have been developed in last twenty years, some of them, based on the simple shadow moirè effect,[1] and useful for qualitative evaluations only, others more precise - using triangulation method to calculate the positions of fixed or projected markers on the surface of the patient back applied in different configurations,[2] others again based on computerised phase evaluation of the image of fringe pattern projected on the object surface.

The last method could be related again to the moirè effect,[1] combined with improving its accuracy phase stepping, or to the analysis of so called fringe pattern with carrier frequency.[1]

One of such a systems built at the Warsaw Technical University, based on the moiré fringe phenomenon has been used as a starting point for our work in construction of the new instrument which is presented in this work.

Schematically, in Figure 1., the accepted measurement geometry of that system is presented.

After having analysed results of clinical tests of the above system and reports describing others, it was possible to establish direction of our research.

In practice the new instrument, maintaining relatively small dimensions and good accuracy, should guarantee:

1. *Short time of the image acquisition, i.e. less than 0.1 sec.;*
2. *Good dynamics;*
3. *Availability of the natural image of the object under test;*
4. *Insensitivity to the external illumination.*

The most important problem, related to the moirè fringe pattern application, is the time consuming operation of the phase stepping, and as a result too long time of images acquisition.

Optical Sensors and Microsystems: New Concepts, Materials, Technologies
Edited by Martellucci et al., Kluwer Academic / Plenum Publishers, New York, 2000.

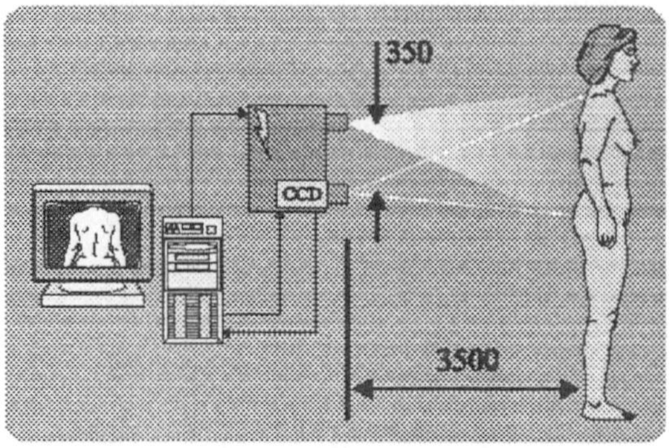

Figure 1. Schema of the applied set-up.

On the other hand, the method of low frequency raster projection, in the form applied till now, could not guarantee sufficient accuracy combined with small dimensions of the instrument.

2. PRINCIPLE OF THE MEASUREMENT

To avoid all these problems we decided to register two images of the object projecting on its surface low frequency fringe pattern (raster), twice, respectively out of phase, as schematically presented in the Figure 2.

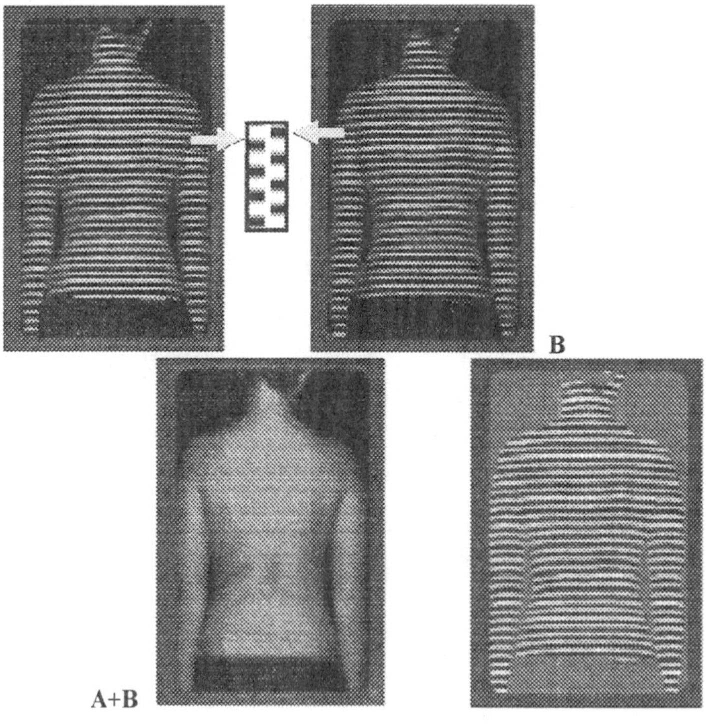

A B

A+B (A-B)/A+B)

Figure 2. Principle of out-of-phase images registration and primary treatment.

Such a type of phase stepping, i.e. of π, allows to calculate the fringe pattern of almost uniform amplitude – independent on the object reflectivity, and of the spatial frequency spectrum, not containing the zero order - as can be seen from Eq. (1-4).

$$I_A(x,y) = \{Object\ Reflectivity(x,y)\} \cdot (1 + \cos(\phi(x,y))), \tag{1}$$

$$I_B(x,y) = \{Object\ Reflectivity(x,y)\} \cdot (1 - \cos(\phi(x,y))), \tag{2}$$

where: x,y pixel co-ordinates in the picture frame,
$\phi(x,y) \Rightarrow$ phase depending on the object height,

$$\{Object\ Reflectivity(x,y)\} = (I_A(x,y) + I_B(x,y)) / 2, \tag{3}$$

$$\{Fringes(x,y)\} = (I_A(x,y) - I_B(x,y)) / (I_A(x,y) + I_B(x,y)). \tag{4}$$

Modulation of fringes described by Eq. (4) is influenced only by the shape of examined object, on the other hand Eq. (3) shows that from this kind of projection it is possible to calculate also the required natural image of the patient – see drawing (A+B) in the Figure 2.

To analyse phase modulation of calculated distribution Eq. (4) we proposed technique based on the same feed back idea which is used in applied in telecommunication Phase Locked Loop method. The new, so called Complex Phase Tracing (CPT) technique, introduces complex local oscillator instead of the real one and not requires the low pass filtering as in a classic PLL schema.

In the Figure 3. block diagrams of both quoted methods are presented.

The input signal to be treated in both cases is a sequence of values in a column of the matrix containing the analysed distribution.

The main advantages of the Complex Phase Tracing (compared to PLL technique) is its elasticity with respect to the analysed frequency variation - it works properly also for low spatial frequencies, and guarantees high accuracy without iterations.

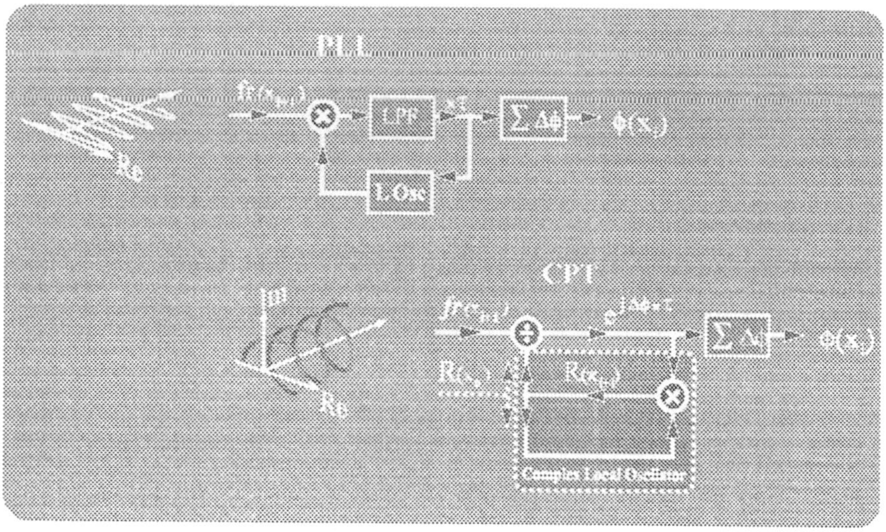

Figure 3. Comparison of the Phase Locked Loop and Complex Phase Tracing method.

The price to be paid applying the CPT method is necessity of the complex input signal - i.e. of a fringe pattern composed of the real and imaginary part, or in others words of sine and cosine of the analysed phase. As it was shown before, the proposed method of π-phase stepping raster projection gives possibility to calculate fringe pattern of very good quality; however this result makes available only the real part of the distribution requested by the Complex Phase Tracing method.

3. COMPLEX PATTERN RECONSTRUCTION AND CORRECTION OF THE PHASE ERROR

In fringe pattern processing there exists well known method of the imaginary component reconstruction, having only real part of the complex distribution. The procedure to be applied is presented schematically in Figure 4.

Unfortunately mentioned process, is generating some errors in the reconstructed complex distribution, especially in proximity of borders of the window under consideration.

The above problem is illustrated in Figures 5a)., 5b). and 5c)., using results of the numerical model.

It can be proved that the obtained real component (see Figure 5b.) is identical with the initial distribution (Figure 5a.), while the imaginary part contains some additive term presented in the diagram (Figure 5c.) with the thick, grey line.

Alteration of one component of the complex number has of course an influence on its phase and modulus what means that the reconstructed complete complex distribution contains an error of both modulus and phase.

This fact was already observed before,[3] curves like those in Figure 6. can be found in many publications.

The first of them presents phase error and the second one modulus of reconstructed complex distribution.

We noted and mathematically proved connection between those dependencies – the derivative of modulus of reconstructed complex distribution is of the same form as that of phase error, to a very good approximation.

Graphical explanation of this fact in the complex plane is presented schematically in Figure 7.

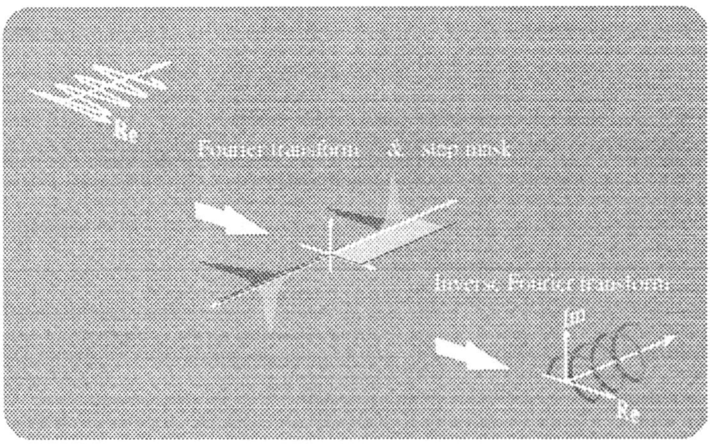

Figure 4. Principle of the complex fringe pattern reconstruction.

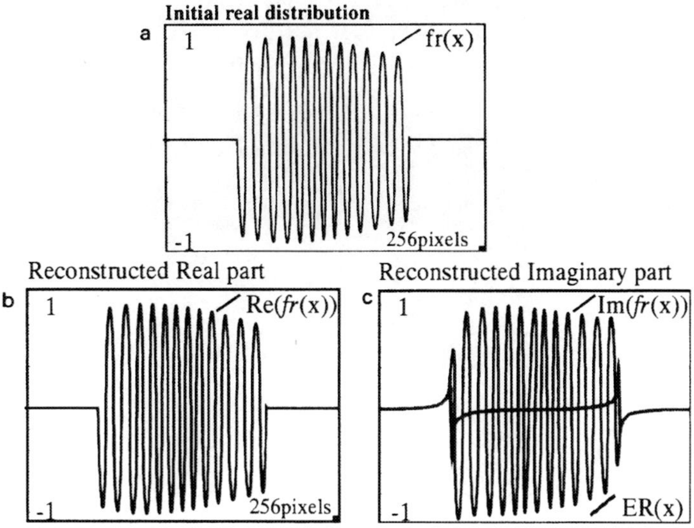

Figure 5. a) initial distribution; b, c) real and imaginary part of the reconstructed complex pattern

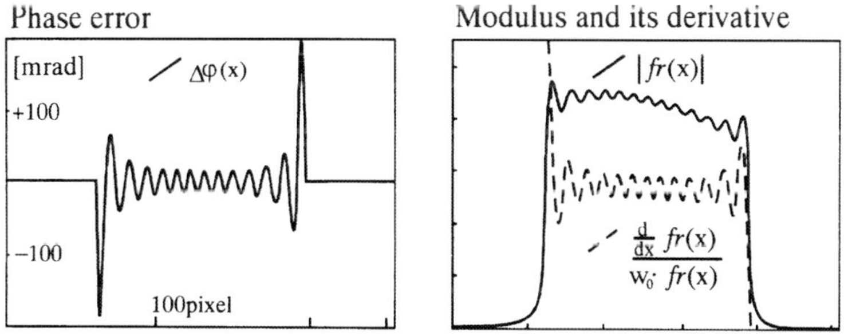

Figure 6. Phase error and modulus oscillation resulting from the applied image processing.

The starting point for this demonstration is the assumption that derivatives of the reconstructed and theoretical, reference distribution are equal in points of the same real components, this can be interpreted in the complex plain as parallelism of the relative tangent vectors.

The only necessary condition to justify the initial assumption is small derivative of the before mentioned imaginary additive term introduced by the process of filtering in the spatial frequency domain.

From geometrical dependencies presented in this figure one can see that, as it was supposed above, the mentioned derivative is equal to the phase difference between reconstructed and reference complex value.

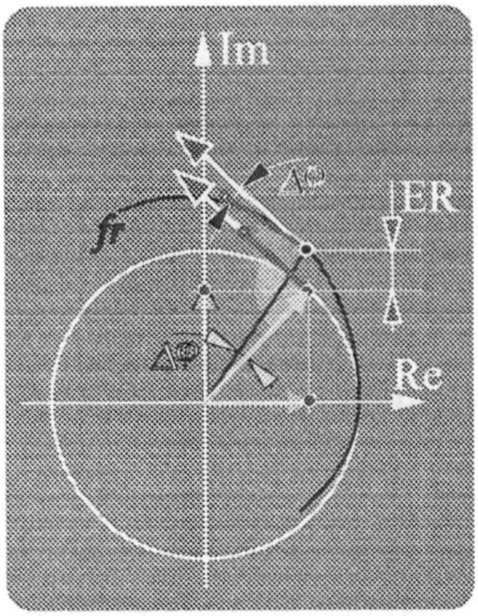

Figure 7. Relation between phase error and derivative of the modulus.

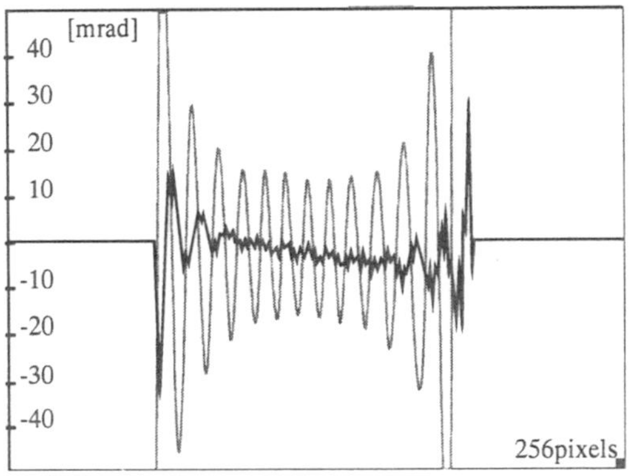

Figure 8. Phase error before and after correction.

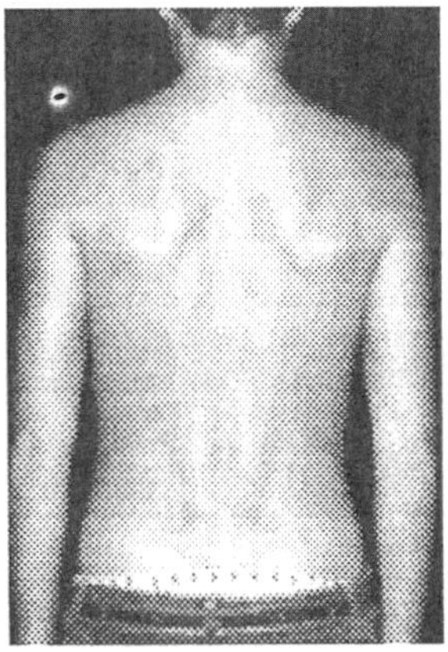

Figure 9. Image of patient.

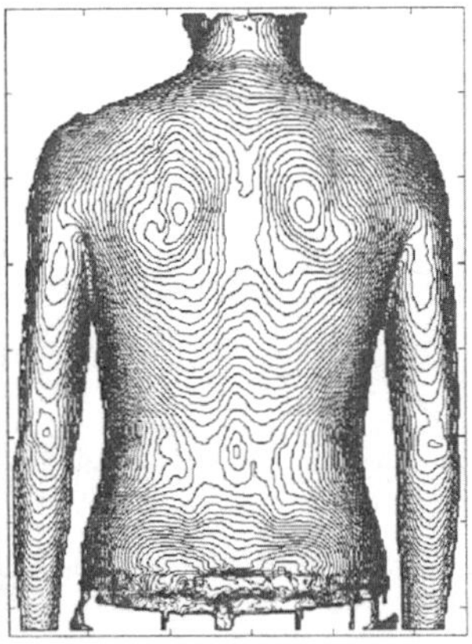

Figure 10. Equilevel lines.

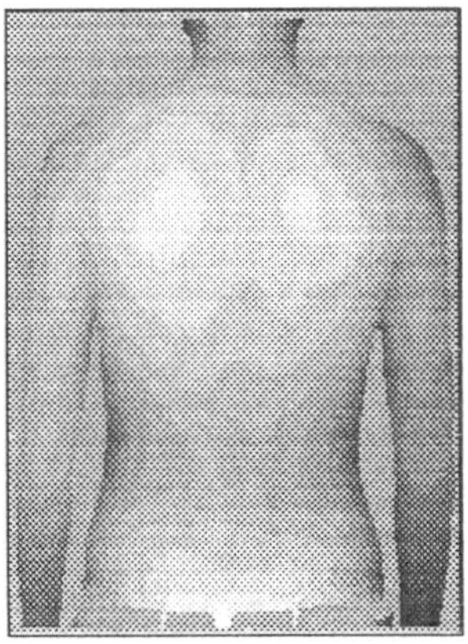

Figure 11. Height as a grey level.

Above relation has been applied in the proposed system to compensate the phase error introduced by presented signal processing.

The obtained final result i.e. curves presenting not compensated and compensated phase error, are shown in Figure 8.

Above method allows to reduce the principle draw back of application of the Fourier transform for the complex signal reconstruction from its real component.

The last part of the data processing to be tackled - called calibration, transforming the recovered phase into the geometrical shape of the patient back, has been realised in two steps – in the first one the optical distortion of used observation system has been measured and compensated by proper geometrical transformation of registered images; and in the second step the non linear re-scaling of calculated phase, using well known geometrical properties of the ideal system, could be applied.

Combining the new idea of the out of phase raster projection, proposed method of correction of the phase error, invented Complex Phase Tracing demodulation technique and presented calibration procedure it was possible to build prototype of the instrument for the human back shape measurement of the following properties:

- *maximum error smaller than 1mm;*
- *measurable slope of over 70 degree;*
- *time of the data acquisition shorter than 0.5sec;*
- *time of the data processing (512x512pixel) in order of 1min.;*
- *the angle between directions of projection and observation smaller than 6 degree, what means no problem with shadowing effect;*
- *the absolute depth measurement in the range of ±70mm;*
  i.e. such a discontinuities (jumps) of the surface under the test are measurable;
- *access to the natural image of the patient;*
- *external dimensions of the prototype of 480x480x150.*

The accuracy tests of the instrument have been done using the plane inclined on different angles and the reference object built for this purpose.

The exemplary result of the human back shape measurements done with presented instrument is shown in Figure 9. as the real image of the patient back, in Figure 10. as contours representing intersection of the examined 3D form with the family of parallel planes distanced 2 mm one from the other, and in Figure 11. as a grey level representation of the object height.

ACKNOWLEDGEMENTS. The authors would like to thank the Regional Government of Sardinia for the financial support.

REFERENCES

1. K.Patorski, *Handbook of the Moiré Fringe Technique,* Elsevier Science Publishers, (1993),
2. *Tridimensional Analysis of Spinal Deformities*, M. D'Amico, A. Merolli, G.C. Santambrogio, ed., IOS Press, Amsterdam (1995).
3. R.J.Green, J.G.Walker, D.W. Robinson. Investigation of the Fourier-transform method of fringe pattern analysis. Optics and Lasers in Eng. 8, 29:44 (1988).

OPTICAL FIBRE CHEMICAL SENSORS FOR ENVIRONMENTAL AND MEDICAL APPLICATIONS

F. Baldini

Istituto di Ricerca sulle Onde Elettromagnetiche "Nello Carrara" - CNR
Via Panciatichi 64
50127 Firenze, Italy

1. INTRODUCTION

There has been a remarkable development of optical fibre chemical sensors in recent years. The first optical fibre chemical sensor, which was a sensor for detecting ammonia, was described in 1976.[1] Since then, investigations have been made of numerous parameters.[2,3,4] This particularly relevant interest is completely justified, because the detection of chemical parameters is extremely important in many industrial and chemical processes, in environmental control and in the biomedical field, and also because optical fibre chemical sensors offer considerable advantages compared to traditional sensors.

In industry the possibility of perfecting remote-detection measurements in a hostile environment and of achieving continuous monitoring of the parameter under investigation is often essential.

In environmental analyses, the possibility of performing continuous in-situ controls without having to resort to drawing samples is of great importance, and is often a winning characteristic for optical fibre sensors.

But it is perhaps in the biomedical field that the detection of chemical parameters by means of optical fibres had its greatest development: their high degree of miniaturization, considerable geometrical versatility, and extreme handiness make it possible to perform a continuous monitoring of numerous parameters, thus enabling performances which are often unique: invasive analyses of numerous parameters present in the blood (such as pH, oxygen partial pressure, carbon dioxide partial pressure, calcium, potassium, glucose); invasive measurement of enterogastric reflux; analysis of enzymes and antibodies.

Before entering the details of sensors for environmental and biomedical applications, a short description of the sensing mechanisms utilized in optical fibre sensors will be given.

Optical Sensors and Microsystems: New Concepts, Materials, Technologies
Edited by Martellucci et al., Kluwer Academic / Plenum Publishers, New York, 2000.

2. SENSING PRINCIPLES

Optical fibre chemical sensors are mainly amplitude-modulation sensors: the intensity of the light transported by the fibre is directly modulated by the parameter being investigated, which has optical properties (spectrophotometric sensors), or by a special reagent connected to the fibre, whose optical properties vary with the variation in the concentration of the parameter under study (transducer sensors). In the latter case, the probe containing the appropriate reagent is called optode. A phase modulation occurs only in a few special cases, since the chemical species being investigated modifies the optical path of the light transported by the fibre.

The main physical phenomena exploited for the realization of chemical sensors are absorption and fluorescence, even if chemical optical fibre sensors have been realized by exploiting other phenomena such as chemical luminescence, Raman scattering and plasmon resonance.

2.1 Absorption

In addition to the substances having their own absorption bands, substances which, by interacting with an appropriate reagent, vary their absorption (e.g. acid-base indicators vary their own absorption depending on the concentration of the hydrogen ions) can also be detected. If the measurement is made by transmission through a solution, the concentration of the parameter being investigated is proportional to absorbance A (Lambert-Beer law), according to the equation:

$$A = \log \frac{I_o}{I} = \varepsilon l c \tag{1}$$

where I_o and I are the light intensities transmitted in the absence and in the presence of the absorbing sample, respectively; ε is the absorption coefficient, l is the optical path and c is the concentration of the absorbing substance. Clearly, this is true if the substance under investigation is the only one absorbing at the considered wavelength; otherwise, the absorption of other substances present in the solution must be considered.

If, instead, the measurement is made by reflectance (e.g. reflection by a solid substrate), a special function (function of Kubelka-Munk) must be introduced which is proportional to the concentration of the substance under examination, according to the Kubelka-Munk theory; concentration c of the absorbing substance can be determined according to the equation:[5]

$$F(R) = \frac{(1 - R)^2}{2R} = k c \tag{2}$$

where R is the reflectance of an infinitely thick sample and k is a constant depending on both absorption and scattering coefficients. If the thickness of the sample can not be considered infinite scattering and transmission through the sample must be taken into account and the relationship between reflectance and concentration of the analyte is much more complex.

In an optode, where the appropriate reagent is immobilised on a substrate, the intensity of the light is partly transmitted, reflected, absorbed and scattered, so that Eq. (1) and Eq. (2) are not followed exactly and a proper algorithm has to be introduced.

Moreover, the two equations are strictly valid only if a monochromatic source is used. If light-emitting diodes (LEDs) are used, as often occurs in optical sensors, a multiwavelength optical beam must be considered; for example Eq. (1) becomes:

$$A' = \log \frac{\int I_o(\lambda)\, d\lambda}{\int I(\lambda)\, d\lambda} \tag{3}$$

where the integral is evaluated on all the wavelengths emitted by the LEDs. Clearly A' is no longer related in linear manner to the concentration of the chemical parameter.

2.2 Fluorescence

Fluorescence can be used in optical sensors for detecting a chemical substance by means of different approaches. Three main cases can be distinguished:

- the substance being investigated is fluorescent;
- the substance is not fluorescent, but can be labelled with a fluorophore;
- the substance interacts with a fluorophore, causing a variation in the emission of fluorescence.

In the first two cases Parker's law is at the basis of fluorescence-based measurements:

$$I(\lambda_{em}) = k\, I(\lambda_{exc})\, \psi(\lambda_{exc})\, \varepsilon(\lambda_{exc})\, l\, c \tag{4}$$

where $I(\lambda_{exc})$ and $I(\lambda_{em})$ are the intensities of excitation and emission radiation, respectively; ψ and ε are the quantum yield and the absorption coefficient; l, the optical path; c, the concentration; and k, a constant depending on the optical set-up and on the configuration of the probe. This equation hypothesizes both low absorbance values by the fluorophore and the absence of inner filter effects, which can be caused by the fluorophore itself (i.e. reabsorption of the emission light) or by other absorbing compounds.

In the latter case, of particular interest is the phenomenon known as fluorescence "quenching", in which the fluorescence intensity decreases as a consequence of the interaction with the substance (quencher) under test, which can thus be detected.[6] This is one of the most used approaches in optical-fibre chemical sensing.

Fluorophore (F) can interact with quencher (Q) at the ground state (static quenching), with a consequent formation of a nonfluorescent complex (FQ):

$$F + Q \leftrightarrow FQ \tag{5}$$

or can interact with it at the excited state (dynamic quenching):

$$F^* + Q \rightarrow F + Q^* \tag{6}$$

and, due to the interaction with the quencher, the fluorophore comes back to the ground state, without the emission of fluorescence.

In both cases, the relationship between fluorescence intensity I and the concentration of quencher [Q] is:

$$\frac{I}{I_o} = \frac{1}{1 + K\,[Q]} \tag{7}$$

where I_o is the fluorescence in the absence of the quencher and K is a constant equal to the dissociation constant of Eq. (5) in the case of static quenching, and is the Stern-Volmer constant* in the case of dynamic quenching.

A decrease in the fluorescence intensity may also be due to an energy transfer from fluorophore F in the excited state to another molecule, acceptor A, whose absorption spectrum, modulated by the chemical species under investigation, overlaps the emission spectrum of the fluorophore. Therefore, fluorescence and absorption can be combined to detect a chemical parameter. In this case, the fluorescence intensity in presence of acceptor, I is given by:

$$\frac{I}{I_o} = 1 - \eta \tag{8}$$

where I_o is the fluorescence intensity in the absence of the acceptor, and η is a term depending on the distance between fluorophore F and acceptor A.

If other chromophores are present in the solution under test, a decrease in the fluorescence, caused by the absorption of the excitation light (primary inner filter effect) or of the emission light (secondary inner filter effect) by these chromophores, can be observed. It is apparent that in this case, the previous equations are no longer valid, but that corrective terms are necessary.

In the case of dynamic quenching, it is more convenient to look at time-dependent decay.[7] In the presence of an interaction with the excited state, the lifetime of the fluorophore is decreased: the higher the concentration of the quencher, the greater the decrease in the lifetime. This is not the case for static quenching, in which the lifetime of the fluorophore is not affected by a change in the concentration of the quencher. Typical fluorescence decay times are in the range between 2 and 20 nsec, while phosphorescence decay times are in the 1 μsec ÷ 10 sec range.

According to Stern and Volmer, the relationship between the decay time and the concentration is:

$$\frac{\tau}{\tau_o} = \frac{1}{1 + K_{sv}[Q]} \tag{9}$$

where τ and τ_o are the decay times of the excited state of the fluorophore in the presence and in the absence of the quencher, respectively.

Lifetime can be measured either in the time domain or in the frequency domain. In the first case, the fluorophore is excited with a narrow pulse and the fluorescence decay is monitored. In the latter case, a modulated excitation is used: the fluorescence emission is still modulated at the same frequency, but is decreased in amplitude and phase shifted. The entity of the amplitude decrease and of the phase shift depends on both the frequency of modulation and on the lifetime of the fluorophore.

Lifetime measurements can be performed not only in the case of reactions involving excited states, but in the case of ground-state reactions. In this case both a reagent and a product must be fluorescent, characterized by different decay-times. For example in the case of pH detection both the basic and acid forms of indicators should be characterized by two different decay times.[8,9] It is apparent that the sensitivity of the method depends on the differences between these two decay times.

The advantage of this approach lies mainly in the fact that there is no more dependence on loading or photobleaching of the chemical transducer fixed at the end of the optical fibres, which is one of the greatest drawbacks of intensity-modulated chemical sensors. Moreover, no problems

---

* The Stern-Volmer constant is equal to the product $k_q \cdot \tau_o$, between the diffusion-controlled rate constant k_q and the fluorescent lifetime τ_o of the excited state F^* in the absence of the quencher.

arise from eventual fluctuations or drift in the source intensity or photodetector sensitivity which, on the contrary, heavily affect intensity of modulated sensors. Furthermore, in the case in which several species interact with the reagent, causing the emission of fluorescences characterized by different decay times,[4] these can be detected simultaneously by using time-resolution instrumentation.

Up to a few years ago, the utilization of this technique was thwarted by the need for expensive and cumbersome optoelectronic instrumentation (laser, very fast detection system, etc.). At present, the advent of fast and powerful light sources, such as emitting diodes and laser diodes, at wavelengths compatible with the fluorophores, makes possible the realization of compact and quite cheap optoelectronic units.[10] This makes this approach one of the most promising for optical fibre chemical detection.

The only drawback, which is intrinsic to the properties of the fibres, is related to the limitation in the length of the optical link due to fibre dispersion.

2.3 Plasmon Resonance

This physical phenomenon is based on the variation in the light reflected by a fine metallic layer as a result of the surface-plasmon resonance.[11] The resonance takes place when the momentum of the photons in the plane of the metallic layer matches that of the surface plasmons, k_{sp}. This momentum is a function of the dielectric constants of the metal, ε_m, and of the external layer, ε, according to the relationship:

$$k_{sp} = \frac{\omega}{c}\left(\frac{1}{\varepsilon} + \frac{1}{\varepsilon_m}\right) \tag{10}$$

If the light is incident to the surface with an angle ϑ, then the wave vector of the light parallel to the surface is:

$$k_x = \frac{\omega}{c}\sqrt{\varepsilon}\,\sin\,\vartheta \tag{11}$$

For angles which satisfy the condition of total reflection, an evanescent wave penetrates the metallic layer and, if the layer is sufficiently thin, interacts with the surface plasmon wave. A definite value of ϑ exists for which the two Eq. (10) and (11) match each other. Experimentally, this resonance can be detected by observing the presence of a minimum in the light reflected in the variation of the angle of incidence on the metal/optical guide interface, which depends also on the refractive index of the external medium ε. Hence, the presence of a chemical species can be detected by following a variation in the refractive index. This technique is not selective by itself but a selectivity can be reached, for example, by covering the metallic surface on the side of the external medium with an appropriate layer, permeable only to the investigated parameter.

2.4 Raman Scattering

In addition to the determination of the concentration of chemical species, the Raman techniques make it possible to obtain important information regarding their structure, in analogous way to information obtained by IR spectroscopy. These techniques utilize wavelengths in the visible band and are therefore useful for analyses in aqueous samples, which are highly absorbent in the IR region, and are also compatible with the use of optical fibres, which are transparent in the visible region. So far, the greatest limitation of Raman scattering has been its rather low sensitivity. On the other hand, in the case of molecules adsorbed on corrugated metallic surfaces, an increase in the Raman signal of several orders of magnitude ($\approx 10^7 \div 10^8$) has recently been observed.[12] This phenomenon, which is known as Surface-Enhanced Raman Scattering, can be attributed to various chemical-physical mechanisms, such as the presence of highly-amplified local electromagnetic fields on the metallic surface and the resonant charge-transfer excitation of the surface/adsorbate complexes. It is clear that, in this case, the optical fibre sensor must be connected to a spectroscopic apparatus, since the Raman spectrum of the sample under examination is obtained as a result of detection. On the other hand, the potential of this technique when used with optical fibres is extremely high.

2.5 Chemiluminescence

In this case, the emission of light as a consequence of a chemical reaction is measured. One of the products of the reaction is found in an excited state, and degenerates into the basic state with the consequent emission of light.

Low detection limits can easily be reached with this approach and a wide dynamic range is obtained. Clearly there is no need of a light source: in some cases, this can be considered an advantage; but since luminous radiation is generated by the chemical reaction and obviously cannot be modulated, as in the case of an external source, interference with ambient light must be carefully avoided. Moreover, in on-line monitoring, a continuous supply of the reagent which is consumed during the reaction is necessary; otherwise, only "single shot" assays are possible.

3. THE PROBE AND THE OPTICAL LINK

Many optical fibre sensors have been prepared on the basis of the principles described above. Particular attention must be devoted to the sensing mechanism in the probe, to the probe design and to the choice of the optical link.

3.1 Sensing Mechanisms

Spectrophotometric-type sensors exploit the absorption band of the associated parameter or the possible emission of fluorescence, and are the simplest to be realized since at most they require the realization of an optimized photometric cell to be connected to the fibre. Examples are the sensors for gases which exploit the vibrational bands or the associated overtones.

In the case of transducer-type sensors, use is made - as already mentioned - of a special reagent whose optical properties vary in accordance with the variation in the concentration of the parameter under examination. In the simplest case, the reaction between reagent and parameter being investigated is direct. Typical examples are oxygen sensors in which the oxygen interacts directly with a fluorophore, causing a decrease in the emission of fluorescence, and pH sensors in which the hydrogen ions react with an acid-base indicator or a fluorophore, causing a variation in the absorption or fluorescence, respectively.

In other cases, the parameter under investigation does not directly modify the optical properties of the reagent connected to the fibre, but reacts chemically, giving rise to a detectable product. For example, the detection of carbon dioxide is based on the detection of the pH of a carbonated solution, since the acidity of the latter depends on the quantity of CO_2 dissolved there:

$$CO_2 + H_2O \Leftrightarrow H_2CO_3 \Leftrightarrow H^+ + HCO_3^- \qquad (12)$$

The detection of ammonia is generally based on the pH variation when the ammonia is dissolved in an aqueous solution:

$$NH_3\ (vapor) + H_2O \Leftrightarrow NH_4^+ + OH^- \qquad (13)$$

Of particular interest are the enzyme sensors which make it possible to detect numerous biological parameters and which, therefore, find important applications for the detection of biological compounds such as pesticides. Detection is based on a selective conversion, catalyzed by a special enzyme, of the parameter under examination in a product which can be optically detected. The concentration of the parameters being analyzed can be linked to the rate of formation of products or to the steady-state concentration of the products.

3.2 The Probe

In transducer-type sensors, the optically-sensitive reagent can be immobilized directly on the fibre at the distal end of the fibre or on a solid external support, and the resulting probe can be indicated as an intrinsic-optode sensor and extrinsic-optode sensor, respectively.

With the intrinsic optode, a compact, highly-miniaturized structure is attained, since the probe is practically the fibre itself. On the other hand the signal levels obtained are generally weak, since the modulation of the optic signal comes from a thin layer of reagent connected to the fibre, thus requiring the use of sophisticated and costly electronic and optical components (laser sources, lock-in, photomultipliers, etc.). By specially treating the tip-end surface of the fibre, enough to increase the sites available for attaching the chromophore, it is possible to obtain a partial improvement in the signal-to-noise ratio.

In general, better results are obtained with extrinsic-optode sensors, since a larger surface is available for attaching the chromophore, even if the realization of a special "envelope" for attaching the support to the tip-end of the fibre is made necessary. Special care must therefore be given to the search for the most appropriate "envelope", since this can heavily affect the performance of the probe, and in particular the response time. In fact, it must be kept in mind that a free exchange for the chemical species being investigated must be guaranteed between the inside of the optode, where the chromophore is located, and the external environment.

In the case of intrinsic optodes, good signal levels can be reached by immobilizing the reagent or chromophore along the fibre. In this configuration the analyte can interact with the optical fibre along the fibre core.[13] The electromagnetic field which propagates along the fibre inside the core extends also in the cladding region. The solution to Maxwell's equations shows that, in the presence of total internal reflection, a standing wave (called evanescent wave) exists in the cladding, propagates in the direction of the fibre axis and decays exponentially in the direction perpendicular to the core/cladding interface. The penetration depth of the evanescent wave is a key parameter for sensing purposes; it is the distance, from the core, at which the amplitude of the electromagnetic is decreased by a factor equal to 1/e and is expressed by the following formula (valid for a step-index fibre):

$$d_p = \frac{\lambda_0}{2\pi n_1} \frac{1}{[\sin^2\vartheta - \sin^2\vartheta_c]^{1/2}} \qquad (14)$$

Typical values of penetration depth are in the order of the utilized wavelength. For example, if n_1=1.5 and n_2=1.33 (aqueous medium) the minimum value of the penetration depth (ϑ= 90°) is about $\lambda/5$ and increases upwards by about 1 wavelength for angles 1° greater than the critical angle. The penetration depth goes to infinity in correspondence of the critical angle. However this fact can be disregarded since, for angles close to the critical angle, the fibre is characterized by losses due to the scattering coming from the surface roughness. In practice, the evanescent wave field is limited to within few microns or less from the core surface.

The efficiency of the approach depends on the fraction r of the optical power carried by the fibre which propagates in the cladding. The optical power carried by the core represent a background since it is not modulated by the analyte: higher this background is, lower is the performance of the evanescent wave sensors. This fraction is clearly high for monomode fibres (r>0.5) and, in multimode fibres, for the modes close to the cut-off condition (the so-called higher modes). In multimode fibres the average power in the cladding, coming from the contribution of all modes, has to be considered. For multimode fibres typical values of r are in the range of 0.01.

Any changes in the microenvironment close to the fibre core which is ascribed to chemical species and modifies the evanescent field distribution can be used in the development of optical sensors. Modification of the refractive index of the cladding due to the penetration of the chemical species in the region close to the core is the sensing mechanism most followed. This approach is not characterized by selectivity, which can be reached by combining the evanescent wave analysis with the absorption or fluorescence coming either directly from the analyte or from the proper chemical transducer, which is located in the proximity of the fibre core.

3.3 The Optical Link

In the design of an optical sensor, attention must be devoted to the transmission properties of the utilized fibres since they must be transparent at the working wavelengths where the reagent changes its optical properties. These wavelengths differ according to whether electronic or vibrational transitions of the reagent are involved. In the former case, the interested region is the UV-visible (e.g. bilirubin λ_{abs}=452 nm, nitrogen dioxide λ_{abs}=496 nm; nitrates $\lambda_{abs}\approx$210 nm; phenols λ_{exc}=266 nm and λ_{em}=270÷400 nm, bromothymol blue λ_{abs}= 616 nm). Instead, in the case of vibrational transitions, the wavelengths involved fall within the mid-IR (e.g. heavy water λ_{abs}=4 μm; propane λ_{abs}=3.3 μm), even if it is possible to operate on the overtones of the absorption bands or on their combinations, which fall within the near-IR where attenuation of the fibres is lower (e.g. methane λ_{abs}=1.33 and 1.66 μm; propane λ_{abs}=1.68 μm).

Table 1. summarizes the working range and the typical losses for the main types of fibres. As can be seen, silica fibres guarantee low attenuations for wavelength values included between 500 nm and 1.9 μm. For λ<500 nm, as the wavelength decreases, the attenuation increases up to values on the order of 3 dB/m for $\lambda\approx$200 nm, making it necessary to use very short fibre lengths.

In the visible band, plastic fibres can also be used, but their attenuation permits utilization only for short distances. What must be considered is their lower cost in comparison with that of other fibres; therefore, in some cases, their utilization can be taken into account, notwithstanding the high attenuation.

Table 1. Working range and attenuation of the most used optical fibres.

Fibres	Working range		Attenuation
Silica fibres	200 nm - 1.9 μm		
	main optical windows	770 nm - 900 nm	3 - 5 dB/Km
		1.05 μm - 1.35 μm	0.5 - 2 dB/Km
		1.45 μm - 1.75 μm	0.2 - 3 dB/Km
Plastic fibres	400 nm - 800 nm		0.3 - 3 dB/m
Fluoride fibres	1.5 μm - 4.5 μm		2 - 20 dB/Km
Chalcogenide fibres	3.0 μm - 11 μm		0.5 - 5 dB/m
Polycrystalline silver-halide fibres	4.0 μm - 20 μm		0.5 - 5 dB/m

For wavelengths longer than 2.0 μm, different fibres must be used, although these are characterized by higher attenuations. This prevents their use if long optical links are necessary.

On the other hand, since all the optoelectronic components (sources, detectors, lenses, etc.) are available at a low price only in the band ranging from 400 nm to 1.9 μm, it is apparent that this is the optical window in which a great many of the proposed optical sensors works.

4. OPTICAL FIBRE SENSORS FOR ENVIRONMENTAL APPLICATIONS

With regard to the environment, pollutants can be analysed using different approaches and different instrumentations, such as:

- portable field instruments for local and not long-term controls;
- fixed field instruments for continuous monitoring, both for ambient quality control (e.g. ambient air quality control) and for the analysis of pollutants which enter the environment in the presence of industrial discharges;
- analytical and laboratory instruments, such as gas- or high pressure liquid-chromatography.

The third approach is required when measurements in field are impossible due to the absence of a transduction process which makes it possible to detect the investigated pollutant or when the available sensors do not offer an adequate sensitivity, selectivity or accuracy.

It is apparent that on-line continuous monitoring of pollutants is preferable, since:

i) it allows a complete view of the trend of the process and avoid errors of evaluation or failures in detection which can exist in the case of measurements limited in time;
ii) it enables in-situ controls without having to resort to drawing samples, and avoids errors arising from a change in the drawn samples during transportation from the field to the laboratory;
iii) it gives a real time response which makes possible, if necessary, an immediate remediation procedure.

New European and American legislation is imposing a continuous control of environmental quality and of industrial discharges which the present sensor technology is still unable to satisfy completely. On the other hand, approval of the new regulations created a very large potential market, a fact that is giving noticeable stimulus to research in this field.

Within this framework, optical fibres play an important role since they are potentially capable of continuous in-situ controls. For this reason, there has been a remarkable development of optical-fibre chemical sensors for environmental applications in recent years. Moreover, the ease in combining the sensing process with an optical network provides an additional advantage for optical fibres: the capacity of interrogating many sensors for different parameters simultaneously and with the same optoelectronic unit.

The analysis of the water quality represents one of the most significant target of pollution monitoring, mainly for the necessity related to the supply of the drinking water. Therefore the detection of harmful pollutants (such as hydrocarbons, pesticides, ammonia, phosphates) is of paramount importance in our society: here below some examples of optical fibre sensors are described.

4.1 Pesticides

The detection of pesticides in river, soil extracts and water wells is becoming more and more important since the highly diffused of pesticides use in agriculture can give rise to strong pollution with a potentially- high problem for drinking water. Their continuous monitoring still represents a challenge for the analyst, and only laboratory test are available at present.

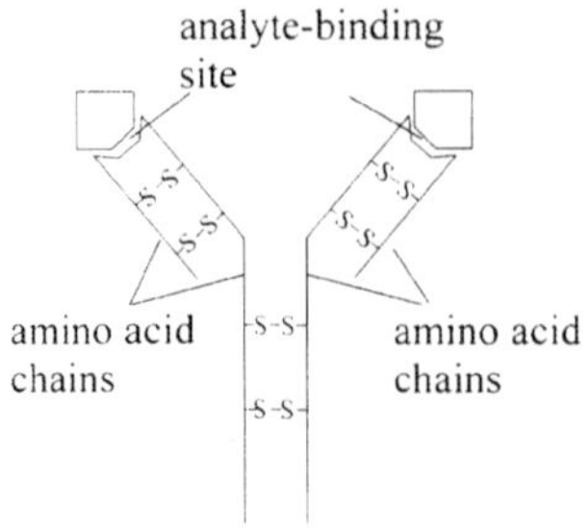

Figure 1. Schematic sketch of an antibody.

The optical fibre approach is based on immunosensing procedures[14,15] or on enzymatic reactions.

Immunosensors are sensors based on antigen-antibody interaction, which offers an excellent degree of selectivity thanks to the process of antibody-antigen recognition.

An antibody is a complex biomolecule belonging to the general class of proteins called immunoglobulins, constituted by hundreds of individual amino-acids properly arranged.

Antibodies are usually depicted as Y-shaped proteins capable of being bound to one-analyte molecule at the tip of each arm (Figure 1.).

They are produced by the immune systems of organisms in presence of an antigen, and are capable of recognizing and combining selectively with such an antigen.

Generally, an antigen is a large protein molecule. Many aspects make a certain molecule capable of evoking a strong antibody response; among these, it is important to recall:

- the fact that the molecule must be "foreign" to the host in which the antibody is made;
- its molecular size: with few exceptions, molecules with a mass < 1000 Da are not able to activate an immunological response;
- its complexity: microorganisms and large proteins are highly complex molecules, and are good immunogens; on the other hand, polymers of a single amino acid, which is a simple molecule, although with a high molecular weight, are not capable of evoking a strong immunological response.

From this point of view, pesticides, which usually have a low molecular mass, are not antigens. On the other hand, many small molecules, called haptens, can induce the production of antibodies if coupled to appropriate proteins as carriers (such as ovalbumin or bovine serum albumin). Frequently, a chemical functionality such as OH, COOH, NH_2 or SH, must be introduced in the molecule to form a hapten. The important aspect is that haptens and the beginning molecule, if structurally closely related to the haptens, react with the specific antibodies produced in response to the hapten-protein conjugate. For example, Figure 2. shows the sequence for the development of antibodies capable of recognizing the pesticide paraquat. The first step is the design of the hapten with the addition of a proper functionality, the second step is the coupling of the hapten with a protein so as to allow the molecule to be able to evoke the antibody response in terms of size and complexity. Finally, the last step is the immunization process, with the injection of the molecule into the body of an animal for the production of antibodies.

An optical immunosensor for the detection of atrazine[16] made use of a monoclonal atrazin antibody immobilized along the core of an optical fibre (core diameter: 1 mm). A competitive binding immunoassay between known solutions of fluorescein-labelled atrazine and samples containing an unknown concentration of atrazine was used as a sensing mechanism. The decrease in the fluorescence excited at 488 nm and detected at 530 nm was related to an increase in the atrazine molecules of the samples, which replaced the labelled molecules (Figure 3.). Detection limit of 100 ng/l was obtained and the comparison with the reading of the optical sensor with the conventional method of HPLC showed a good agreement.

A pesticide sensor based on an enzymatic reaction has recently been proposed.[17] It is based on the measurement of activity of the enzyme acetylcholine esterase (AChE) which is inhibited by the presence of the pesticide. The activity of the enzyme is measured by exploiting its capacity in converting a yellow synthetic enzyme substrate (2-(2-acetoxy-5-methyl-phenylazo)-N-methyl-1,3-thiazolium methosulfate) into a hydrolized blue product: the enzyme substrate and its hydrolized product are characterized by different absorption spectra. Absorbance measured at 580 nm makes it possible to measure the concentration of the hydrolized product, which is clearly correlated to the activity of the enzyme and, therefore, to the concentration of the inhibitors. Measurements were performed with a flow system fed by a peristaltic pump capable of mixing substrate, buffer and inhibitor solutions. The mixed solution passed through an enzyme column, constituted by the enzyme covalently bound on controlled pore glasses, and reached a flow cell connected with an optical fibre spectrophotometer. Its disadvantage in comparison with the immunosensing approach is the lack of selectivity due to the existence of numerous inhibitors of AchE: the system is sensitive not only to different species of pesticides (e.g organophosphorus compound and carbamate), but also to heavy-metal ions. Therefore the system can be seen as an alarm system to be used in on-line monitoring, capable of warning if a more accurate test, such as HPLC, is necessary.

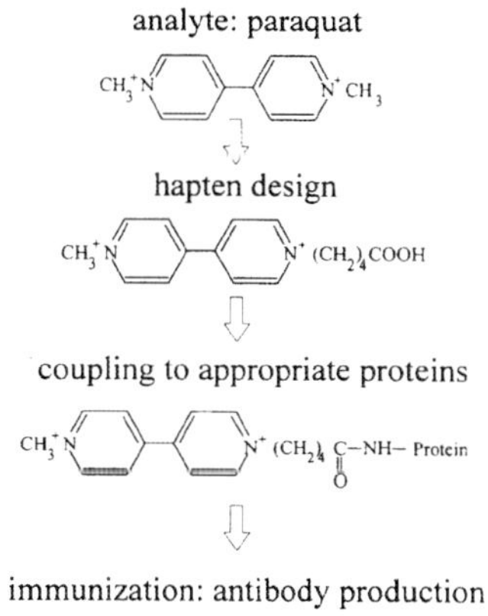

Figure 2. Sequence for the production of antibodies for the recognition of paraquat.

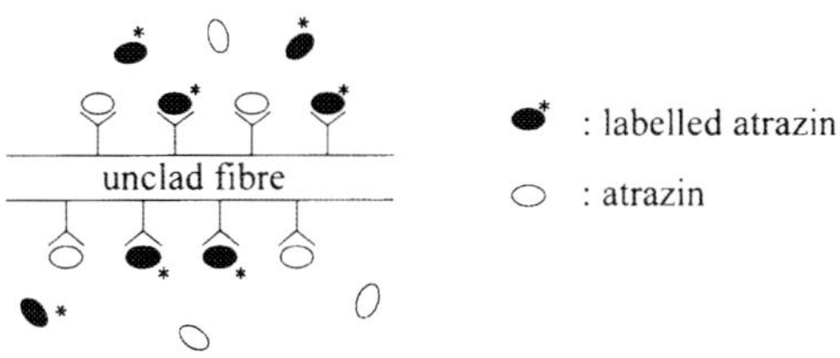

Figure 3. Competitive immunoassay between labelled and unlabelled atrazine.

4.2 Hydrocarbons and Related Derivatives

The continuous monitoring of hydrocarbons and their derivatives (such as chlorinated hydrocarbons) is a very important task since they are among the main potential pollutants of groundwater coming from leakages and spreading from waste sites. The analysis of drawn samples of ground water with analytical methods is a time-consuming job and is also an expensive procedure if we consider the enormous number of sites which should be monitored. The importance and effectiveness of continuous and on-line monitoring is, in this case, apparent.

An absorption-based optical fibre sensor has been proposed[18] which makes use of the colourless reagent pyridine, which reacts with chlorinated hydrocarbons such as chloroform and trichloroethylene (TCE), giving an intense red product characterized by a strong absorption in the 530-570 nm range.

Due to the irreversibility of the reaction, continuous delivery of the reagent is necessary; the sketch of the probe is shown in Figure 4.; single optical fibres are used to carry the light from the source, a filtered incandescent lamp, to the probe and from the probe to the photodetector. Fill and drain capillaries are used to transport the reagent in and out of the probe. A membrane permeable to organochlorides permits the entrance of the chemicals in the analyzed volume.

The probe was tested in real conditions by inserting it in a penetrometer cone. The probe was located at different depths, up to 41 metres, and TCE was the target molecule. The working range from 2 ppm by volume to 120 ppm by volume and an accuracy of about 1 ppm were obtained.

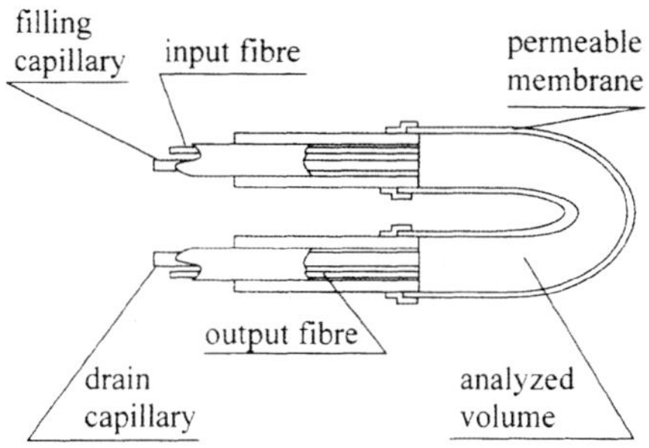

Figure 4. Sketch of the probe for chlorinated hydrocarbons.

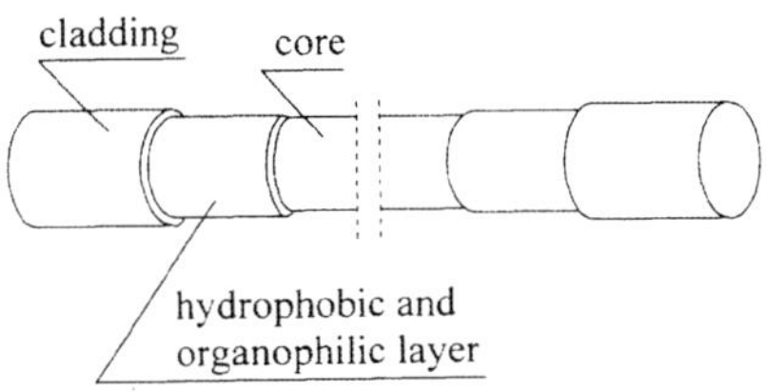

Figure 5. Modified fibre for hydrocarbon detection by means of evanescent wave technique.

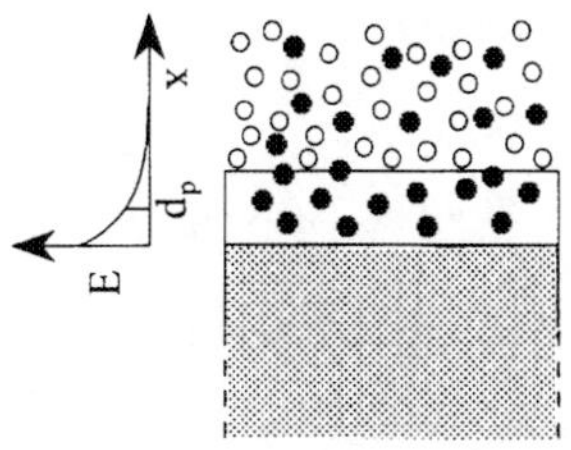

Figure 6. Adsorption of the analyte in the modified cladding; on the left side the exponential decay of the electric field outside the fibre core along with the penetration depth is shown.

A different approach is based on evanescent wave technique: the core of the probe is covered for a definite length with an hydrophobic and organophilic layer, which becomes the new cladding of the sensitive fibre (Figure 5.), provided that it has a refractive index smaller than the refractive index of the fibre core. Since the interaction between the light carried in the fibres and the external environment occurs by means of the evanescent field, the sensing area is limited to a few microns from the core/cladding interface. If the new cladding has a thickness greater than the penetration depth, the investigated volume is inside the cladding; therefore, the effect of interfering elements which may be found in the investigated sample can be eliminated through the choice of an appropriate layer. The choice of a proper hydrophobic and organophilic layer can prevent the diffusion, close to the fibre core, of water molecules (Figure 6.). These cannot interfere with the measurement, which makes it possible to detect the dissolved species, which is adsorbed inside the layer.

Hydrocarbons, if present, are adsorbed on this layer and modulate the light carried by the fibre. The modulation can be induced in two ways:

- change in the refractive index of the cladding after the adsorption: the sensor is not selective but responds to all the adsorbed hydrocarbons; its advantage lies in the simplicity of the optoelectronic system, since no requirements are given on the choice of the wavelength;
- light absorption in correspondence with the absorption bands of the hydrocarbons: the selectivity can be reached, but a multiwavelength system must be used either in the NIR region, if the overtones of chemical compounds are utilized, or in the infrared region, if the vibrational bands are exploited. Clearly, the organophilic layer should be carefully chosen so as to avoid the overlapping between its absorption peaks and the absorption peaks of the hydrocarbons.

Notwithstanding the lack of selectivity, the first approach can be efficiently used for the detection of leakage from storage tanks and remediation efforts. Silylating agents[19] or heteropolysiloxane polymers[20] were used as organophilic and hydrophobic layers; the Petrosense system, distributed by Whessoe Varec,[21] is used for the analysis of petroleum hydrocarbons.[22]

The spectrophotometric investigations in the IR region make it possible to obtain high selectivity coupled with high sensitivity. The 2-15 μm infrared spectral region, is the most informative for detection of the various molecules.[23,24,25] It is known as the "fingerprint" region, because it covers the majority of the absorption bands of the fundamental molecular vibrations. The disadvantages are represented by the impossibility of utilizing long optical link, due to the high attenuation of chalcogenide fibres and the very expensive optoelectronic system (e.g., tunable lasers coupled with highly-efficient detection system or an IR spectrophotometer coupled to the fibres).

Also the near-infrared spectral region 0.8-1.8 μm provides some useful information, since at these shorter wavelengths, overtones or combination bands occur, presenting however a much weaker intensity than the fundamental bands in the infrared. This disadvantage is counter-balanced by the advantages coming from the fact that, at these wavelengths, quartz fibres are characterized by low attenuation. This means:

- very long optical link can be used (up to many kilometres);
- optical fibre networks can be easily developed, and multiple detection in different sensing points is feasible.

Commercially-available silica fibres with a siloxane cladding were used and tested for the first time by DeGrandpre and Burgess.[26]

The same approach was then followed by Ache and coworkers for the detection of hydrocarbons.[27,28] The fibres (core diameter ranging from 200 μm to 500 μm) were coupled with a spectrophotometer and an halogen lamp was used as optical source. Good detection limits were reached (0.4 ppm for p-xylene and chlorobenzene, 0.9 ppm for toluene), proving the effectiveness of the approach.

4.3 Biological Oxygen Demand

A knowledge of the oxygen content in water is a very important parameter for establishing water quality: the number of microbes and their activity are generally proportional to the quantity of oxygen consumed and clearly, the richer of organic substances the water, the faster the decrease of the oxygen content. The biological oxygen demand (BOD) is defined as the number of milligrams of oxygen which is required by the microbial flora for the degradation of the organic matter per litre. It is apparent that this parameter is very important in aerobic sewage treatment, since it is capable of evaluating the effectiveness of the water purification process.

The present approach is to measure BOD over 5 days, under specified standard conditions, the so-called BOD_5. Clearly, this test is not appropriate for on-line monitoring, and alternative methods have been established, by exploiting the immobilization of microorganisms on a porous membrane in contact with an oxygen electrode. The advantage of the use of fluorescent-based optical sensors is the absence of oxygen consumption by the transducer during its detection, which is not a negligible aspect particularly in the case of samples with a low oxygen content.

The proposed optical fibre sensor[29] makes use of a yeast, Trichosporon cutaneum, as microorganism, immobilized in a poly(vinylalcohol) layer spread over the sensitive membrane, which contains a ruthenium complex, tris(4,7-diphenyl-1,10-phenantroline)ruthenium(II) perchlorate, the fluorescence of which is quenched by oxygen. A polyester film acts as an inert solid support, while a polycarbonate cover, permeable to the dissolved organic matter present in the sample, is capable of retaining the microorganism. A sketch of the optode is shown in Figure 7.: an additional layer of charcoal, permeable to oxygen, serves as an optical isolator.

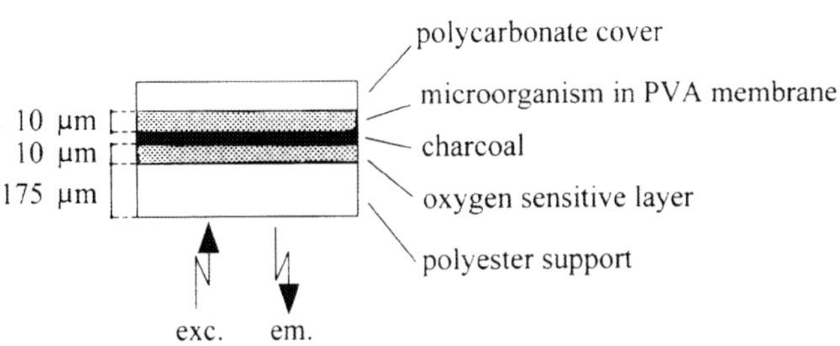

Figure 7. Sketch of the optode for the BOD detection.

The diffusion of organic compounds in the optode stimulates the metabolism of the immobilized microbial cells which, therefore, consume oxygen. As a result of this process, the quantity of oxygen which diffuses through the oxygen-sensitive membrane decreases, with a consequent increase in the fluorescent signal. The optode is fixed at the end of an optical fibre bundle, connected with a light source and a photomultiplier. The excitation light from a Xenon lamp is selected by means of a 480 nm interference filter, while the emitted fluorescence, characterized by a maximum at 610 nm, is isolated with a 560 nm long-pass filter placed in front of the photomultiplier.

It is evident that the thicknesses of the different layers in the optode characterize its performance. A thin layer of the microbial layer would mean a low loading of the microorganism with a consequent low sensitivity (low signal change in presence of a low BOD); but, on the other hand, a thick layer would mean a very slow response, so that a compromise between high sensitivity and fast response time must be reached. The oxygen-sensitive layer must be as thin as possible in order to have a fast oxygen diffusion. Upper limits of detection equal to 110 mg/l BOD with a response time of 3-10 minutes have been obtained. The values of BOD obtained with the optical sensor were successfully compared with the values of BOD_5 obtained by the conventional 5-day method.

5. OPTICAL FIBRE SENSORS FOR BIOMEDICAL APPLICATIONS

An important application field for optical fibre chemical sensors is the biomedical one which seems to have good future development perspectives as also indicated by market forecasts at European level, which have foreseen in the last five years of this century an increase of about 300%.

The first optical fibre chemical sensors was a pH sensor for blood analysis developed at National Institute of Health of Bethesda (Washington, USA) in 1980.[30] Since then, optical fibre sensors have found increasing application and rapid progress due to their peculiar advantages such as miniaturization, geometrical versatility, electromagnetic immunity and absence of electrical contacts. These advantages make them highly competitive in invasive applications for the continuous monitoring of many chemical parameters. Although the tendency of physicians is to use non-invasive sensors, the necessity of continuous monitoring of many biological compounds for long periods entails the use of invasive sensors. Here below some examples of invasive sensors are given and their performances are discussed.

5.1 Bile

Optical fibre sensors play a fundamental role for the detection of enterogastric and non-acid gastro-oesophageal refluxes,[31] which are considered contributing factors to the development of several pathological conditions such as gastric ulcer, "chemical" gastritis, upper dyspeptic syndromes, and severe oesophagitis. Under certain conditions, the enterogastric reflux may also increase the risk of gastric cancer.

Optical detection is based on the optical properties of bile, which is always present in such refluxes.[32] Bilitec 2000 is the commercially available sensor, commercialized by Medtronic-Synectics Medical AB.[33] Basically, it utilizes two light emitting diodes, as sources, at λ=465 nm and 570 nm (reference) and an optical fibre bundle to transport the light from the sources to the probe (which is actually a miniaturized spectrophotometric cell, of 3 mm external diameter) and the returning light from the probe to the detector. The instrument evaluates the logarithm of the ratio between the light intensities collected by the detection system. According to the Lambert-Beer law, the difference in the logarithms measured in the sample and those measured in a pure water reference is proportional to the bilirubin concentration. This is related to bile-containing reflux in the stomach and/or oesophagus. The method was validated on numerous patients, by inserting the optical fibre bundle into the stomach or oesophagus via the nasal cavity.[34] The sensitivity of the sensor was 2.5 µmol/L (bilirubin concentration) and the working range was 0÷100 µmol/L. This range fits well with that encountered in the stomach or oesophagus (although bilirubin concentration in pure bile can be up to 10 mmol/L, it is progressively diluted to its final concentration in the refluxate by pancreatic enzymes, duodenal secretion and, finally, by gastric content). These characteristics clearly refer to in vitro tests, whereas, in the case of in vivo measurements the nonhomogeneity between the gastric content and the mucus and the solid particles in suspension represents a serious impediment. In such case, although absorbance values could numerically express bilirubin concentration, they can only allow an approximate quantitative assessment of the overall bile-reflux concentration. However the sensor is able to accurately measure the contact time between the refluxate and the gastric and/or oesophageal mucosa.

5.2 pH

pH is a very important parameter in biomedical applications; its knowledge is strictly related to the diagnosis of good working of many organs and systems in the human body. It is generally detected by a chromophore which changes its optical spectrum as a function of the pH; absorption-based indicators or fluorophores are usually used.

Real-time monitoring of the blood pH, together with the detection of the blood oxygen (pO_2) and carbon dioxide (pCO_2) partial pressures is of paramount importance in operating rooms and intensive care units in order to determine the quantity of oxygen delivered to the tissues and the quality of the perfusion. All these three parameters are conventionally measured by benchtop blood-gas analysers on hand-drawn blood samples. However significant changes can occur in blood samples after the removal from the body, before the measurements are carried out by the blood gas analyser. Due to their ability to provide continuous monitoring, optical fibre sensors represent a welcome and significant improvement in patient management.

The first pH sensor was developed at NIH (Bethesda, Maryland) and made use of phenol red as acid-base indicator, covalently bound to polyacrylamide microspheres[30]; such microspheres are contained inside a cellulose dialysis tubing (internal diameter 0.3 mm) connected to a 250 µm plastic fibre (Figure 8.). The probe was inserted into either the tissue or the blood vessel through a 22-gauge hypodermic needle. The probe was tested in vivo on animals for the detection of extracellular acidosis during regional ischemia in dog hearts,[35] and for the measurement of transmural pH gradients in canine myocardial ischemia,[36] and of conjunctival pH.[37]

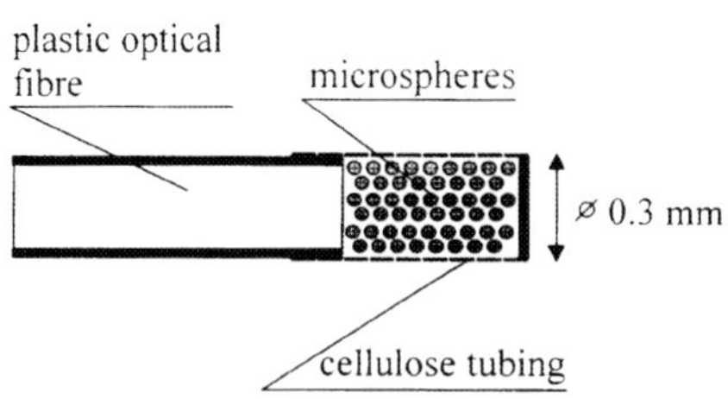

Figure 8. Optode with phenol red bound to polyacrylamide microsphere for blood pH monitoring.

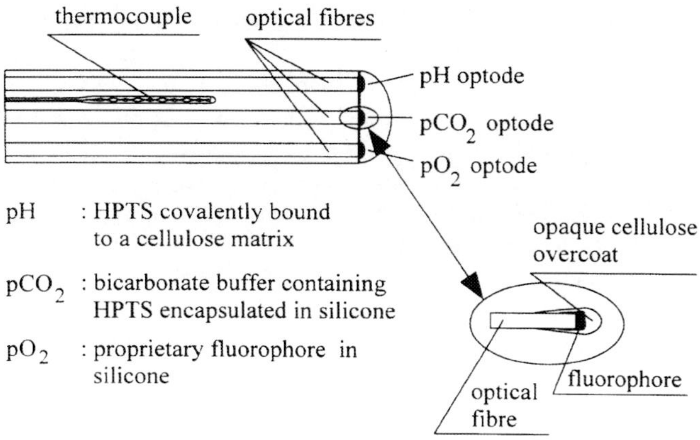

Figure 9. Sketch of the probe for the invasive detection of pH, pO_2, and , pCO_2 in human blood.

The first intravascular sensor for simultaneous and continuous monitoring of the pH, pO_2, and pCO_2 was developed by CDI-3M Health Care (Tustin CA)[38] based on a system designed and tested by Gehrich et al..[39] Three optical fibres (core diameter = 125μm) are encapsulated in a polymer enclosure, along with a thermocouple embedded for temperature monitoring (Figure 9.). pH measurement is carried out by means of a fluorophore, hydroxypyrene trisulfonic acid (HTPS), covalently bonded to a matrix of cellulose, attached to the fibre tip. Both the acidic (λ_{exc}=410 nm) and alkaline (λ_{exc}=460 nm) excitation bands of the fluorophore are used, since their emission bands are centred on the same wavelength (λ_{em}=520 nm). The ratio of the fluorescence intensity for the two excitations is measured, to render the sensor relatively insensitive to fluctuations of optical intensity. Measurements of partial pressure of oxygen (pO_2) and of carbon dioxide (pCO_2) are described in the following sections.

The probe (OD=0.6 mm) has been tested in-vivo on animals,[40,41] and has shown satisfactory correlation with data obtained ex-vivo from electrochemical blood gas analysers.

On the other hand some problems regarding the intravascular use of this sensor have emerged during clinical trials on volunteers in critical care and on surgical patients:[42]

- a blood flow decrease due to peripheral vasoconstriction lasting for several hours after surgical operations; this can give rise to a contamination by flush solutions which can seriously affect the measurements;
- the so-called "wall effect" which primarily affects the oxygen count (if the fibre tip is very close to or touches the arterial wall, it measures the tissue oxygen, which is lower than the arterial blood oxygen);
- the formation of a thrombus (clot) around the sensor tip which alters the value of all the analyte values.

Other intravascular-probe systems have been proposed by Abbott (Mountain View, CA)[43] and by Optex Biomedical (Woodlands, TX),[44] where the structure of the probe is essentially similar to the CDI one previously described: i.e. three different multimode fibres, each of which is associated with the specific chemistry and charged with the detection of a single measurand. In the Optex Biomedical approach, the configuration is modified slightly, as each single fibre is bent, and a side-window sample chamber is built up to contain the appropriate chemistry (Figure 10.). The use of plastic fibres assures that the bundle does not break during insertion, routine patient manipulations and removal. There are basically three advantages in this lateral configuration: a) the real optode does not suffer any mechanical stress during insertion through the arterial catheter, b) the "wall effect" can be avoided by rotating the probe into areas where the blood flow is good, and c) the washability of the sensing-element in the presence of a thrombus or other fouling phenomena, is improved.

Notwithstanding these technical improvements, invasive sensors for blood gas measurements remain at the research level and the fundamental drawbacks appear difficult to be overcome to the exacting standards necessary.

These problems are clearly avoided in a system working in an extracorporeal blood circuit, developed by CDI-3M, which has been commercially available since 1984. A disposable probe which uses the same chemistry as the previously described intravascular optode, is inserted on line in the blood circuit on one side and connected to the fibre bundle on the other. The system is currently employed in open heart measurements, with more than 10,000 disposable probes produced monthly by CDI-3M.

A different approach has recently been described:[45] a fibre optic blood gas and pH monitoring system, capable of performing "paracorporeal" measurements for use at the patient's bedside, has been developed. The developed probe, consisting of three different fluorescent based sensors for pH, pCO_2 and pO_2 respectively, is placed in series with a standard arterial line. Measurements have been performed on withdrawn samples of arterial blood which return to the patient after the detection.

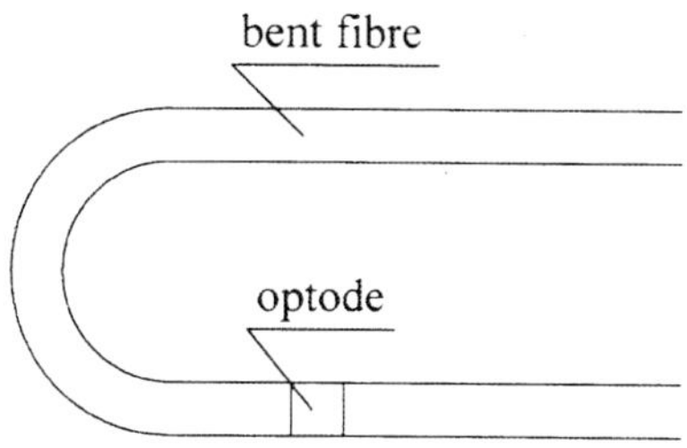

Figure 10. Sketch of the bent fibre and side-window sample chamber used in the intravascular probe developed by Optex Biomedical.

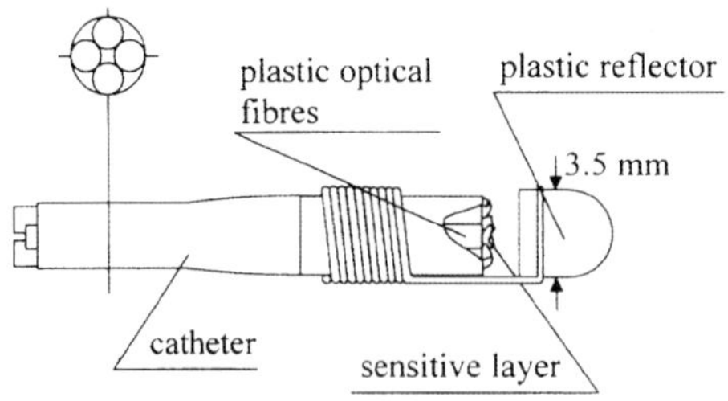

Figure 11. Optical probe for the measurement of gastric pH.

Though this procedure does not allow continuos monitoring, samples can be withdrawn continuously without any risk of inducing anaemia (which may otherwise occur with the use of benchtop analysers), since no blood loss results from such paracorporeal system. Such an approach seems very promising, since it is capable of quasi-continuous monitoring, without the problems affecting intravascular sensors.

Optical fibres sensors have been also proposed for the detection of the acidity in the stomach and in the oesophagus. Monitoring gastric pH for long periods (for example, 24 hours) serves to analyse the physiological pattern of acidity. It provides information regarding changes in the course of the peptic ulcer, and enables assessment of the effect of gastric antisecretory drugs. In the oesophagus, the gastro-oesophageal reflux, which causes a pH decrease in the oesophagus contents from 7 to 2, can determine oesophagitis with possible strictures and Barrett's oesophagus, which is considered a preneoplastic lesion. In addition, in measuring the bile-containing reflux, the bile (generally slightly alkaline) and pH should be measured simultaneously, since, due to a shift in the bilirubin absorption peak to lower wavelengths, the accuracy of the bile measurement decreases by about 30% for values of pH <3.5.[46]

The current practice is to insert a miniaturized glass electrode mounted in a flexible catheter, in the stomach or oesophagus through the nostrils. This electrical system is, however, impractical, owing to the size and rigidity of the glass electrode, and has the added drawback of being subject to electromagnetic interference. Optical fibre sensors overcome these problems, although the broad range of interest (from 1 to 8 pH units) requires the use of more than one chromophore, thereby complicating the design and construction of the optode. This is probably the reason why almost all the pH sensors developed for biomedical applications have been proposed for blood pH detection, with only a few intended for the detection of gastric or oesophageal pH.[47,48,49]

The first sensor proposed for detecting gastric and oesophageal pH[46], made use of two fluorophores, fluorescein and eosin, immobilized onto fibrous particles of amino-ethyl cellulose, fixed on polyester foils. Only tested in vitro, the sensor reveals a satisfactory response time of around 20 seconds. In vivo tests have been reported very recently,[47,50,51] but none of the proposed pH sensors appears completely satisfactory.

A sensor proposed by Peterson et al.[50] is based on two absorbance dyes, meta-cresol purple and bromophenol blue, bound to polyacrylamide microspheres. The configuration of the probe is similar to that one shown in Figure 8. The laboratory optical system arrangement was composed of a lamp plus filters, a fibre coupled probe, a CCD spectrometer and a personal computer. The sensor was tested on samples of human gastric fluid and was also tested in vivo after inserting the optical probe into the stomach of a dog. The accuracy (better than 0.1 pH units) satisfies clinical requirements, but the response time to each pH step was longer than desired, ranging between 1 and 6 minutes. Such a long response time would prevent the detection of fast changes in pH, and makes the sensor useless for the detection of gastro-oesophageal reflux, where pH changes are usually extremely rapid (less than 1 minute).

Another sensor makes use of two dyes, bromophenol blue (BPB) and thymol blue (TB), to cover the range of interest.[49] The chromophores, immobilized on controlled pore glasses, are fixed at the end of plastic optical fibres. The distal end of the fibres is then heated and the CPGs form a very thin pH-sensitive layer on the fibre tips. The probe has four fibres (two for each chromophore) and its sketch is shown in Figure 11. A Teflon diffuse reflector was held in front of the fibres, using a small fine steel wire, in order to improve the return coupling of the modulated light. An optoelectronic unit, similar to that used for bilimetric monitoring, was developed. It consists of two identical channels, separately connected to each of the two fibres carrying TB and BPB, for the detection of pH in the ranges 1 to 3.5 and 3.5 to 7.5, respectively. The use of LEDs as sources, photodetectors and an internal microprocessor make it a truly portable, battery-powered sensor.[52] Response time was less than 1 minute for every pH step. Although the in vitro accuracy was 0.05 pH units, satisfactory in vivo accuracy has not been attained, since, in some cases, a step of some tenths of pH is present between the response of the optical sensor and of the pH electrode. The reason for this behaviour could be ascribed to colour/turbidity interference, since the optical fibres see not only the colour of CPGs, but also the colour /turbidity existing between the fibres and the reflector. This drawback could be removed by measuring the absorption coming from the sample at the working wavelengths of TB and BPB, by means of other fibres without the immobilized CPGs, but this would imply the manufacture of a very complicate probe and optoelectronic unit. The use of only one indicator to cover the whole range would simplify the whole system and recent studies indicate methyl red as feasible indicator for gastric pH measurement.[53]

5.3 Oxygen

Together with pH, oxygen is surely the chemical parameter most investigated for biomedical applications, as its knowledge and continuous monitoring are very important in many fields, such as those of circulatory and respiratory gas analysis.

A knowledge of the oxygen content in blood is essential in order to know how the cardiovascular and cardiopulmonary systems work. This measurement can be performed either spectroscopically by exploiting the optical properties of haemoglobin, the oxygen-carrying pigment of erythrocytes, or by using an appropriate fluorophore, the fluorescence of which is quenched by oxygen.

Oxygen saturation, i.e. the amount of oxygen carried by the haemoglobin (Hb) in the erythrocytes in relation to its maximum capacity, was the first quantity measured with optical fibres.[54] This parameter is measured optically by exploiting the different absorption spectra of the Hb and the oxyhemoglobin (OxyHb) in the visible/near infrared region. Numerous artery and vein insertion models are now commercially available, for example the instruments made by Oximetric Inc., Mountain View CA; BTI, Boulder CO; Abbott Critical Care, Mountain View, CA. In the simpler version, reflected or absorbed light is collected at two different wavelengths and the oxygen saturation is calculated via the ratio technique on the basis of the isosbestic regions of Hb and OxyHb absorption. On the other hand, the presence of other haemoglobin derivates, such as carboxyhemoglobin, carbon monoxide haemoglobin, methemoglobin and sulfhemoglobin, makes preferable the use of multiple wavelengths or of the whole spectrum, which allow their discrimination.[55] Noninvasive optical oximeters, which calculate oxygen saturation via the light transmitted through the earlobes, toes, or fingertips have also been developed, primarily for neonatal care. In this case, particular attention has to be paid to differentiating between the light absorption due to arterial blood and that due to all other tissues and blood in the light path. This implies the use of multiple wavelengths, such as the eight-wavelength Hewlett Packard ear oximeter. Such a drawback can be avoided by using a pulse oximeter. This original approach is based on the assumption that a change in the light absorbed by tissue during systole is caused primarily by the arterial blood. By an appropriate choice of two wavelengths, it is possible to measure noninvasively the oxygen saturation by analysing the pulsatile, rather than the absolute transmitted or reflected, light intensity.[56,57]

Spectrophotometric measurements performed directly on the skin tissue and on the organ surface provide essential information on the microcirculation in tissue and skin and on the metabolism of an organ, respectively. For example, on-line monitoring of the oxygen supply in peripheral organs has considerable importance. It is apparent that a perfect and adequate perfusion of all the organism is basic to the safety of the patient. In the presence of pathological changes in the oxygen transport chain, the organism, by itself, decreases the perfusion in peripheral organs (e.g. skin, skeletal muscles, gut) in favour of central organs such as the brain and the heart (centralization). This mechanism is one of the most effective and important ones during different shock forms. Spectra from biological tissues are able to detect the beginning of centralization before any external, physical and more dangerous symptoms become visible. On the other hand, during the early stages of shock, such an alteration of oxygen transport does not occur homogeneously on the tissue surface: therefore, only fiber optics offers a sufficient spatial resolution for immediate detection. A special algorithm is generally used, since the spectra of the Hb and its derivatives are unevenly distributed in a highly scattering medium, and thus are notably altered.

With the spectrophotometer analyser developed by BGT (Überlingen, Germany),[58] important parameters such as intracapillary haemoglobin oxygenation, intracapillary haemoglobin concentration, local oxygen uptake rate, local capillary blood flow, changes in subcellular particle sizes, and capillary wall permeability (via the injection of exogenous dyes) can be measured in real time. Light from a xenon arc lamp illuminates the tissue via a bifurcated fiber bundle. The back-scattered light, filtered by a monochromator, impinges a photomultiplier connected to a computer that records the spectra resulting from the biological tissue. Thanks to the use of fibres, only small volumes of tissue are investigated, thus making possible the resolution of spatial heterogeneities. The instrument is able to record spectra of high quality even in moving organs.

The disadvantage of utilizing haemoglobin as an indirect indicator for the measurement of oxygen is its full saturation at $\approx$100 Torr: this fact prevents, for example, the use of this method in the case of the respiration of gas mixtures with an O_2 content larger than 20% as routinely used in anaesthesia. Therefore the use of a chemical transducer becomes necessary in some cases. The first optode-based oxygen sensor was described by John Peterson, and makes use of a fluorophore, perylene-dibutyrate, the fluorescence of which is efficiently quenched by oxygen.[59] The dye, adsorbed on amberlite resin beads, was fixed at the distal end of two plastic optical fibres with a hydrophobic membrane permeable to oxygen. The probe described was tested in-vivo for the measurement of the arterial pO_2 level in dog eyes.[60]

Other optodes have been developed and tested in-vivo, all of them using a fluorophore, the fluorescence of which is quenched by oxygen. In the intravascular sensor developed by CDI, previously described, a specially synthesized fluorophore, a modified decacyclene (λ_{exc}=385 nm, λ_{em}=515 nm), is combined with a second reference-fluorophore that is insensitive to oxygen, and is incorporated into a hydrophobic silicon membrane that is permeable to oxygen.

A new type of noninvasive sensor has been proposed to measure the local oxygen uptake through the skin. The direct measurement of the oxygen flow on the skin surface provides information regarding the oxygen flow inside the tissue, which can help physicians to diagnose circulatory disturbances and their consequences. Two optodes, which make use of a ruthenium complex, measure the difference in oxygen pressure across a membrane placed in contact with the skin. In-vivo tests performed on the left lower forearm of a patient gave good results.[61]

A knowledge of the concentration of oxygen, as well as of many other gases, in exhalation analysis is very important, since it may provide important information on the correct metabolism of the human body. An optode for the simultaneous detection of oxygen and carbon dioxide, potentially suitable for respiratory gas analysis, has recently been proposed.[62] It makes use of two fluorescent dyes dissolved in a very thin layer (1-3 μm) of a plasticised hydrophobic polymer which is fixed at the distal end of optical fibres. A wavelength discrimination by appropriate interference filters makes possible the simultaneous monitoring of O_2 and CO_2, as the emission wavelengths of the two fluorophores are sufficiently separated.

5.4 Carbon Dioxide

As for oxygen, the measurement of CO_2 is capable of providing important information in regard to the working of the circulatory and respiratory systems. Its detection is based on the detection of the pH of a carbonate solution, since its acidity depends on the quantity of CO_2 dissolved therein. Therefore, all optodes developed for blood CO_2 make use of the same dye utilized for blood pH detection, fixed at the end of the fiber and covered by a hydrophobic membrane permeable to CO_2. In the intravascular CDI system the fluorophore, hydroxypyrene trisulfonic acid (the same one used in the pH optode), is dissolved in a bicarbonate buffer solution which is encapsulated in a hydrophobic silicon membrane permeable to CO_2 and attached to the fiber tip.

6. CONCLUSIONS

The detection of chemical parameters by means of optical fibres has had a decisive boost in recent years, thanks to the advantages which these sensors are capable of offering in comparison with traditional sensors. So far, some difficulties still exist, for these sensors, to be present on the market. This is mainly related to the realization of the optodes, which is characterized by significant problems concerning both the structural aspect (compactness, guarantee of a rapid and efficient exchange with the outside environment) and the chemical aspect (search for the most suitable optically-sensitive reagent and its application to the fibres) to which considerable attention must be dedicated to obtain very high performance from the probe in terms of response time, reversibility and stability. In spite of these difficulties, the first sensors have already appeared on the international market: it is conceivable, therefore that we can expect in the future an industrial conversion of numerous prototypes that until now have been realized only on research level.

Moreover optical fibre sensors for environmental applications are finding a continuous stimulus due to the regulations which impose a continuous monitoring of many analytes in aqueous or gaseous environment. It is one of the few cases where legislation came before the results of the research; the great many of on-line sensors, not only based on optical fibres, still suffer from poor stability, cross-sensitivity and problems with packaging and protection. This is the main reason for which many sensors for several analytes were proposed at laboratory level but only some few are available on the market. On the other hand the great effort made by the research world already led to a noticeable progress in the field and the presence on the market of many other sensors in the forthcoming years is foreseeable. Optical fibre approach present some winning features in comparisons with the other approaches, such as electromagnetic and electric isolation, network of sensors for multipoint detection, which make optical fibre sensors for the environment very attractive.

In biomedical field, especially for invasive measurements, optical fibre sensors have given good results in invasive testing. Their utilization is increasing continuously, and this fact makes it feasible to imagine their more and more widespread diffusion for invasive monitoring. The possibility, in some case already exploited, of monitoring several parameters with a single instrument[63,64] makes them still more competitive in comparison with the other techniques and is continuously encouraged by physicians, who greatly appreciate the possibility of having a multitest portable unit with low-cost disposable probes, that can be easily managed by both doctors and patients.

REFERENCES

1. D.J. David, M.C. Willson, and D.S. Ruffin, Direct measurement of ammonia in ambient air, *Anal. Lett.* 9:389 (1976).
2. A.M. Scheggi and F. Baldini, Chemical sensing with optical fibres, *Intern. J. Optoelectr.* 8 (2):133 (1993).
3. *Fiber Optic Chemical Sensors and Biosensors*, O.S. Wolfbeis ed., CRC Press, Boca Raton, Fl, USA, Voll. I-II (1991).
4. O.S. Wolfbeis, Chemical sensing using indicator dyes, in *Optical fiber sensors. Application, analysis, and future trends*, J. Dakin and B. Culshaw eds, Artech House, Norwood, MA, Vol. IV (1997).
5. A.J. Guthrie, R. Narayanaswamy, and D.A. Russell, Application of Kubelka-Munk diffuse reflectance theory to optical fibre sensors, *Analyst* 11:457 (1988).

6. P. Yuan and D.R. Walt, Calculation for fluorescence modulation by absorbing species and its application to measurements using optical fibres, *Anal. Chem.* 59: 2391 (1987).
7. H. Szmacinski and J.R. Lakowicz, Fluorescence lifetime-based sensing and imaging, *Sensors and Actuat.B* 29:16 (1995).
8. M.E. Lippitsch, S. Draxler, and M.J.P. Leiner, Time -domain fluorescence methods as applied to pH sensing, *Proc. SPIE* 1796:202 (1992).
9. S. Draxler and M.E. Lippitsch, pH sensors using fluorescence decay time, *Sensors and Actuat.B* 29:199 (1995).
10. D. Kieslinger, S. Draxler, K. Trznadel, and M.E. Lippitsch, Lifetime-based capillary waveguide sensor instrumentation, *Sensors and Actuat.B* 39:300 (1997).
11. B. Liedberg, I. Lundstrom, and E. Stenberg, Principles of biosensing with an extended coupling matrix and surface plasmon resonance, *Sensors and Actuat.B* 11:63 (1993).
12. J.M. Bello and T. Vo-Dinh, Surface-enhanced Raman scattering fiber-optic sensor, *Appl. Spectr.* 44 (1):63 (1990).
13. B.D. MacCraith, Enhanced evanescent wave sensors based on sol-gel-derived porous glass coatings, *Sensors and Actuat.B* 11:29 (1993).
14. J.M. Van Emon and V. Lopez-Avila, Immunochemical methods for environmental analysis, *Anal. Chem.* 64 (2):79A (1992).
15. T. Vo-Dinh, G.D. Griffin, and M.J. Sepaniak, Fiberoptic immunosensors, in *Fiber Optic Chemical Sensors and Biosensors*, O.S. Wolfbeis ed., CRC Press, Boca Raton, Fl, USA (1991).
16. P. Oroszlan, G.L. Duveneck, M. Ehrat, and H.M. Widmer, Fiber-optic atrazine immunosensor, *Sensors and Actuat.B* 11:301 (1993).
17. W. Trettnak, F. Reininger, E. Zinterl, and O.S. Wolfbeis, Fiber-optic remote detection of pesticides and related inhibitors of the enzyme acetylcholine esterase, *Sensors and Actuat.B* 11:87 (1993).
18. F.P. Milanovich and S.B. Brown, A fiber-optic sensor system for remote, long-term monitoring of soil and groundwater contamination, *Proc. SPIE* 2360:98 (1994).
19. F.K. Kawahara, R.A. Fiutem, H.S. Silvus, F.M. Newman, and J.H. Frazar, Development of a novel method for monitoring oils in water, *Anal. Chim. Acta* 151:315 (1983).
20. C. Ronot, M. Archenault, H. Gagnaire, J.P. Goure, N. Jaffrezic-Renault, and T. Pichery, Detection of chemical vapours with a specifically coated optical-fibre sensor, *Sensors and Actuat.B* 11:375 (1993).
21. Whessoe Varec, Inc., 10800 Valley View Street, Cypress, California 90630.
22. S.M. Klainer, J.R. Thomas, and J.C. Francis, Fiber-optic chemical sensors offer a realistic solution to environmental monitoring needs, *Sensors and Actuat.B* 11:81 (1993).
23. R. Gobel, R. Krska, R. Kellner, and A. Katzir, Development of protective polymer coatings for silver halide fibres and their application as threshold level sensors for chlorinated hydrocarbons in sea water, *Fresen. Z. Anal. Chem.* 348:780 (1994).
24. R. Gobel, R. Krska, S. Neal, and R. Kellner, Performance studies of an IR fiber optic sensor for chlorinated hydrocarbons in water, *Fresen. Z. Anal. Chem.* 350:514 (1994).
25. B. Mizaikoff, R. Gobel, R. Krska, K. Taga, R. Kellner, M. Tacke, and A. Katzir, Infrared fiber-optical chemical sensors with reactive surface coatings, *Sensors and Actuat.B* 29:58 (1995).
26. M.D. DeGrandpre and L.W. Burgess, A fiber optic FT-NIR evanescent field absorbance sensor, *Appl. Spectr.* 44:273 (1990).
27. J. Burck, J.-P. Conzen, and H.J. Ache, A fiber optic evanescent field absorption sensor for monitoring organic contaminants in water, *Fresen. J. Anal. Chem.* 342:394 (1992).
28. J.-P. Conzen, J. Burck, and H.J. Ache, Characterization of a fiber-optic evanescent wave absorbance sensor for nonpolar organic compounds, *Appl. Spectr.* 47:753 (1993).
29. C. Preininger, I. Klimant, and O.S. Wolfbeis, Optical Fiber Sensor for Biological Oxygen Demand, *Anal. Chem.* 66:1841 (1994).
30. J.I.Peterson, S.R. Goldstein, R.V. Fitzgerald, and D.K.Buckhold, Fiber optic pH probe for physiological use, *Anal. Chem.* 52 (6):884 (1980).
31. R. Falciai, A.M. Scheggi, F. Baldini, and P. Bechi, U.S.A. Patent 4,976,265 (1990).
32. R. Falciai, F. Baldini, F. Cosi, P. Bechi, and F. Pucciani, Bile enterogastric reflux sensor using plastic optical fibres, *Fiber and Integr. Opt.* 12:215 (1993).
33. Medtronic-Synectics Medical, Renstiernas Gata 12, S-116 28 Stockholm, Sweden.
34. P. Bechi, F. Pucciani, F. Baldini, F. Cosi, R. Falciai, R. Mazzanti, A. Castagnoli, A. Passeri, and S. Boscherini, Long-term ambulatory enterogastric reflux monitoring. Validation of a new fiberoptic technique, *Digest. Dis. Sci.* 38:1297 (1993).
35. G.A. Tait, R.B. Young, G.J. Wilson, D.J. Steward, and D.A. Mac Gregor, Myocardial pH during regional ischemia: evaluation of a fiber-optic photometric probe, *Am. J. Physiol.* 243:H1027 (1982).
36. R.M. Watson, D.R. Markle, Y.M. Ro, S.R. Goldstein, D.A. McGuire, J.I. Peterson, and R.E. Patterson, Transmural pH gradient in canine myocardial ischemia, *Am. J. Physiol.* 246:H232 (1984).

37. E. Abraham, S.E. Fink, D.R. Markle, G. Plinholster, and M. Tsang, Continuous monitoring of tissue pH with a fiberoptic conjunctival sensor, *Am. Energ. Med.* 14:840 (1985).
38. CDI-3M Health Care, 1311 Valencia Avenue, Tustin CA 92680.
39. J.L. Gehrich, D.W. Lübbers, N. Opitz, D.R. Hansmann, W.W. Miller, K.K. Tusa, and M. Yafuso, Optical fluorescence and its application to an intravascular blood gas monitoring system, *IEEE Trans. Biomed. Eng.* BME-33:117 (1986).
40. W.W. Miller, M. Yafuso, C.F. Yan, H.K. Hui, and S. Arick, Performance of an *in-vivo*, continuous blood-gas monitor with disposable probe, *Clin. Chem.* 33:1358 (1987).
41. D.R. Hansmann and G.L. Gehrich, Practical perspectives on the *in-vitro* and *in-vivo* evaluation of a fiber optic blood gas sensor, *Proc. SPIE* 906:4 (1988).
42. G. Khalil and S.F. Malin, *In-vivo* blood gases: promises and realities, *Proc. SPIE* 2388:502 (1995).
43. G. Khalil, J. Yim, and G.G. Vurek, *In-vivo* blood gases: problems and solutions, *Proc. SPIE* 2131:437 (1994).
44. L. Schlain and S. Spar, Continuous arterial blood gas monitoring with transmitted light sensors and LED light sources, *Proc. SPIE* 2131:452 (1994).
45. R.C. Martin, S.F. Malin, D.J. Bartnik, A.M. Schilling, and S.C. Furlong, Performance and use of paracorporeal fiber optic blood gas sensors, *Proc. SPIE* 2131:426 (1994).
46. M.F. Vaezi, R.G. Lacamera, and J.E. Richter, Validation study of Bilitec 2000: an ambulatory duodenogastric reflux monitoring system, *Amer. J. Physiol.* G1050 (1994).
47. H.E. Posch, M.J.P. Leiner, and O.S. Wolfbeis, Towards a gastric pH-sensor: an optrode for the pH 0-7 range, *Fres.Z. Anal. Chem.* 334:162 (1989).
48. F. Baldini, S. Bracci, F. Cosi, P. Bechi, and F. Pucciani, CPG embedded in plastic optical fibres for gastric pH sensing purposes, *Appl. Spectrosc.* 480:549 (1994).
49. E.J. Netto, J.I. Peterson, and B. Wang, Fiber-optic pH sensor for gastric measurements - preliminary results, *Proc. SPIE* 1886:109 (1993).
50. F. Baldini, P. Bechi, S. Bracci, F. Cosi, and F. Pucciani, *In-vivo* optical fibre pH sensor for gastro-oesophageal measurements, *Sensors and Actuat.B* 29:164 (1995).
51. E.J. Netto, J.I. Peterson, M. McShane, and V. Hampshire, A fiber-optic broad-range pH sensor system for gastric measurements *Sensors and Actuat.B* 29:157 (1995).
52. Prodotec, Viadotto Indiano 50145, Firenze, Italy.
53. See the Chapter of F. Baldini and A. Falai, Characterization of an optical fibre pH sensor with methyl red as optical indicator, in these proceedings.
54. N.S. Kapany and N. Silbertrust, Fiber optics spectrophotometer for in-vivo oximetry, *Nature* 208:138 (1964).
55. D.W. Lubbers, Chemical *in vivo* monitoring by optical sensors in medicine, *Sensors and Actuat.B* 11:253 (1993).
56. Y. Mendelson and B.D. Ochs, Noninvasive pulse oximetry utilizing skin reflectance photoplethysmography, *IEEE Trans. Biomed. Eng.* BME-35:798 (1988).
57. W. Cui, L.E. Ostrander, and B.Y. Lee, B.Y. *In vivo* reflectance of blood and tissue as a function of wavelength, *IEEE Trans. Biomed. Eng.* BME-37:632 (1990).
58. K.H. Frank, M. Kessler, K. Appelbaum, and W. Dümmler, The Erlangen micro-lightguide spectrophotometer Empho I, *Phys. Med. Biol.* 34:1883 (1989).
59. J.I. Peterson, R.V. Fitzgerald, and D.K. Buckhold, Fiber-optic probe for *in vivo* measurement of oxygen partial pressure, *Anal. Chem.* 56:62 (1984).
60. E. Stefansson, J.I. Peterson, and Y.H. Wang, Intraocular oxygen tension measured with a fiber optic sensor in normal and diabetic dogs, *Am. J. Phys.* 256:H1127 (1989).
61. G.A. Holst, T. Köster, E. Voges, and D.W. Lübbers, FLOX-an oxygen-flux-measuring system using a phase modulation method to evaluate the oxygen dependent fluorescence lifetime, *Sensors and Actuat.B* 29:231 (1995).
62. O.S. Wolfbeis, L.J. Weis, M.J.P. Leiner, and W.E. Ziegler, Fiber-optic fluorosensor for oxygen and carbon dioxide, *Anal. Chem.* 60:2028 (1988).
63. C.D. Anderson, D. Vokovich, and M.T. Wlodarczyk, Fiber optic sensor for simultaneous oxygen saturation and blood pressure measurement, *Proc. SPIE* 1648:116 (1992).
64. M. Sun and A. Kamal, A small single sensor for temperature, flow, and pressure measurement, *Proc. SPIE* 1420:44 (1991).

INTRODUCTION TO THE MULTICOMPONENT ANALYSIS WITH ARRAYS OF NON-SELECTIVE CHEMICAL SENSORS

C. Di Natale, and A. D'Amico

University of Rome Tor Vergata
Department of Electronic Engineering
Via di Tor Vergata
00133 Roma, Italy

1. INTRODUCTION

Sensor science has been mainly driven since the beginning by the chase of sensitivity, stability and selectivity.[1] But while sensitivity and stability are straightforward concepts, selectivity deserves to be more deeply discussed. Actually a sensor output can always be considered as the result of a process of synthesis of the whole features of the environment, to which the sensor is exposed. The mechanism is evident, for instance when optical sensors are considered, in this case the whole spectral intensity of a luminous source is converted in an electrical signal through the sensor spectral responsivity. The process can be thought to take place also in chemical sensors (see Figure 1.). The environment, in this case, can be represented as a chemical pattern (discrete), in which all the chemical species, each at a certain concentration level and contemporaneously present in the environment, are shown as a histogram. In the same way, sensor selectivity can be represented as a pattern of sensitivities.

In chemical sensing those sensors showing a selectivity spectra where one of the sensitivities is much greater than the others are called specific. It is interesting to note that currently specificity can be achieved only for few kinds of sensors, as an example those based on immunoreactions, such as the antigene-antibody interactions.[2] Anyway considering the huge amount of chemical compounds it is not foreseen to develop immunosensors for each of them.

Another way, currently pursued, to enhance the selectivity is that of supramolecular chemistry. In principle it is possible to design molecules characterized by specific key-lock interactions toward certain species. As an example cage compounds, such as calixarenes or resorcinarenes, able to bind only molecules having a certain shape, have been successfully synthesized.[3] Nevertheless when these compounds are utilized (e.g. as coating of quartz microbalances) the effective selectivity of sensor turns out to be wide, this because if it is true that cages are selective the rests of the molecules are rather non selective and Van der Waals interactions with many other analytes can occur.

Optical Sensors and Microsystems: New Concepts, Materials, Technologies
Edited by Martellucci et al., Kluwer Academic / Plenum Publishers, New York, 2000.

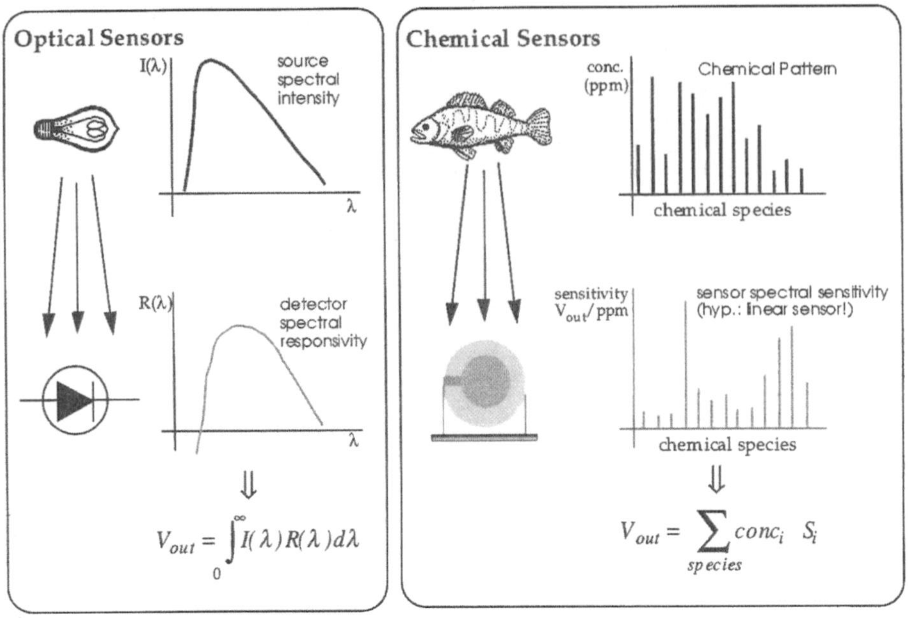

Figure 1. The convolution between the environment and the sensor selectivity is the basic operation describing the functionality of any sensor. This mechanism, straightforward in optical sensing, can be extended also to model the interactions occurring in chemical sensors.

After years of chase of selectivity it is possible to conclude that many families of solid-state chemical sensors (among them mass transducers and metal-oxide semiconductors) are intrinsically non specific, it is then mandatory to develop some strategy allowing their utilization in practical applications.

In this context optics can offer another useful analogy. It is indeed common practice in this field to use a number of broad-band sensors to infer some information about the nature of luminous sources, for instance it is usual in astronomy to classify the stars according to their photometric responses in selected bands. In the same way with a number of chemical sensors it is possible to get information about the nature of the chemistry of an environment.

From this point of view Nature provides us with an important paradigm, indeed the chemical senses (olfaction and tongue) do not make use of specific sensors but rather they are based on a great number of non-selective broad-band sensors which give to the living beings the capability to smell and taste a great variety of compounds being able to classify, recognize and discriminate different aromas and tastes, also, in some animals, with great resolutions.[4] Due to this analogy with the natural sensor systems arrays of non selective sensors are currently denominated *electronic noses*. Arrays of such non-selective sensors can then be utilized both for quantitative purposes (to estimate the concentrations of some chemical species, or for qualitative analysis to classify samples according to some general feature.

For both the procedures data analysis is the basic issue allowing the extraction from the sensors outputs of the required information. Its usual role is to allow a correlation between sensors outputs pattern and the properties of the environment which are the object of the measurement strategy.

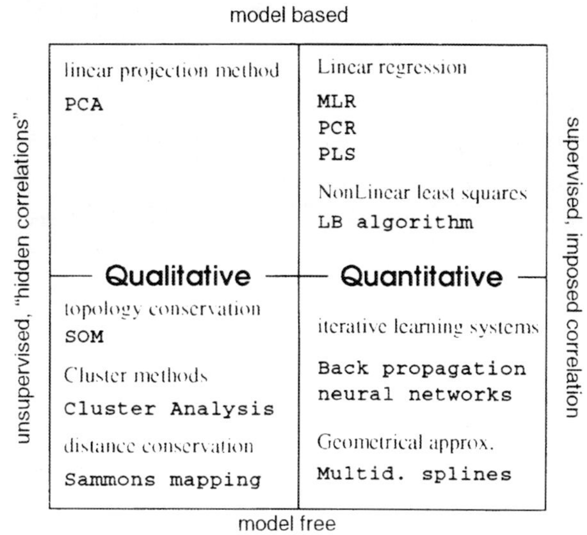

Figure 2. In figure a taxonomy of the data analysis techniques utilized with sensor arrays is shown. On the left side methods for electronic nose applications are shown while on the right side methods for quantitative analysis are listed. On the upper side of the table are grouped the models which require a a-priori defined model between the data while in the bottom part model-free methods are shown.

Data analysis has been the main concern of many research papers since the first development of the field. They have been dealing with the application of several techniques, borrowed from other application fields. Chemometrics and neural networks are those disciplines from which the most utilized techniques are derived.[5]

In Figure 2. a taxonomy of the techniques which are mostly utilized is shown. A main division among them concerns the quantitative and qualitative character of the analysis. In the following some specific definitions for both these aspects of sensor arrays will be introduced.

2. QUANTITATIVE ANALYSIS

A sensor array can be defined as a vectorial organization of the output of single sensors, so that the contribution of each array element is a component of a multidimensional general problem. The mathematical scenario of a sensor array is drawn in Figure 3.

Let us consider an array of m sensors whose outputs depend on n state variables characterizing the environment on which the sensor is exposed. In the following matrices will be indicated with capital boldface letters (A, B,...) and vectors with small cap boldface letters (a, b,...). Let z_i be the output of the i-th sensor and z the output vector ($z \in \Re^m$), w_j the value of the j-th state variable and w the state vector ($w \in \Re^n$).

A generic sensor array can be represented through a non linear transformation f: W_{op} fi Z where W_{op} is an open sub-set of the state space W identifying the domain of variability of the state variables and Z is the sensor output space. Let us also indicate with $f_i(w)$ the I/O characteristic of the i-th sensor.

With these positions the array is described by the following array equation:

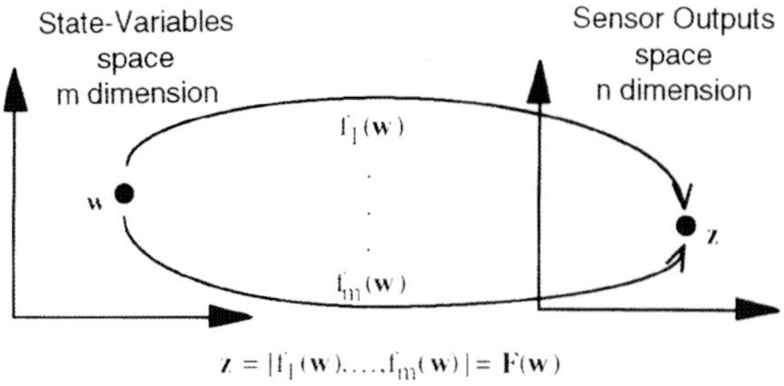

Figure 3. Mathematical scenario for quantitative analysis.

$$z = f(w) \text{ where } f(w) = \left[f_1(w), \ldots, f_m(w)\right] \tag{1}$$

The inversion of the Eq. (1) is the procedure that allows the array to accomplish the measurement finding out, from the values of the sensor outputs, the values of the state variables. This operation resides in a numerical deconvolution procedure.

This operation can be accomplished by using many different techniques, as displayed in the right side of the classification scheme proposed in Figure 2. Among them those derived from chemometrics and those based on the use of artificial neural networks have been mostly utilized. Neural networks are treated in dozens of excellent textbooks (for instance ref. 6 offers a good introduction and ref. 7 gives a wide review of applications) while although chemometrics is investigated since the beginning of 70's it is not sufficiently diffused among the sensor community. Chemometrics is a powerful approach to data analysis and it shows how, starting from a very simple linear approach is possible to get performances that in some cases are comparable with those obtained with more costly non-linear methods, such as neural networks.

2.1 Chemometrics

Chemometrics is a discipline aiming at providing analytical chemists with a number of algorithms for the extraction, from multichannels instruments (such as spectro-photometers and gas-chromatography), of information on the samples under analysis.[8] The development of chemometrics started at the beginning of the seventies, so that when, at the middle of the eighties, the arrays of chemical sensors started to be object of research and development a number of chemometrics methods were yet part of the background of the analytical chemistry community.

Many chemometrics methods are rather simple (from conceptual and often practical point of view) and beside the interpretation of the results is often straightforward. The major limitations of these methods come from their linear character, so that in some cases the results are affected by non tolerable errors.

Concentration estimate is based on the assumption that a linear relation between sensor response and species concentrations hold:

$$z = \sum_{i=1}^{N} s_i c_i \tag{2}$$

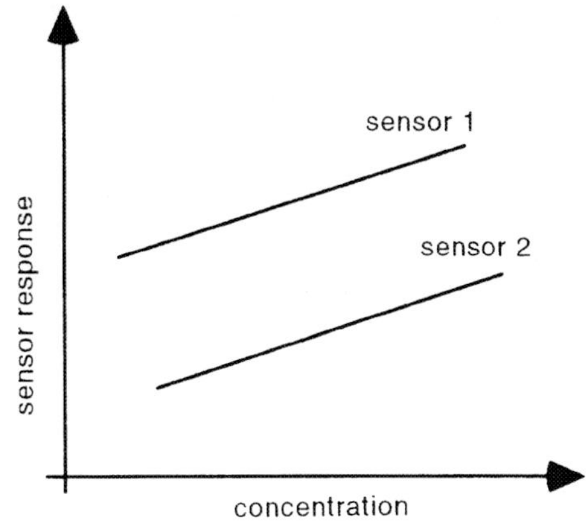

Figure 4. The evident colinearity of sensors puts in evidence that colinearity is related with the sensitivity of the sensors namely with the slope of the characteristic curve.

where z indicates the response of a generic sensor, c_i is the concentration of the i-th species and S_i is the sensitivity of the sensor to the i-th species.

When more sensors are gathered together to form an array, the whole array can then be represented, writing Eq. (2) in matricial form, as:

$$z = \boldsymbol{S} \cdot \boldsymbol{c} \tag{3}$$

From the point of view of sensor arrays, chemometrics provides a number of tools for the solution of the Eq. (3). These methods go from the trivial solution, based on the pseudoinversion, to the sophisticated utilization of latent variables. All these techniques originates from the necessity to overcome the problem of colinearity.

Colinearity is a crucial concept in sensor array. Two sensors are said to be colinear when their responses are proportional one each other. It can be easily understood that, from the point of view of their information content, these sensors behave as one. From a mathematical point of view colinearity means that two rows of the matrix S are linearly dependent, therefore the matrix cannot be inverted. It is worth while to remark that when sensors are quasi-colinear the inversion is still possible in principle, but considerable errors are introduced in calculus.

It is important to remark that two sensor becomes colinear when their sensitivities are equal, so it can be the case that the amplitude of the responses are different but the slopes of the responses are equal as shown in Figure 3. Colinearity problem is solved by chemometrics reducing Eq. (3) only to those directions that bring effective information.

The scope of the quantitative analysis is to use the calibration data to determine a regression matrix B which can be utilized to calculate, from the sensor array output, an estimate of the concentrations according to:

$$\boldsymbol{c} = \boldsymbol{B} \cdot z \tag{4}$$

Chemometrics offers many ways to solve this problem the many utilized are discussed in the forthcoming sections.

2.1.1 Multiple Linear Regression: In this routine matrix B is carried out through a direct inversion of Eq. (2) as an application of the Gauss-Markov theorem for the solution of least squares problems, making use of the operation of pseudoinversion:

$$\boldsymbol{B} = z^{+} \cdot c \qquad (5)$$

where z^{+} is the pseudoinverse matrix of z.[9]

2.1.2 Principal Component Regression: In this routine the concept of Principal Component is applied to the solution of Eq. (2).

PCR is based on the operation of singular value decomposition which states that given a matrix zmxn it can be written as a product of three matrices Umxm, Snxn, and Vnxm, SVD gives also the possibility to express the matrix z^{+} as combination of U, S, and V according to:

$$z^{+} = V^{T}_{m,n} \cdot S^{T}_{m,m} \cdot U_{m,m} \qquad (6)$$

The dimensions of z^{+} are given by the first and the last index in Eq. (6), so that the internal index m can be exchanged with an index q, said factor. Index q spans all the dimensions of the sensor output space, so that for q=m all the dimensions are considered, while for q<m some dimensions are neglected, it is important, at this point, to note that the dimensions of sensor output space are ordered according to their content of information by the diagonal matrix S. A suitable selection of q gives the possibility to eliminate from the array all the colinear or quasi-colinear components giving a reduction of errors in determining the matrix B. For each factor q different performances are obtained, in order to choice the best q it is necessary to have a factor of merit to quantify the performances. As an example, relative average error (RAE) can be chosen, it is defined on a test data set as:

$$RAE = \frac{1}{N} \sum_{N} \left| \frac{estimate - true}{true} \right| \qquad (7)$$

where true is the true value of concentration, estimate value estimated by PCR at factor q and N is the number of test data. The value of q in correspondence of which the RAE is minimized, is the chosen for the PCR model of the array. The regression matrix is then given by:

$$B_{n,t} = V^{T}_{n,q} \cdot S^{T}_{q,q} \cdot U_{q,m} c_{m,t} \qquad (8)$$

PCR capabilities can be improved considering a different data decomposition (the so-called Lanczos Bidiagonalization) which takes into account also the concentration matrix. This method called *Partial Least Squares* is one the most adopted chemometrics tools.[10]

3. QUALITATIVE ANALYSIS (ELECTRONIC NOSE)

Electronic noses are instrumental apparatus based on the utilization of a chemical sensor array where each sensor is characterized by its own degree of selectivity.[11] This last feature is the key property on which electronic noses found their working principle. Electronic noses are commonly utilized for classificatory purposes. In fact they can distinguish among samples according to some classification. Typical examples are found in the field of food analysis where sometimes very intuitive classes are adopted according to accepted categories such as freshness or edibility.[12]

In an electronic nose the chemical patterns occurring in a certain environment are "translated" by the sensors into a response pattern. With respect to the chemical pattern, the response pattern is characterized by having less dimensions; basically it is a combination of all the components which form the chemical pattern. The rule of combination, generally non-linear, is given by the selectivities and the sensitivities of each single sensor.

In each application optimal performances are achieved when the pattern translation process preserves those features allowing the discrimination among those classes which are relevant to the particular case. This procedure based on a reduction of dimensions in the patterns has, as a consequence, a reduction of the information content.

Data analysis is an important issue in Electronic noses. Its usual role is to allow a correlation between sensors outputs pattern and the properties of the environment which are the object of the measurement strategy. Data analysis has been the main concern of many research papers since the first development of the field. They have been dealing with the application of several techniques, borrowed from other application fields.[13]

Data analysis techniques can be presented as belonging to two different categories: explorative (unsupervised) and regressive (supervised). Explorative methods deal with the structure of a set of data and are employed to find-out the relations between the data and any eventual occurrence of clusters. The knowledge of this intrinsic classification is of primary importance to understand the property of the Electronic noses.

Regression methods are those aiming at establishing correlations between sets of input and output data. These methods are naturally utilized for quantitative analysis where the scope of the measurement is to find-out, from the sensor outputs, estimations about the concentrations of a given number of species. Nevertheless the same methods can also be utilized to establish correlations between sensor outputs and a pre-defined set of classes which are typical of the environment under test.

A detailed description of the methods can be found in ref. 5, in the next section the attention will be focused on one of these methods: the self organizing map. It is a neural network which allow a complete study of the properties of a sensor array, in particular for electronic noses it provides classification, clustering and an evaluation of the contribution of each sensor to the whole array.[14]

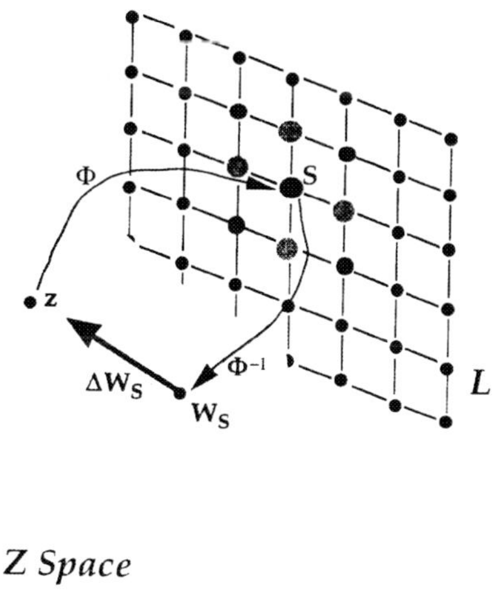

Figure 5. Schematic representation of the SOM learning algorithm.

4. SELF ORGANIZING MAP (SOM)

SOM is one of the most important neural models. It belongs to the category of competitive learning methods and it is based on unsupervised learning. This last aspect means that the SOM algorithm does not require any additional information but the sensors outputs. As an example a SOM can discover the clustering properties of a set of data without the knowledge about their class membership. SOM's were introduced, at the beginning of the eighties, by T. Kohonen, and due to its great flexibility it has been utilized in many different fields, from speech recognition to the process control in industrial plants. A comprehensive introduction to the basic principles and many examples of applications can be found in the monograph of T. Kohonen.[15]

Hereafter a brief introduction to the SOM is given from the point of view of multisensor applications. SOM is a network formed by N neurons arranged as the nodes of a planar grid. Each neuron is identified by a vector r, whose components are the node coordinates in the grid. The neurons are logic elements with two possible states; they have m input (vector z) and one output. An input is a real value vector, while the output is either active (value 1) or inactive (value 0).

Each single neuron is characterized by a m-component codebook vector w_r, which represents the neuron in the input space. In our sensor system this logical structure accepts inputs from the sensor array: due to each input the codebook vectors of neurons are modified by a learning algorithm, called "Kohonen algorithm"; it aims at constructing the whole set of codebook vectors $\{w_r\}$ of the grid as a discrete approximation of all the supplied z vectors, a pictorial view of the algorithm is shown in Figure 5.

Once a new z is provided, the learning algorithm prescribes two stages:

1- Response. Determination of the index s from the condition:

$$\|z - w_s\| \leq \|z - w_r\| \qquad \text{for all } r \tag{9}$$

Namely the neuron whose codebook vector is closest to the input z is selected.

2- Adaptation. Variation of the codebook vectors of all the neurons according to:

$$w_r^{new} = w_r^{old} + \alpha \cdot h_{rs}\left(z - w_r^{old}\right) \qquad \text{for all } r \tag{10}$$

where h_{rs} (neighbor function) can have, as an example, the following form:

$$h_{rs} = \exp\left(\frac{-\|r - s\|^2}{2\sigma^2}\right) \tag{11}$$

The function h_{rs}, defines an area, around of the s neuron, involving those neurons participating in the adaptation stage. The parameter s is the length scale of the proximity of neuron s.

From a practical point of view a calibration data-set is used to train the SOM and, at each learning step a data, randomly selected, is presented to the SOM. The process takes place until the network converges, namely the codebook vectors do not change more than a negligible quantity. In order to ensure the convergence, the parameter α appearing in Eq. (10) is not constant but it is a decreasing function of the time operating along the learning process. Various kinds of decreasing laws (linear and the hyperbolic are the most common) can be imposed to α.

Once the network reaches a convergence the codebook vectors of the SOM neurons contain the model of the phenomena of which the calibration data-set is a sample. Obviously the more the sample is representative of the phenomena, the more reliable is the model encoded in the SOM. In order to read-out, from the codebook vectors, information about the sensor array, it is necessary to properly read the codebook vectors of the neurons.

4.1 Tools for Sensor Array Modeling and Data Analysis

4.1.1 Data classification: The most immediate way to utilize a SOM after the training is to identify the neurons of the SOM which respond to the calibration data and to eventual test data. This operation allows to draw, onto the map plane, regions which are pertinent to different qualitative classes occurring in data. It has to be in evidence that SOM lattice is not a metric space, in the sense that the proximity between neurons is not the same everywhere in the map. The proximity between neurons can be evaluated by the difference between their coedebook vectors. SOM provide a discretization of the sensor output space and neurons are in some sense the "pixels" of this process. The volume of the sensor space represented by each neuron is not a constant but it depends on the statistical distribution of the calibration data. If the data are very few, and/or if they are very compact in some regions, it happens that only few neurons of the map are interested by the data while the others are utilized to fit the distances between the sparse data.

4.1.2 Evaluation of single sensors contribution: In sensor arrays there is the problem to evaluate the effective contribution that each single sensor brings to the whole array. One of the most utilized method consists in calculating the loading in a PCA representation of a data-set. For this methods the same consideration made previously holds. PCA also in this case provide a results which is, in some sense, averaged all over the domain of the data. Another possibility makes use of analytical functions describing the behavior of the sensors towards the species characterizing an environment under analysis, this methods, also if exact in principle, brings a lot of numerical problems that makes it very hard from a practical point of view. SOM gives an easy opportunity to evaluate the behavior of each sensor simply studying the components of the codebook vectors. Indeed each sensor is a coordinate in the SOM input space and therefore is a component of the codebook vector. It is then very easy to represent graphically this information onto the SOM plane or using a colored map (as in the clustering problem) or a 3D shape. Both the choice give the possibility to monitor the influence of a sensor on the whole domain of the data and the comparison of the representation reveals, by similarity, if two or more sensors are correlate one each other.

4.1.3 Sensor drift effects: A reading of the map gives also the opportunity to foresee the effects of sensors drift on the performances of the multisensor system. Drift is one of the major drawbacks affecting chemical sensors, although there are a plenty of explanations for drift in many kinds of sensors none of them allow correct predictions. SOM allows to predict the variations in sensor system response due to sensor drift and also to estimate the drift levels which are acceptable for any specific application. This result can be obtained very simply considering a particular working point and simulating a drift changing the response of, as an example, one of the sensors. Each point can be displayed onto the SOM plane so that the drift can be followed as a track on the map.

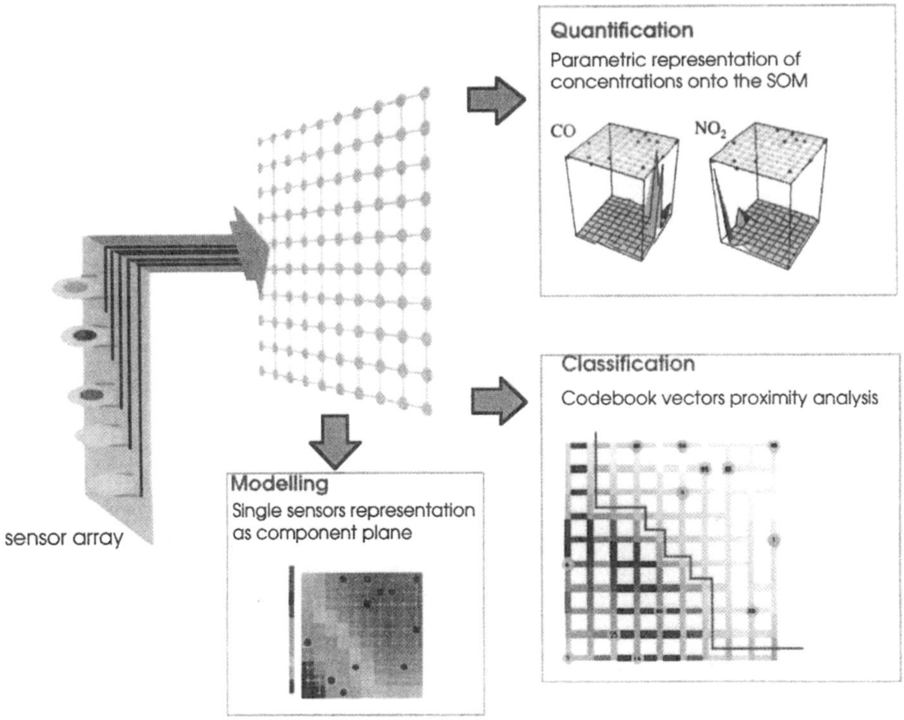

Figure 6. A SOM can be utilized to provide information about the classification of sensor array data (electronic nose), quantification of concentrations working as a sort of *look-up* table (quantitative analysis) and to give an evaluation of the single sensors contribution to the whole array (array modeling).

REFERENCES

1. W. Göpel, T.A. Jones, M. Kleitz (eds.); *Sensors Vol. 2: Chemical and biochemical sensors*, VCH (Weinheim, Germany) 1991.
2. D.J. Newmann, Y. Olabiran, C.P. Price; Bioaffinity agents for sensing systems, *in Handbook of Biosensors and Electronic Noses*, E. Kress-Rogers (eds.), CRC Press (Boca Raton, USA) 1997.
3. D.J. Cram; The design of molecular hosts, guests and their complexes, *Science*, 240: 760 (1990).
4. T.V. Getchell; Functional properties of vertebrate olfactory receptor neurons, *Physiol. Rev.*, 66: 772 (1986).
5. A. Hierlemann, M. Schweizer, U. Weimar, W. Göpel; Pattern recognition and multicomponent analysis, in *Sensors update Vol. 2*, W. Göpel, J. Hesse, H. Baltes (eds.), VCH (Weinheim, Germany) 1995.
6. D. W. Patterson; *Artificial Neural Networks: Theory and Applications* (Prentice Hall, Singapore) 1996.
7. A.J. Maren (ed.); *Handbook of Neural Computing Applications*, J. Wiley and sons (London, UK) 1991.
8. D. Massart, B. Vandeginste, S. Deming, Y. Michotte, L. Kaufmann; *Data Handling in Science and Technology: Chemometrics a Textbook*, Elsevier (Amsterdam, The Netherlands) 1988.
9. S.L. Campbell, C.D. Meyer; *Generalized Inverse of Linear transformation*, Pitman (London, UK) 1976.
10. P. Geladi and B. Kowalski; Partial least squares regression: a tutorial, *Anal. Chim. Acta*, 195: 1 (1986).
11. E. Kress-Rogers (ed.); *Handbook of Biosensors and Electronic Noses*, CRC Press (Boca Raton, USA) 1997.
12. C. Di Natale, A. Macagnano, R. Paolesse, A. D'Amico, T. Boschi, M. Faccio, G. Ferri; An electronic nose for food analysis, *Sens. and Act. B*, 44: 521 (1997).
13. C. Di Natale, F. Davide, A. D'Amico; Pattern recognition in gas sensing: well assessed techniques and advances, *Sens. and Act. B*, 23: 111 (1995).
14. C. Di Natale, A. Macagnano, F. Davide, A. D'Amico; Data analysis and modelling of electronic nose with self organizing neural networks; *Meas. Sci. and Techn.* 8: 1236 (1997).
15. T. Kohonen; *Self Organizing Map*, 2nd edition, Springer Verlag (Berlin, Germany) 1997.

HIGH SENSITIVITY TRACE GAS MONITORING USING SEMICONDUCTOR DIODE LASERS

C. Corsi[†], and M. Inguscio[††]

[†]Università degli Studi di Firenze
Dipartimento di Neuroscienze
Viale Pieraccini 6
50139 Firenze, Italy

[††]European Laboratory for Nonlinear Spectroscopy (LENS) and
INFM-Istituto Nazionale Fisica della Materia
Largo E. Fermi 2
50125 Firenze, Italy

1. INTRODUCTION

In recent years semiconductor diode lasers in the visible near-infrared have been applied to high sensitivity gas detection for a variety of environmental, medical and industrial applications.

The advantage of laser-based gas sensor with respect to conventional electro-chemical and semiconductor point sensors resides in their characteristics of non-intrusiveness, high gas selectivity, high detection speed and low cost.

In particular, great attention has been attracted by InGaAs-InP Distributed Feed-Back (DFB) diode lasers, operating at room temperature. Thanks to the knowledge and technology developed for telecommunication diode lasers emitting around 1.3μm and 1.5μm, these lasers can be easily designed to emit single mode almost anywhere in the region between 1μm and 2μm.

This spectral region is of particular interest because of the presence of molecular overtone vibrational bands for many important gases, like for instance, CO_2, CO, H_2S, HCl, HF, NH_3, CH_4, H_2O, NO, N_2O.

Although the transition strengths are at least one order of magnitude weaker than those for the fundamental bands in the mid-infrared, this problem is compensated, in real applications, by the advantages coming from the reduced opacity of the atmosphere and from the possibility of connecting the lasers with optical fiber systems, developed for telecommunication purposes.

Furthermore diode lasers are particularly suited for high-sensitivity absorption spectroscopy because they show much smaller amplitude noise than most other laser sources. In addition, diode lasers can easily be modulated at frequencies up to several GHz.

Optical Sensors and Microsystems: New Concepts, Materials, Technologies
Edited by Martellucci et al., Kluwer Academic / Plenum Publishers, New York, 2000.

This allows to apply various detection techniques[1,2,3] such as low-wavelength modulation spectroscopy (LWM) and frequency modulation spectroscopy (FM)[4,5] depending on the sensitivity level necessary to be reached.

In principle, sensitivities near to the fundamental quantum limit (shot-noise limit) are possible. In practice, quantum-limited sensitivity has proven to be difficult to obtain, due to a number of technical noise sources.

The noise sources can be divided in detector noise, noise due to amplitude fluctuations of the laser field and optical noise due to interference fringes.

1.1 Detector noise

There are three major components in a photodetector noise: Johnson (thermal) noise, detector shot noise, and 1/f-noise.

Johnson noise is due to the thermal fluctuations of the charge carrier density within the resistor itself. The thermal noise current can be expressed as:

$$i_T = \left(\frac{4KT\Delta\nu}{R}\right)^{\frac{1}{2}} \tag{1}$$

where K is the Boltzmann constant, T the temperature, $\Delta\nu$ the detection bandwidth and R the detector system resistance. The Johnson noise has a white frequency spectrum and can be reduced by cooling the detector.

Shot noise is due to quantum fluctuations of the radiation field. They give rise to fluctuations of the detected current in a photodetector and can be expressed by:

$$i_S = \left(2e\cdot\frac{eP\eta}{h\nu}\cdot\Delta\nu\right)^{\frac{1}{2}} \tag{2}$$

where e is the electronic charge, η is the quantum efficiency of the detector, P is the incident power and ν is the photon energy. Shot noise is also white and independent of the modulation frequency but is proportional to the square root of the laser power. For optical powers in the mW range the output noise-voltage of the photodetector is dominated by shot noise rather than by Johnson noise.

Often devices show various sources of other noise mechanisms. In many cases this additional noise shows a 1/f dependence, but unfortunately no theoretical analysis are available. An empirical expression for the 1/f noise current is given by:[6]

$$i_{1/f} = \left(\frac{C\cdot\Delta\nu}{f^a}\right)^{\frac{1}{2}}\left(\frac{eP^b\eta}{h\nu}\right) \tag{3}$$

where C is a proportional factor, a and b are constants close to unity. The 1/f noise depends on the manufacturing processes, in particular on electric contacts and surfaces. It dominates the detector noise for frequencies below 1 kHz and drops below the Johnson and shot noise levels at higher frequencies. The rms 1/f noise current shows an approximately linear dependence on the photocurrent and therefore on the light intensity.Since all three detector noise sources depend on the detection bandwidth Δν a reduction of the bandwidth results in a noise-reduction. It is also convenient to work at higher frequencies where the 1/f-detector noise has become lower than the shot noise.

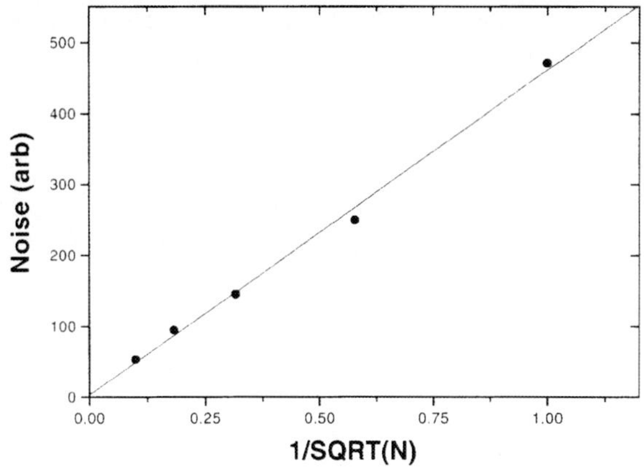

Figure 1. Improving the signal to noise ratio decreasing the detection bandwidth, by means of signal averaging. In figure the signal noise is plot versus the inverse of the square root of the number of averages, which is proportional to the detection bandwidth, for a DFB laser at 1578 nm.

1.2 Laser Excess Noise

In practice, the sensitivity of absorption measurements is often limited by excess-noise which is due to fluctuations of the laser power. The fluctuations are generated by external effects like current and temperature instabilities, mechanical vibrations, or optical feedback as well as by intrinsic noise sources, such like photon and carrier density fluctuations and partition noise. The external noise sources can be minimized by battery-driven or highly stabilized current sources together with proper alignment, ar-coatings, and optical isolators. The intrinsic noise depends mostly on the manufacturing process and design and on the operating conditions, like temperature and current.

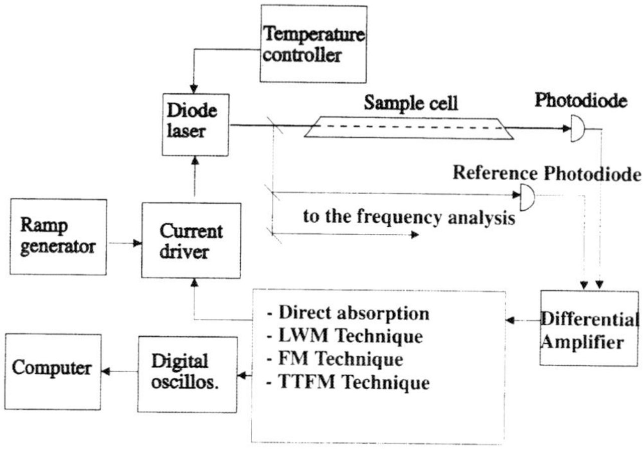

Figure 2. Schematic view of a laboratory set-up for a semiconductor diode laser spectrometer.

Several investigations of amplitude-noise characteristics of lead-salt lasers[7] and DFB and DBR-lasers[8] have shown that in most cases the laser excess-noise exhibit a 1/f dependence with cut-off frequencies ranging between 1 and 100 Mhz. The rms current at the detector caused by the laser-excess noise can be expressed as:

$$i_{ex} = \left(\frac{\Delta\nu}{f^{a}}\right)^{\frac{1}{2}}\left(\frac{eP_{ex}\eta}{h\nu}\right) \tag{4}$$

where a defines the frequency dependence of the laser-excess noise and ranges between 0.8 and 1.5, P_{ex} defines the magnitude of the laser power fluctuations at 1 Hz in a 1 Hz detection bandwidth. P_{ex} is approximately proportional to the laser power and depends on the intrinsic noise of the diode laser and on the external effects of the particular measurement system. As in the case of detector noise, laser excess noise is detection bandwidth dependent and can be reduced with appropriate techniques.

1.3 Residual Amplitude Modulation

Frequency modulation of diode lasers results in a simultaneous amplitude modulation, since not only the wavelength depends on the current but also the laser power. This residual amplitude modulation (RAM) becomes often the main noise source in high-sensitive absorption measurements. The RAM can be reduced using a dual beam subtraction method, in which the laser beam is split in a reference beam and a probe beam, which are detected by the same kind of photodiodes.

It is important that the amplifier have the same quantum efficiency, the same gain and the same frequency response. Therefore all electronic components have to be selected with great care to avoid manufactural differences of photodiodes and electric components, such as resistance, operational amplifiers and capacities.

1.4 Interference Fringes

Every transmitting element in the optical path, such as beamsplitters, lenses, absorption cell windows or the laser collimator itself can create Fabry-Perot fringes. These optical fringes often exhibit a free spectral range (FSR) comparable to the linewidth of absorption lines when the laser is scanned across the lines. They appear as periodic oscillations with sufficient amplitudes to obscure weak absorptions signals. Interference fringes arise from reflections between parallel surfaces in the optical path and the transmission depend on the laser wavelength. Any scanning of the wavelength results in an amplitude variation if resonant structures appear in the path. The free spectral range between two fringes can be expressed as:

$$FSR = \frac{c}{2nl} \tag{5}$$

where n is the refractive index of air and l is the distance between the optical surfaces. The distance l ranges typically between 1 m and 5 mm, which corresponds to resonances at a distance of 150 MHz and 30 GHz, respectively.

There are several possibilities to minimize these etalon effects. For example, if the free spectral range of the fringes is very different compared to the width of the absorption feature they can be removed by filtering the detected signal. However, for spacings comparable to the linewidths the fringes can only be removed by a careful design of the experimental set-up.

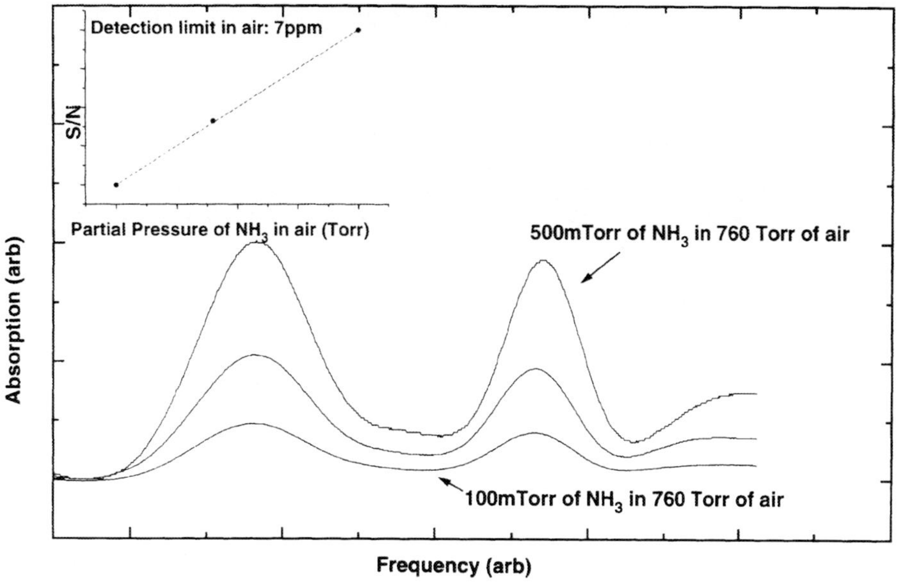

Figure 3. Second derivative profiles of different pressures of NH_3 in air: a detection limit of 7 ppm×m in air is extrapolated. The combination band of ammonia at 1.5μm is investigated, with a laser at 1494 nm.

Whenever possible, transmissive optics should be avoided or anti-reflection coated optics should be used.

It is important to note that only detector 1/f noise and laser excess noise depend on the detection frequency, while all noise sources discussed so far depend on the detection bandwidth $\Delta\nu$. Therefore, high sensitivity detection can be achieved by increasing the detection frequency (to a value for which the laser excess noise is lower than the shot- and Johnson noise) and by reducing the detection bandwidth.

1.5 Detection Techniques

In general, in a direct absorption measurements, the small changes in the transmitted amplitude, arising from gas traces, have to be distinguished on a large background, resulting in a poor sensitivity.

Frequency modulation techniques are based on a fast modulation of the laser emission in the frequency domain resulting in an amplitude modulation of the ligth intensity. By means of phase sensitive detection is then possible to extract only the absorbed signal from the background of the transmitted power. An additional increase in sensitivity is due to the fact that the detection frequency can be moved to larger values, in order to reduce the laser amplitude fluctuations.

The modulation techniques are divided according to the modulation frequency.

In low wavelength modulation spectroscopy (LWM), the laser frequency is modulated at a relatively low frequency (hundreds of kHz), which is small compared to the width of the line to be probed. The observed signal arises from the difference in the absorption of different sidebands which probe simultaneously the absorption line. The signal is then demodulated, at n times the modulation frequency. Usually, first (n=1) and second (n=2) derivative detection are used.

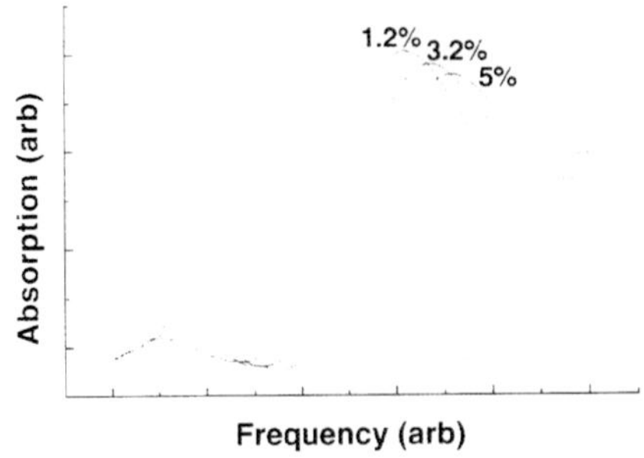

Figure 4. Second derivative absorption signal of CO at the exit of a catalytic exhaust with a DFB laser at 1578 nm.

In frequency modulation spectroscopy (FM), the laser is modulated at much higher frequencies, that range usually between 100 MHz and several GHz, comparable with the width of the absorption features, and produce sidebands which are widely spaced in frequency. In this technique, only one sideband is absorbed at a time, giving rise to a heterodyne beat signal at the modulation frequency. The detection frequency is moved in a region where the laser excess noise presents its minimum value and, in addition, the selective absorption of the sidebands results in a larger detected signal.

In order to retain a similar sensitivity for atmospheric pressure broadened absorption profiles, the laser must be modulated at frequencies in excess of one Ghz. Electronics for heterodyne detection at these frequencies can be complicated. The problem can be simplified by using two-tone frequency modulation spectroscopy (TTFM).[9] The laser emission is simultaneously modulated at two distinct but closely spaced frequencies, $\nu_1+\nu_2$ and $\nu_1-\nu_2$, once more comparable to the line-width of interest (ν_1~a few GHz for pressure broadened profiles and ν_2~few MHz). The heterodyne beat signal is then obtained at the much lower frequency $2\nu_2$, eliminating the need of high speed detectors and electronics. Such detection frequency can be anyway maintained sufficiently large (a few MHz) to avoid the 1/f excess noise and reach the same noise level as in the FM technique.

It has to be noted that such high modulation frequency can be obtained only in semiconductor diode laser by means of the injection current. Electro-Optical modulators (EOM) could be a possible solution for other laser sources, but efficient modulation at frequencies above 1 GHz are not very easy to be reached.

Further improvement in sensitivity can be obtained by means of a double beam configuration, in which amplitude fluctuations can be canceled out, achieving a shot-noise limited detection.

1.6 Bandwidth reduction

Another important parameter that greatly influences the sensitivity of the apparatus is the electronic detection bandwidth of the signal. Indeed, many noise sources (shot-noise, Johnson noise, laser excess noise) depend on it, in particular they decrease proportionally to the square root of the bandwidth.

For example, an appropriate filtering of the signal can result in an enhancement of the sensitivity. In addition, by means of a signal averaging, the noise can be further reduced, as if the bandwidth was decrease according to the relation:[7]

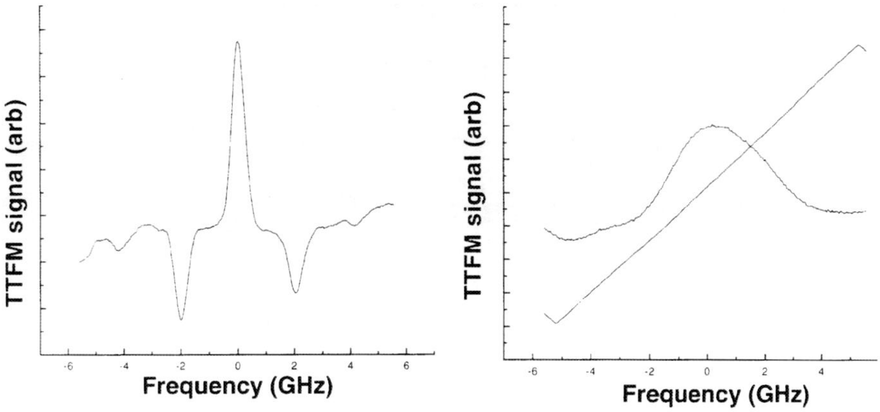

Figure 5. Two-tone signal from 90 mTorr of pure H_2S and b) in an atmosphere of air (1Hz detection bandwidth), with a DFB laser at 1578 nm.

$$\Delta \nu_{eff} = \frac{\Delta \nu}{n} \qquad (6)$$

where n is the number of averages. This effect has been verified experimentally with a DFB diode laser at 1578 nm, as can be seen in Figure 1.

2. EXPERIMENTAL SET-UP

In Figure 2. an experimental laboratory setup of a typical tunable diode laser sensor is shown. Different diode lasers are used, depending on the gas under study. The emission frequency is selected with an appropriate control of laser temperature and current in order to match the chosen molecular line. A current ramp of about 10 mA amplitude is added to the injection current, leading to continuous sweep of the laser frequency of several GHz, which allows to record the whole selected line with one current scan.

The output beam of the laser is collimated and splitted into three parts: one is used for the frequency control, the second one is used to detect the unabsorbed light intensity and the third one is used to detect the transmitted light intensity.

The absorbed and reference signals are amplified and subtracted from each other to avoid the background slope due to the amplitude modulation produced by the current scan.

The sample cell, a 1.5 m long Pyrex tube, is pumped out to 10^{-4} Torr and filled with the gas sample.

This simple experimental set-up is used to extract atmospheric relevant broadening parameters with the advantage of a straightforward interpretation of the broadened absorption profile.[10]

For high sensitivities measurements LWM and TTFM techniques are used.

In the low wavelength modulation apparatus an ac component at ω_0=3.5 kHz is added to the injection current, the absorbed signal is demodulated at a frequency $2\omega_0$ by means of a lock-in amplifier.

In the two-tone frequency modulation scheme the signal from a synthesizer at ν_1=2.04 GHz is mixed with that at ν_2=5.35 MHz of a function generator.

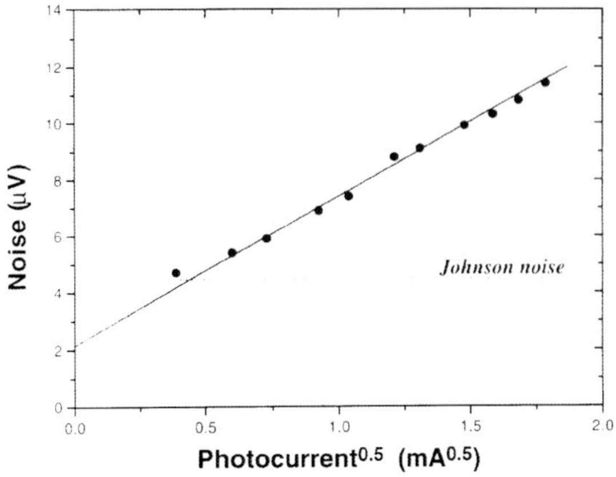

Figure 6. Noise power as a function of the optical power incident on the photo-detector. The linear dependence indicates that the sensitivity is shot-noise limited, since other noisesources depend in a different way on the optical power.

The output, at ν_{mod}=2 Ghz±5.35 MHz is coupled to the laser diode current by means of a high pass filter. The signal of the function generator is frequency doubled and is mixed in a phase detector with the filtered and amplified absorption signal coming from an InGaAs-PIN photo-diode. The output of the phase-detector, which is proportional to the Fourier component of the signal from the photo-diode at ν_{det}, is low pass filtered (100 Hz), amplified and recorded on a digital oscilloscope.

To perform a quantum limited detection a careful attention to experimental details like amplifiers has to be paid. In particular, for the balanced dual-beam set-up used in our application, it is of fundamental importance that the probe signal and the reference signal are detected with the same kind of photo-diodes and amplified with the same kind of amplifiers.

Table 1. Comparison between minimum detectable concentrations in air obtained with different detection techniques.

	λ	Detection limit in air		
	nm	ppm×m		
NH_3	1494.35		7 (LWM)	
CO	1579.74	500 (AD)	170 (LWM)	7 (TTFM)
CO_2	1579.57	800 (AD)	260 (LWM)	10 (TTFM)
H_2S	1577.32	400 (AD)	130 (LWM)	4 (TTFM)
O_2	760.89		1000 (LWM)	
HCl	1742.3		0.5 (LWM)	

To avoid manufactural differences of resistances or capacitance they have to be tested and chosen carefully. For that reason a low-noise pre-amplifier circuit for the photo-diodes was especially designed to work for detection frequencies between 1 and 40 MHz.

The preamplified output signals of the photo-detector are combined in a 180° rf hybrid, which shifts one signal by 180° before adding it to the other one, eliminating in this way all coherent fluctuations of the two beams. A quantitative description of our quantum limited dual beam spectrometer can be found elsewhere.[11]

With all the detection techniques listed before a series of signal to noise ratio measurements at different total and partial gas pressures are performed. The minimum detectable gas pressure is then obtained by a nonlinear fitting procedure, extrapolating the value at which the signal to noise ratio becomes unity.

3. RESULTS AND DISCUSSION

The sensitivity limits from the three detection techniques are reported using the absorption lines of some interesting molecules in the 1μm-2μm region (Table 1.).

3.1 Ammonia NH_3

The combination band $\nu_1+2\nu_4$ at 1.5μm has been investigated. Low wavelength modulation spectroscopy has been performed on the strongest component at λ = 1494.35 nm using a DFB diode laser (Fig.3). A minimum detectable concentration of 7 ppm×m of ammonia in air has been achieved.This low detection limit is important for monitoring the NH_3 concentration in various applications. Selective Catalytic Reduction (SCR) of NO_x with ammonia represents the most effective technology currently available for deep NO_x removal. Furthermore, alteration in NH_3 concentration in the breath can indicate severe epatic failure.

3.2 Carbon Monoxide CO, Carbon Dioxide CO_2 and Hydrogen Sulphide H_2S

These three molecules have, in the region around 1578 nm, overlapping overtone bands: 3ν (CO), $2\nu_1+2\nu_2+\nu_3$ (CO_2)[12] and $\nu_1+\nu_2+\nu_3$ (H_2S)[13]. By using one single semiconductor diode laser and a TTFM scheme, a minimum detectable absorption of 5×10^{-7} is achieved. This corresponds at a minimum detectable concentration in air of 7 ppm×m for CO, 10 ppm×m for CO_2 and 4 ppm×m for H_2S.

CO and CO_2 are important in combustion processes and exhaust gases control (Figure 4.), while H_2S detection is fundamental for security on sour-oil-rigs (Figure 5.).

Furthermore a TTFM spectrometer is a powerful tool for non-invasive diagnostic of human breath, where detection of the isotopic ratio $^{12}CO_2/^{13}CO_2$ can give information of the presence of Helicobacter Pylori in the stomach.

3.3 Molecular Oxygen O_2

A magnetic dipole transition of the $b^1\Sigma_g^+$ (v'=0)$\leftarrow$$X^3\Sigma_g^-$ (v"=0)--band at 761nm is studied. A low wavelength modulation scheme is used, but a sensitivity of only 1000 ppm×m is achieved, because the transitions' line-strengths in this forbidden band are more than one order of magnitude weaker than those of the other molecules investigated.

Table 2. Self-broadening and Air-broadening coefficients for the gases under study (* measured at 1.65 μm).

	λ	Self-broadening	Air-broadening
	(nm)	(MHz/Torr)	(MHz/Torr)
NH_3	1494.35	37(3)	4.0(2)*
CO	1579.74	3.95(9)	3.67(8)
CO_2	1579.57	4.1(3)	3.42(7)
H_2S	1577.32	6.7(2)	3.09(9)
O_2	760.89	2.04(2)	1.92(5)
HCl	1742.3	8.6(4)	2.9(5)

3.4 Hydrogen Chloride HCl

The first overtone band (2ν) at 1.77μm is investigated[14] with a low wavelength modulation and two-tone frequency modulation apparatus. Sensitivity of 0.5 ppm×m is reached. Monitoring of HCl is very important in the emitted gases of waste incinerator.

For the shot-noise limited detection the absorption spectrometer has been tested using H_2S as a target gas, once the noise sources are minimized and the optimal working conditions are found. We have measured the rejection ratio using the difference method.

We have demonstrated that the used dual beam configuration can suppress laser noise by more than 25 dB, in the region of interest near 10 MHz. Then we have measured the difference of the two channels, using the 180° combiner, and the sum using the 0° combiner, that is only 5 dB over the difference. So, taking in account the rejection ratio we concluded that we have reached the shot-noise level. The detection limit of H_2S in air was found to be approximately 500 ppb over 1 meter path-length.

As an additional check to the reached shot-noise limit, the light noise was measured as a function of the square root of the optical power incident on the photo-detector (see Figure 6.). In addition, pressure broadening parameter are measured for all these molecules. Results are listed in Table 2.

4. CONCLUSION

The achieved quantum limited sensitivity demonstrate that using distributed feedback diode lasers in combination with two-tone frequency modulation spectroscopy is a powerful technique for gas detection. It is possible to detect traces gases in air with a sensitivity on a ppm-level over a 1-meter path-length for some molecules of industrial, medical and environmental interest.

The natural combination with fiber optics technology make these sensors attractive for a large variety of "in situ" measurements.

ACKNOWLEDGMENTS. Many collegues and visitors have contributed to the progress in diode laser spectroscopy at Lens. We are indebted to them all: R. Benedetti, F. D'Amato, P. De

Natale, M. De Rosa, K. Ernst, M. Gabrysch, K. Giulietti, L. Lorini, F. Marin, G. Modugno, F.S. Pavone, N. Pique, M. Prevedelli, M. Snels.

We want to thank Prof. M. Rosa-Clot and SIT (Science Industry and Technlogy) for stimulating help in the transfer of our knowledge to industry.

This work has been supported by ASI contract ARS96107, ECC contract ERB FMGE CT 950017 and the Fire & Gas project N. OG-0269-95.

REFERENCES

1. F.S. Pavone, "Diode lasers and their applications in spectroscopy", *Rivista del Nuovo Cimento*, **Vol 19**: 1-42(1996)
2. F.S. Pavone, M. Inguscio, "Frequency- and Wavelength-Modulation Spectroscopies: Comparison of Experimental Methods Using an AlGaAs Diode Laser", *Appl. Phys. B*, **56**:118-122 (1993)
3. M. Gabrysch, "High-Sensitivity Spectroscopy Using Semiconductor Diode Laser in the Visible and Near Infrared Spectral Region", Ph.D. Thesis, Heidelberg 1997
4. G.C. Bjorklund, "Frequency-modulation spectroscopy: a new method for measuring weak absorptions and dispersions", *Opt. Lett.* **5**, no.1: 15-17 (1980)
5. J.A. Silver, "Frequency-modulation spectroscopy for trace species detection: theory and comparison among experimental methods", *App. Opt.*, **31** no. 6: 707-717 (1992)
6. H.I. Schiff, G.I. Mackay, J. Bechara "Air Monitoring by Spectroscopic Techniques", John Wiley & Sons (1994)
7. P. Werle, F. Slemr, M. Gehrtz and C. Brauchle, "Quantum limited FM-spectroscopy with a lead salt diode laser", *Appl. Phys.* **B 49**: 99 (1989)
8. W.H. Richardson, Y. Yamamoto, "Quantum correletion between the junction-voltage fluctuation and the photon-number fluctuation in a semiconductor laser", *Phys. Rev. Lett.* **66**, 1963 (1991)
9. G.R. Janik, C.B. Carlisle, T.F. Gallagher, "Two-tone frequency-modulation spectroscopy", *J. Opt. Soc. Am.*, **B3**: 1070-1074 (1986)
10. C. Corsi, M. Gabrysch, M. Inguscio, "Detection of molecular oxygen at high temperature using a DFB-diode-laser at 761 nm", *Opt. Comm.*, **128**: 35-40 (1996)
11. C. Corsi, M. Gabrysch, F. Marin, G. Modugno "Quantum noise limited detection with semiconductor diode laser", *App. Phys. B*, special issue on "Environmental Trace Gas Detection Using Laser Spectroscopy" (1998)
12. M. Gabrysch, C. Corsi, F.S. Pavone, M. Inguscio, "Simultaneous detection of CO and CO_2 using a semiconductor DFB diode laser at 1.578 µm", *App. Phys. B*, **B65**: 75-79 (1997)
13. G. Modugno, C. Corsi, M. Gabrysch and M. Inguscio, "Detection of H_2S at the ppm level using a telecommunication diode laser", *Opt. Comm.* in press
14. C. Corsi, S. Czudzynsky, F. D'Amato, M. De Rosa, K. Ernst, M. Inguscio, "Detection of HCl on the first and second overtones using semiconductor diode lasers at 1.7 µm and 1.2 µm", in press.

OPTICAL FIBER SENSORS FOR THE NUCLEAR ENVIRONMENT

P. Ferdinand, S. Magne, O. Roy, V. Dewynter Marty, S. Rougeault, and M. Bugaud

LETI (CEA - Technologies Avancées)
DEIN/SPE – CE Saclay
91191 Gif-sur-Yvette Cedex, France

1. INTRODUCTION

Half a century after it was discovered, nuclear power produces about 17 % of the electricity in the world, 30 % in the EU, and over 75 % in France. In countries where a large part of the energy production is based on nuclear power industry (France, USA, Japan, ...), the instrumentation has ever been one of the essential aspects of the R&D in this sector due to the crucial need to improve both safety of operations and the monitoring of equipments in Nuclear Power Plants (NPP).[1] Among the physical parameters needed to be determined, temperature, pressure, strain, electromagnetic field, radiation doses and gas concentrations are the most prevalently quoted.

In the eighties, Optical Fiber Sensors based on already known but unexplored Physics laws began to appear due to the extraordinary enthusiasm shown by R&D laboratories in developing sensors and systems. These sensors can be used separately or together according to many topologies such as : serial, parallel, star, hybrid,... and so form Optical Fiber Sensor Networks (OFSN).

2. POTENTIAL NEEDS FOR OFS(N) IN NUCLEAR POWER PLANTS

2.1 The Nuclear Fuel Cycle : From Mining to Waste Conditioning

The Nuclear Fuel Cycle involves Mining, Ore Concentration and Conversion, Enrichment, Fuel Loading, Reprocessing, Recycling and Waste Conditioning. LETI-CEA Saclay is mostly involved in Mining, Waste Conditioning and Supervision of nuclear structures integrity.

2.2 Nuclear Power Plant instrumentation improvement

In a Nuclear Power Plant, although the cost devoted to the instrumentation is relatively small (< 5 %) by comparison to the total cost of the plant, the nuclear power sector is continuously looking into safety improvement of its plants and operations, decreasing the risks for the environment, reducing the waste and effluents, as well as increasing their efficiency. As nuclear facilities are sometimes difficult to access, relatively complex, corrosive or radioactive in some restrictive areas, they are more and more subject to control, monitoring and intervention most often scheduled (predictive maintenance) but sometimes in emergency. All these activities are concerned with sensing techniques and equipments for surveys, monitoring, controls and interventions.

In a NPP, problems related to "measurements" are mainly linked to ON/OFF sensors, pressure, level, temperature and flow sensors, gains of productivity, safety and maintenance could certainly be achieved with more reliable sensors and detection systems. Moreover, new functionalities given by new technologies are on the way to upgrade instrumentation performances.

In countries involved in an important Electronuclear program, formal and experimental studies have been done or are still currently under investigation to evaluate the impacts of these new technologies both for retrofitting and for designing new NPP programs (CEA-EDF-Framatome in France,[2] Electrical Power Research Institute-Tennessee Valley Authority (EPRI-TVA) in the USA[3,4,5] and the Center Research Institute of Electric Power Industry (CRIEPI) in Japan.[6] Moreover, TVA strongly insists about the necessity for field trial demonstrations.

In France, the same factors and advances in the fiber-optic technology have prompted the LETI-CEA to push its activities in this field of fully innovative instrumentation. In this framework, the LETI-CEA together with the end-user EDF (Electricité de France) and the nuclear firm FRAMATOME wanted to review possible applications for utility systems and to evaluate this new technology of instrumentation from an industrial point of view.

Thus, a task force named CORA 2000, was set-up in 1991 to initiate a collaboration action between the "nuclear energy actors" and the "fibre optic community". The primary goal was to determine which kind of OFS(N) are currently being developed or are actually available on the market, and also what are the main operating configurations for which OFS(N) can provide a substantial return-on-investment (technical and economical) either to retrofit the actual NPPs or for designing the new generation of NPPs.

The main applications selected by the CORA2000 Task Force[2] are similar to the EPRI Working Group conclusions are closed.[3] Some common orientations remain between these two studies that were started in 1991 and 1987 respectively, *i.e.* detection of abnormal conditions and safety improvement (as explained later on). Very recently, a relevant final report has been prepared by United Technology Research Center (Hartford CT-USA) on behalf of EPRI.[7] This report investigates the use of FBGs as sensor transducers elements for electric utility applications and shows the benefits of this new technology.

The applications for OFS and OFSN listed below, which are the main applications selected by the Task Force CORA 2000 and EPRI Working Group, deal with the improvement of monitoring related to preventive maintenance, *i.e.* the monitoring of «structures» and measurements devoted to increase power plant safety (Optical monitoring of structures, particularly the main containment shield, H_2 risk, Fire hazards, Nuclear power plant and disposal dosimetry, ...).

3. ANALYSIS & EXPERIMENTS OF SOME OFS & OFSN FOR NPP

The French REP nuclear containment building is the ultimate barrier. It is designed to withstand an hypothetical hydrogen explosion which could occur in case of a nuclear accident. The ultimate pressure level is about 10 bar to 12 bar for 900 MW NPP (single shield) and 7 bar to 9 bar for 1300 MW NPP (double shield). An integrity test is periodically performed on each building. Nevertheless, a networking approach would considerably enhance reliability, safety and save maintenance costs.

3.1 Nuclear shield monitoring

3.1.1 Bragg gratings for structure monitoring. Fiber Bragg gratings (FBG) are now recognized as very important optoelectronic components for guided-wave optics owing to the large number of device functions they can facilitate.[8,9] They are of widespread use either in telecommunications, in instrumentation and sensors for the measurement of strain, temperature and hydrostatic pressure as well as many other measurands *via* appropriate transducing mechanisms and have unique advantages over classical electrical strain gauges (*e.g.* in smart structures).

These advantages are at first conveyed by OFS intrinsic features such as electromagnetic interference (EMI) immunity, light weight and small size, high temperature and radiation tolerance, flexibility, stability and durability against harsh environments. FBGs have the advantage of being absolute, linear in response, interrupt-immune and of very low insertion loss so that they can be multiplexed in series along a single monomode fiber. Also, any specific network (star, series, fish-bone, ...) can be implemented and modified a long time after the setting-up, thus increasing the return on investments. As the spectral signature renders the measurement free from intensity fluctuations, it guarantees reproducible measurements despite optical losses (bending, ageing of connectors) or even under high radiation environments (darkening of fibers).[10,11]

Moreover, FBGs may be easily embedded into materials (*e.g.* composite materials) to provide local damage detection as well as internal strain field mapping with high localization, resolution in strain and large measurement range. The FBG is therefore a major component for the development of smart structure technology. It offers the promise of undertaking 'real-time' structural measurements with built-in sensor systems expected to be cost-effective when the number of sensors to be multiplexed is large.

The advantages for nuclear shield monitoring are: passive measurement (no need for electricity or energy at the measurement points), large multiplexing capabilities (and reduction of the cost per measuring point), high measurement range (larger than that of conventional extensometers) (Figure 2.).

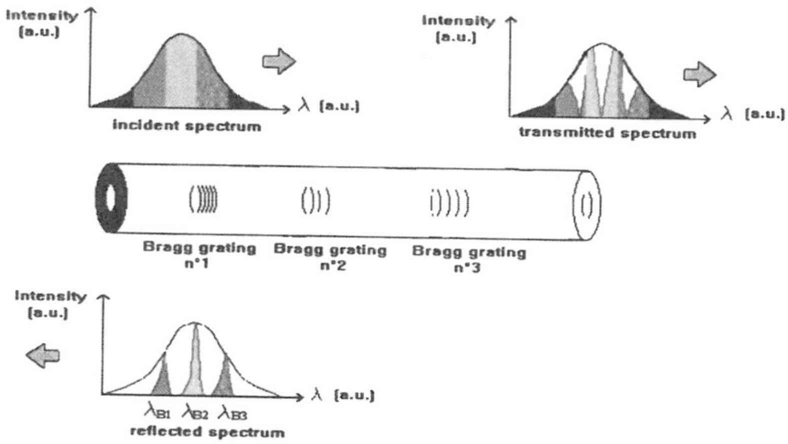

Figure 1. Bragg grating effect in a single mode fibre.

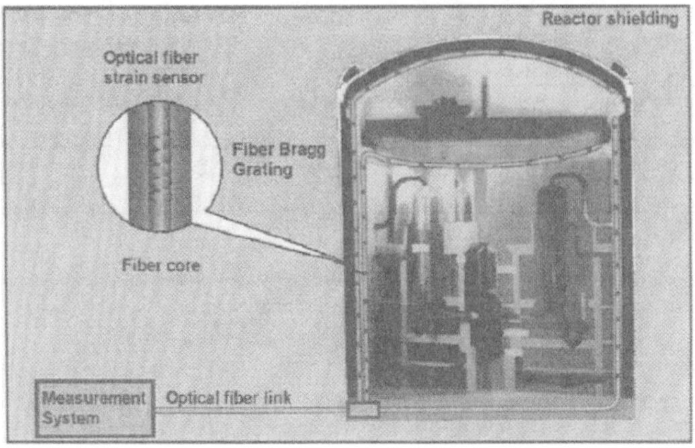

Figure 2. Containment building monitoring concept based on in-FBGs.

They are many way to achieve multiplexing and demultiplexing. The wavelength analysis for both sensing and multiplexing seems to be the simplest and the more pragmatic one, and presents also many advantages (absolute, self-referencing measurement, interrupt-immune, measurement free from intensity fluctuations, ...). This last point is crucial for NPP applications where nuclear radiations darken fibres. Moreover, remote sensing, up to kilometres, with any separation length between sensors (ranging from centimeters up to hundred meters) can be achieved.

3.1.2 Sensitivity of Bragg gratings to main physical parameters. FBGs are wavelength-selective reflectors at the Bragg wavelength : $\lambda_B = 2\ n.\Lambda$, where Λ is the grating period and n is the effective index of the propagating mode[9]. A linear response is obtained with limited change in temperature T, pressure P and strain ($\varepsilon = \Delta L/L$) : $\Delta\lambda_B = (10\ pm/K).\Delta T + (-\ 5\ pm/MPa).\ \Delta P + (1\ pm/\mu strain).\varepsilon$ at $\lambda = 1.3\ \mu m$.

3.1.3 Bragg grating behaviour under γ-ray irradiation. It is now clear that FBG are very attractive for sensing in NPP and especially for structure monitoring (*in-situ* measurements with embedded sensors for instance). Nevertheless, before proposing such a new sensing technology and related equipments to the nuclear end-users, it is of prime importance to anticipate the behaviour under irradiation not only of the optical fibres but also of the sensing elements (FBGs), both being exposed to an hostile environment during normal operations,[12] and even more in case of a nuclear accident.[13,14] During the last decade, a lot of papers have been published devoted to silica-based fibre behaviour during short or long-term steady-state γ-ray exposures at low or high dose rates.

When fibres are exposed to nuclear radiations, color center formation in the optical core and cladding severely degrades their transmission.[10,11,15,16,17,18] A very wide variation in radiation responses of fibres occurs according to their compositions and fibres with pure silica core have demonstrated the best performances (Dainichi or Heraeus Fluosil for instance). In the visible and near IR domains, the radiation-induced attenuation decreases as the wavelength increases. 1.3 µm and 1.5 µm are thus the more useful wavelengths from the standpoint of both intrinsic and radiation-induced attenuations.[10] In many cases, this latter damage is virtually permanent although a recovery is possible (mostly for pure silica core fibres) which kinetics strongly depends on temperature (several minutes at 80°C for instance).[10,11,15,16] Because of this temperature-dependent recovery phenomenon, the induced loss growth is resolved into two terms: a dose-dependent and a dose-rate dependent one. In normal operation (low dose rate), the dose rate dependent term is negligible while in nuclear accident or extreme conditions (high dose rate), both dose and dose-rate dependent terms are of importance, the first one being predominant at temperatures higher than about 100°C.[16]

It is now well established that the best results are obtained with pure silica-core fibres, especially step-index polymer-clad fibres transmitting 0.85 μm, graded-index fluorine-clad multimode fibres and pure silica single-mode fibres (both under normal and accidental operations) at 1.3 μm.[18] Unfortunately, up to now, grating growth has not been reported in fibres with pure silica core. However, photosensitivity has been demonstrated in a several silica-based fibres doped with germanium,[9,19] europium[20] or cerium.[21,22] Various considerations show that the germanium-doped core fibres are, at this time, the most interesting photosensitive fibres.[23] Indeed, massive amounts of these fibres are produced at low cost for telecommunications. Writing a grating within these fibers does not induce significant excess loss at wavelengths of interest for optical sensor applications. Moreover, numerous methods of photosensitization of germanium-doped core fibres have been reported.[24] For example, hydrogen loading is able to induce large change in refractive index even in fibres which otherwise show poor photosensitivity.[25] Furthermore, long life span of gratings written in germanosilicate fibres has been reported by many groups.[24,26] However, although researches on radiation response of optical fibres doped with germanium have been undertaken for a long time, little was known about the effects of ionising radiations on gratings written within these fibers.

The behaviour of FBG written in different kind of Ge-doped silica fiber has been tested in our laboratory under low and high γ dose rate, in order to check the independence of its sensitivity. Moreover, thermal and strain sensitivities of five silica fibers of different composition have also been determined to test their reliability under nuclear environments.

Our results prove that sensitivities and spectral behaviour of FBG are not affected by γ-ray.[27,28] This means that a FBG-based instrumentation can be used without any spectral recalibration, within an hostile nuclear environment; for instance in a NPP during normal operations until installation dismantling, and also in case of a nuclear accident.

3.1.4 Bragg grating extensometer' experiments with concrete. The main aspects considered for the development of this extensometer are the fidelity to the strain transduction and its robustness, specially for rough civil engineering handling. The sensor is composed of a central metallic rod used for transduction. The two ends are used to anchor the sensor to the concrete surface or when embedded. The extensometers were designed with a 10 and 20 cm base length.

3.1.4.1 Experimental results. Real time strain acquisitions concerns both surface and *in situ* measurements. Two kinds of concrete samples, cylindrical for compressive tests or flooring for tensile measurements and crack detection were used.

a) Experiments on concrete surface: cylindrical samples (25 cm diameter and 50 cm high)

The characteristics of the concrete sample were determined 29 days after fabrication to be 49.6 MPa for the compression strength, 4.14 MPa for the traction strength and 37.9 GPa for the Young modulus. Each cylinder is equipped with 3 different kinds of sensors: 2 FBG extensometers, 2 resistive strain gauges, and 2 inductive sensors (C.A.D.I.), as depicted on Figure 3. The concrete sample is placed in a press (accuracy ± 1.5 % of the applied force) and submitted to a charge in compression from 0 N to 1200 kN by step of 100 kN.

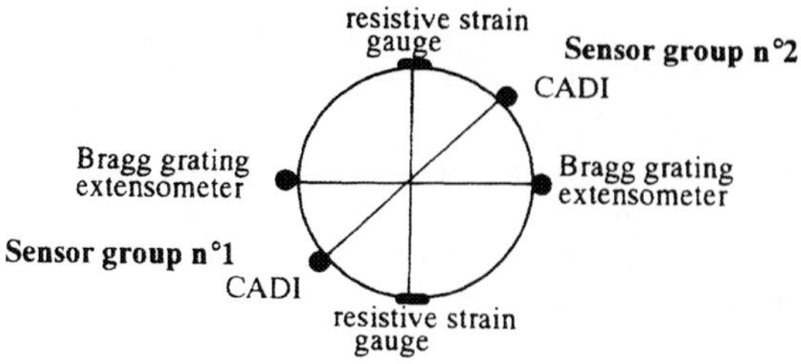

Figure 3. Sensors around the concrete cylinder.

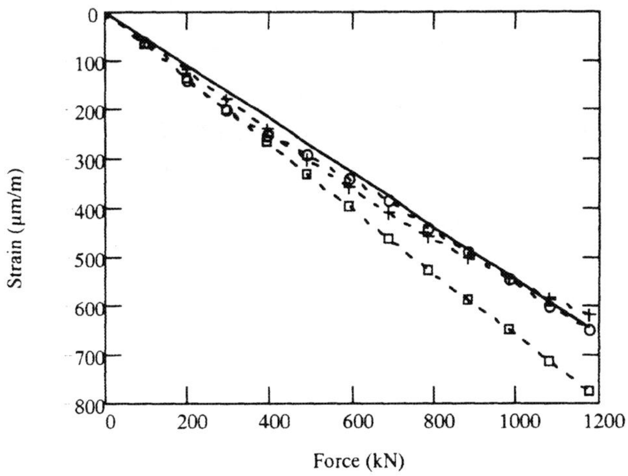

Figure 4. Strain for the 3 sensors of group n°1.

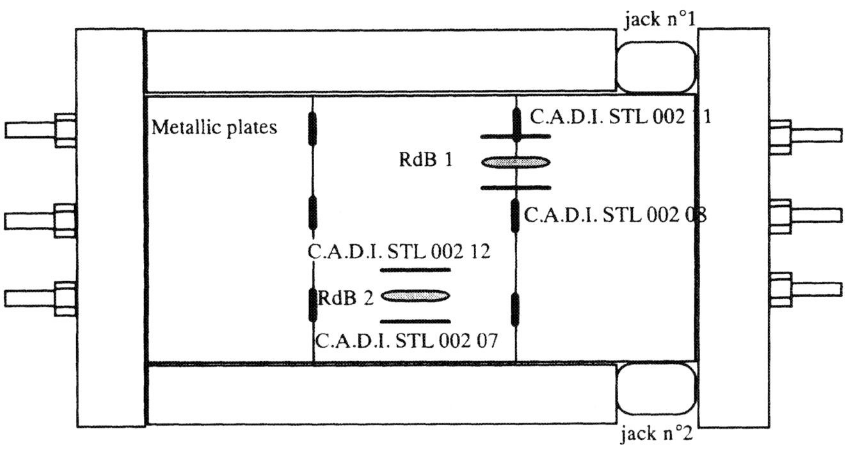

Figure 5. Sensors, and metallic plates in the flooring.

A typical result of a compression test for one of the sensors is shown on Figure 4. We could note the good linearity of the BG extensometers response in compression up to - 700 µstrain. A second type of experiment consists in repetitive charge cycles from 0 to 1000 kN with a fast return to zero. Lack of hysteresis has been checked (within 1 % FS) for both sensors and a good repeatability of the measurements under load is obtained.

b) Experiments on concrete surface: flooring (1.5 m length, 1 m wide and 0.26 m high)

One flooring was fabricated in concrete (compressive strength after 29 days was 37.5 MPa and Young modulus was 70.6 GPa). During the fabrication, some metallic plates were included in the concrete so as to initiate preliminary cracks (Figure 5.).

One FBG extensometer and two inductive extensometers (C.A.D.I.) are positioned above the plates and a second sensor group is placed above a free metallic plate area. The flooring is installed in a metallic frame stressed by two hydraulic jacks.

The traditional extensometer (CADI 002 11) actually measures higher strains than the other two sensors do, because the crack begins at the flooring periphery (time '105 mn' on Figure 6.). When the load increases, the crack widens. At the end of experiment (F = 1 MN), the strain reaches 2500 µstrain which represents a crack of 500 µm wide. On Figure 7., a second crack event appears at the time '180 mn', which induces a slope decrease of the strain (typically 400 µm wide).

c) Experiments in situ: cylindrical samples

Each concrete sample (compressive strength : 56.6 MPa, traction strength: 4.53 MPa and Young modulus : 35.5 GPa) contains two embedded extensometers (one FBG and one Telemac vibrating wire, C110 or F2), and three inductive sensors fixed outside at 120° with respect to each other.

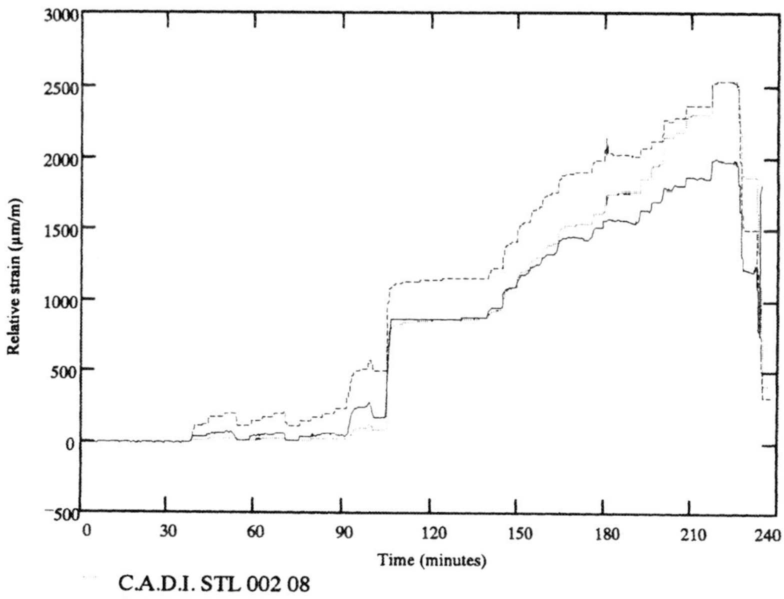

Figure 6. Kinetics of sensor response (metallic plate zone).

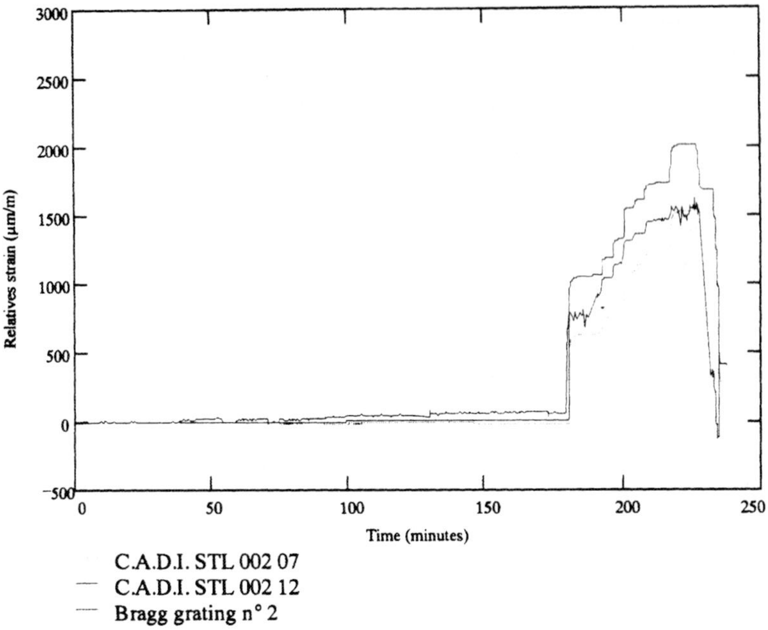

Figure 7. Kinetics of sensor response above a free area.

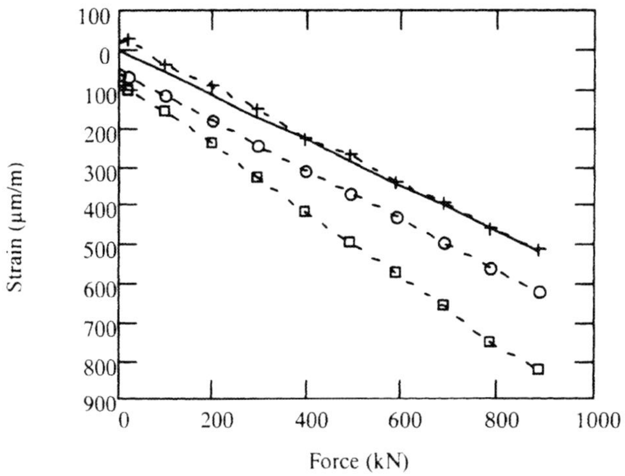

Figure 8. The 3 different sensors' responses (2nd cycle).

All of the concrete samples are submitted to several compression tests. A typical experimental result obtained with the FBG extensometer (the second of three cycles from 0 N to 1000 kN by step of 100 kN) is shown on Figure 8.

Since the first cycle, the embedded FBG presents a good linearity < ± 2 % FS, nevertheless the best fit to the theoretical curve appears for the 2nd or 3rd cycle.

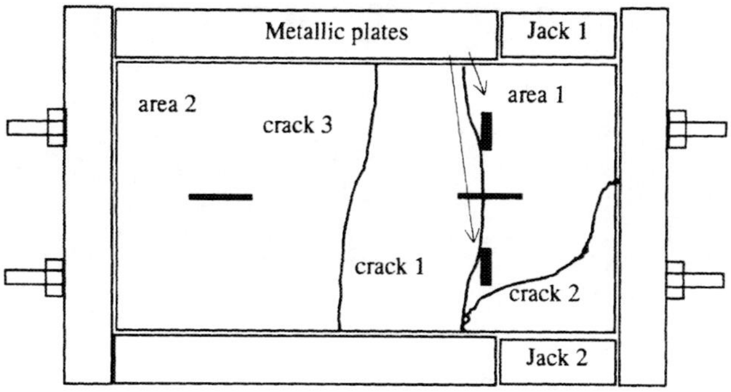

Figure 9. Flooring scheme (top view).

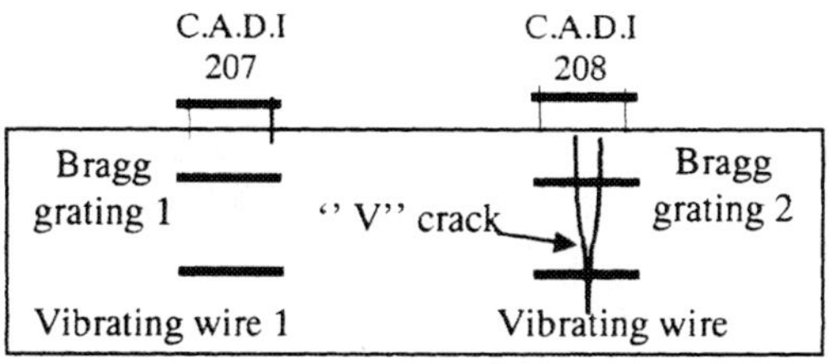

Figure 10. Flooring scheme (side view).

It is interesting to note that the first return-to-zero shows a residual offset for all of the sensors. These offsets disappear since the 2nd cycle.

As previously stated, during rapid repetitive forces (from 0 N to 1000 kN), the embedded sensors show a very good return-to-zero with an error of 5 μstrain and a reproducibility of the measurements of 4 μstrain at 1000 kN.

d) Experiments in situ: flooring

A second rectangular concrete flooring (compressive resistance of 42.5 MPa and Young modulus of 35.5 MPa) is used for embedding extensometers: two Bragg gratings and two vibrating wire extensometers are embedded. Moreover, two inductive sensors are fixed on the surface (Figure 9. and 10.).

At a load of 350 kN, two cracks (n°1 and n°2) appear suddenly at the same time, one of them in the middle of the extensometers (Figure 9. and 10.). The first one is immediately detected by all the sensors of area 1 (Figure 11. and 12.): the C.A.D.I measures 423 μstrain, the FBG 228 μstrain (45.6 μm wide) and the wire 52 μstrain.

At the same time, a strain relaxation (compression) appears around the free zone (crack 3 on Figure 9.) and is detected by the others three sensors. We have observed that the crack propagates (as for surface experiments) from the edge to the center of the flooring. A third crack appears few tens of seconds later in the middle of the flooring which induces a strain relaxation in the area 1 (Figure 11.). After 600 kN (at time labelled ~ '100 mn') the applied force decreases by step of 100 kN to 0 N (Figure 11. and 12.).

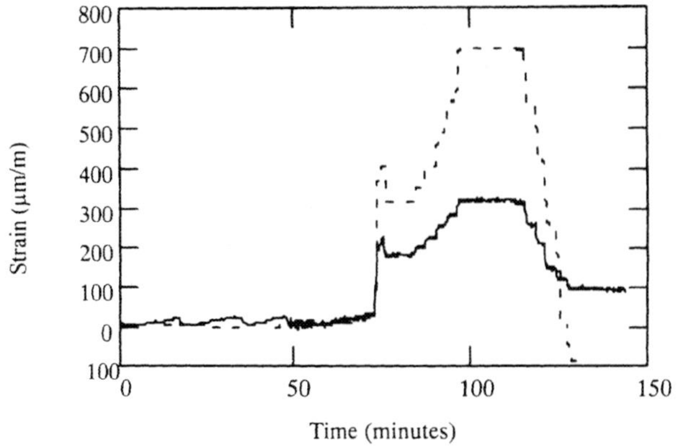

Figure 11. Sensors' responses above the metallic plate zone.

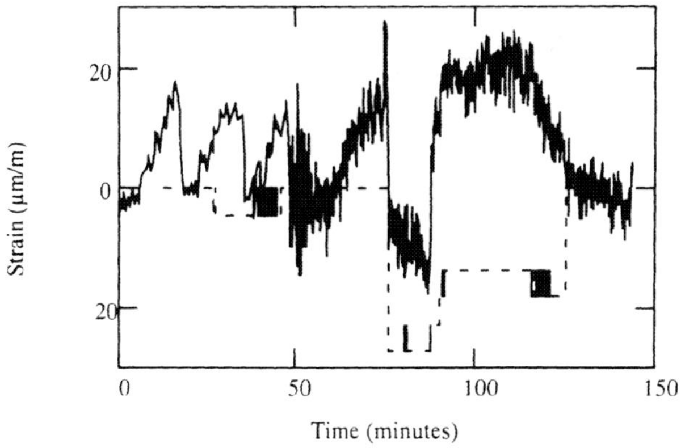

Figure 12. Sensors' responses above the free zone.

3.1.4.2 Conclusion. The FBG extensometer, specially designed for civil engineering (ruggedized and waterproof), convenient for surface strain measurements on buildings as well as for *in situ* measurements, is able to measure both traction and compression in the range of ± 2500 μstrain.

When connected to our system, global specifications are : measurement range: - 2500 μstrain to + 250 μstrain (up to + 2500 μstrain for cracks), low strain detection threshold: 2 μstrain, repeatability : 0.2 % FS, linearity: ± 2% FS, instrumentation repeatability: 0.17 % FS @ 0.5 Hz bandwidth.

Strain monitoring applied to concrete part of a nuclear building includes some specificities of the nuclear industries. Nevertheless, it seems obvious that this kind of approach could be applied to many "structures" in the power plant itself (concrete support for the generator, cooling tower, ...), in the civil engineering and public works domains (dams, mines, underground constructions, bridges, tunnels, skyscrapers, ...), and everywhere else where structure failures might jeopardize safety of people or ecology (plane, train, ship, off shore, pipe line, ...).

3.2 Hydrogen risk

3.2.1 Motivations for safety. In case of a nuclear accident, oxidation of zirconium sheaths which protect any nuclear fuel rods releases hydrogen according to the chemical reaction : $Zr + 2H_2O \rightarrow ZrO_2 + 2H_2$. Moreover, in case of LOCA (Loss of Coolant Accident), water steam is still produced in the nuclear building. Detonation or deflagration may occur depending on the ternary mixture "H_2 - H_2O - air" (Cf. the Three Mile Island accident in Pennsylvania dated March 28, 1979 and the more recent Tchernobyl accident in Ukraine dated April 26, 1986). The Shapiro's diagram (Figure 13.) can be split in three zones, a safe zone (no risk), a deflagration zone (flame) and a detonation zone (explosion). Deflagration occurs for a concentration of 4 % hydrogen in dry air. The aim of the H_2 risk assessment is therefore to provide reactor supervisors with an instrumentation able to locate in the Shapiro's diagram at any area in the containment building by measuring water vapor, O_2 and N_2 concentrations as well as the temperature and pressure.

3.2.2 "H_2 risk" monitoring system specifications. Post-Accidental (PA) specifications have been evaluated by the CEA-DRN as follows: Pressure up to 0.7 MPa or 0.9 MPa, a peak temperature of 170 °C (no vessel failure) and saturated water vapor atmosphere. Hydrogen must be detected in the range: 0.5 % to 20 % with a typical 0.5 % resolution. The response time of the detection system should be of the same order than that of the hydrogen production rate (related to the number of oxidized zirconium sheaths) and also compatible with the delay for action , *i.e.* about one minute. Finally, the measurement system must be able to stay in "stand-by" during an entire NPP lifetime (up to 40 years) and to be operational at any time in the above hostile conditions, may be without any power supply available inside the containment shield.

Taking into account these conditions, the remote distributed measurement concept (sensors located in the containment area while the system is outside) is an interesting and elegant possibility given by optical fibers techniques. Fibers used are pure silica core fibers that withstand the total γ dose of nearly 100 Mrad (normal dose rate x 40 years added with the dose arising from a hypothetical nuclear accident) and the detection method free of intensity level variations.

3.2.3 System development and experiments. The advocated technique is Raman Spectroscopy. Previous devices have been already realized in the past (*e.g.* a bifurcated fiber bundle for *in-situ* remote chemical analysis[29,30]). The so-called RLFO technique (Raman Laser Fiber Optics)[31] has proved to be efficient in designing versatile, efficient, multipoint, remote, *in-situ* chemical sensors.

Another device based on spontaneous Raman scattering has been designed for in-line sensing of H_2 in a mobile launch platform of NASA[32] (high-vibration environment) and makes use of a multiple-pass cell. If most complex polyatomic molecules are often both Raman and Infrared active, diatomic molecules (H_2, N_2, O_2,..) are infrared inactive[33]. Infrared absorption techniques cannot be used to detect such molecules and only Raman Spectroscopy technique can be. Since fibres are not transparent in U.V., fluorescence techniques are merely not achievable.

The principle of Raman effect is as follows : when light is incident on transparent solid, liquid or gas, most of it is transmitted without change, but, in addition, some scattering of the radiation occurs affecting a tyre-like dipole radiation pattern. In the spectrum of the scattered light, there will be observed to be present not only the incident radiation wavelengths (Rayleigh scattering) but also two associated different Raman bands. Raman band at wavelength greater than the incident wavelengths (*i.e.*, lower energies) is called Stokes band and Raman band at wavelength less than the incident wavelength (*i.e.*, higher energies) is called anti-Stokes band (Figure 14.). Stokes shifts of different molecules are detailed in Table 1. As this Stokes shift increases, the anti-Stokes bands is much lower in intensity than the Stokes one so that Stokes bands are the most quantified in Raman spectroscopy of gas molecules.

3.2.4 Experimental. Since absence of interference between species and detection limits are the most relevant factors for the development of such sensor, a low-cost, integrated spectrometer of high throughput has been chosen for demonstration. The typical spectral resolution of this Raman set-up is around 2 nm. As can be seen in Raman spectra, a higher spectral resolution is unnecessary for such application since no interference occurs between corresponding Raman bands. The experimental set-up is depicted in Figure 15.

An argon ion laser (single line at 514.5 nm) has been used for demonstration. Incident laser light is focused into the sample cell so as to excite a cylindrical volume.

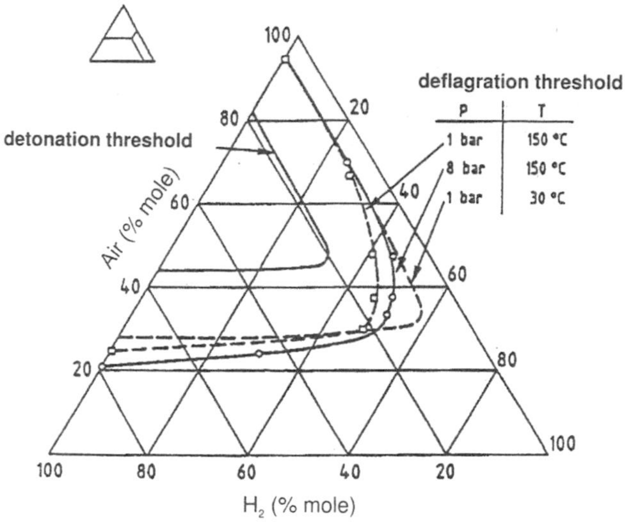

Figure 13. Shapiro's diagram (H_2 - H_2O - air mixture).

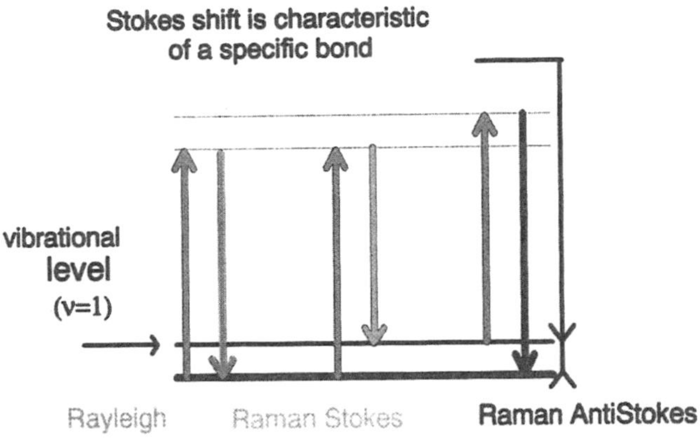

Figure 14. Basic interaction schemes of Rayleigh and Raman scattering.

Table 1. Raman Stokes shifts of many gases of interest for P.A. application; corresponding relative cross-sections relative to that of N_2.

molecule	$\Delta\sigma$ (cm^{-1})	$\Delta\lambda$ (nm) [excitation 514.5 nm]	λ (nm)	$\frac{\sigma}{\sigma_{N_2}}$
H_2	4161	140.2	654.7	2.4
N_2	2331	70.1	584.6	1
O_2	1556	44.8	559.3	1.3
H_20	3652	119	633.5	3.8
CO_2	1388	39.6	554.1	1.4
CO	2145	63.8	578.3	1

Scattered light is simultaneously observed in a direction orthogonal to the polarisation of incident light and propagation vector and is collected by two doublets and a spherical mirror.

For the purpose of demonstration, an hydrogen container and a water flask was connected to the cell *via* a by-pass and a primary pump. Collected scattered light is focused on an Oriel 77400 imaging spectrometer equipped with a cooled CCD. At - 10 °C, saturation occurs at 65536 counts after about 10 minutes (maximum exposure time). High-pass filters are used to attenuate stray light induced by Rayleigh scattering (Schott OG 530).

We obtained Raman spectra of hydrogen at 760 torr and 50 torr (Figure 16.), of air (21 % oxygen, 78 % nitrogen) (Figure 17.). At room temperature, water vapor pressure is 17 torr (22 mBar). The Raman spectra of air has been recorded after desorption of hydrogen.

3.2.5 Conclusion. Raman Spectroscopy-based chemical sensors are inherently selective and versatile. They fully meet the challenge of nuclear Post-Accident supervision because they totally comply with the PA specifications. Minimum concentrations of 10 to 100 ppm can be detected with corresponding exposure times ranging from tens of seconds to ten minutes.

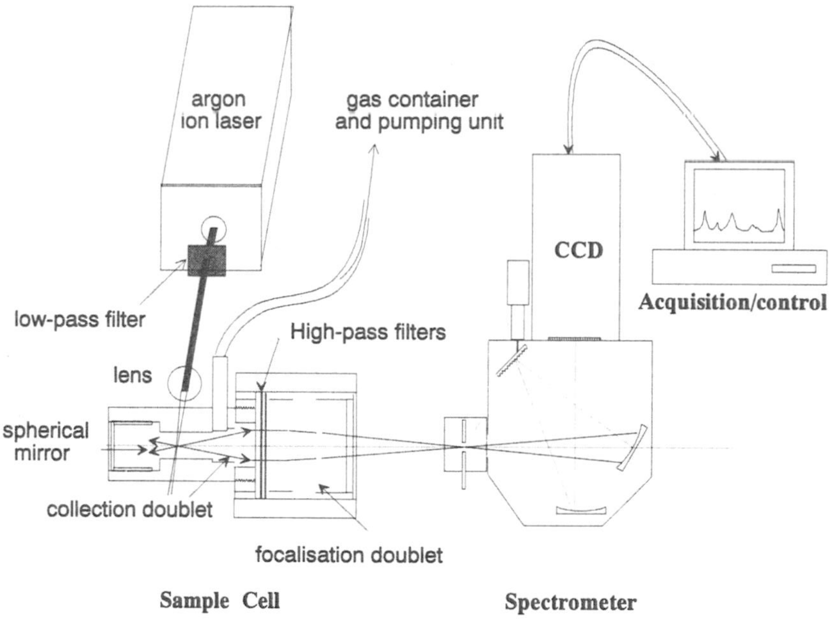

Figure 15. Raman experimental set-up.

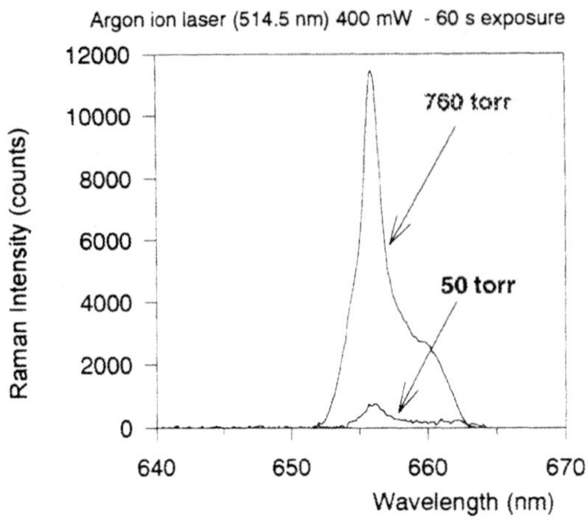

Figure 16. Raman spectrum of hydrogen.

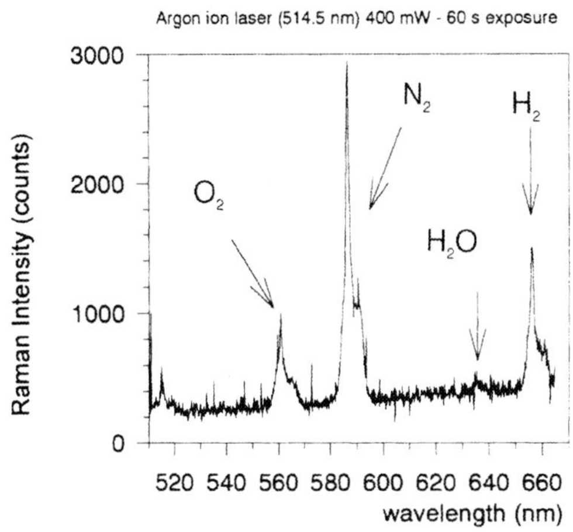

Figure 17. Raman spectrum of air and hydrogen.

An important feature of such system is the absence of calibration since a reference gas such as nitrogen can be used for self-referencing quantitative measurements. An integrated solid-state laser (for instance, a diode-pumped doubled YAG laser or CW diode laser bar) may advantageously be used instead of the argon ion laser used for demonstration purpose.

The use of an imaging spectrometer enables to achieve remote, multi-point and multi-gas monitoring with a unique acquisition unit (*multi-track technique*). Subsequently, the cost per measurement point and gas to be identified is dramatically lowered (about 2000 US \$ to about 4000 US \$ per cell performing real-time and simultaneous measurement of several gases).

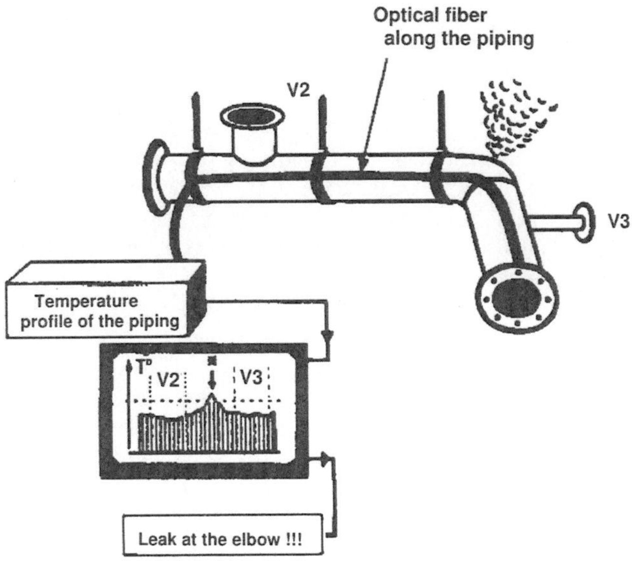

Figure 18. Steam pipe monitoring with (quasi-) distributed optical fiber sensing measurement.

This is also a very interesting feature of Raman Spectroscopy based sensors compared to other sensors (conductance, catalytic, ...) that are not amenable to remote, multiplexed and versatile measurements, such sensors can be easily reconfigured when needed (change in the gases to be detected). Sample cells may be rugged and cost-effective in a large scale manufacturing process. In an industrial context, a background subtraction signal analysis can be performed.

3.3 Steam pipe monitoring

In a PWR power plant, the primary circuit (including the reactor vessel and three or four loops) is not integrated in an unique and specified shield. So, breaks could occur between elements (vessel, pump, pressurizer, steam generator). The reasons for the potential risk are: a) circuit complexity (geometry), b) large number of solder joints ageing under thermomechanical cycling, c) relative vessel-steam generator movements in case of earthquakes,...

Nevertheless, safety factors of pipe materials are very large and safety rules take into account a complete breakage of a pipe (*i.e.* the LOCA event). So, an early detection of leakage is an important issue for safety. Early location of small cracks on steam pipes (primary and secondary circuits) is of prime importance to avoid any fissuration or even wide opening or pipe breaks enabling an emergency situation in the plant. Improved methods of measuring strain on high pressure components, piping (and welds) could thus be of great benefits in assessing remaining life (small leaks of borated water could cause faster corrosion) and detecting incipient failure. A steam or over-heated water leak induces an increase of temperature of the pipe insulation. For instance, experiments realised by the CEA have shown a steam panache and an increase in the range 40°C to 100°C under the thermal insulation in case of 100 liters/h water leak from a pressurised water pipe in normal REP conditions (155 bar, 325°C).[12] The rise time of the phenomena is quite short and an early detection is strongly linked to the sensor location (1 s on the top of the pipe; 30 s on the pipe bottom). Moreover, in case of large water leakage (230 l/h), the increase reaches 100°C both for the primary and the secondary water circuits outside the thermal insulation, while 140°C could be reached for steam in the secondary circuit.

Obviously, classical approaches exist to detect such kind of leak (global water balance, dose level measurement around the pipe (only for primary circuit), ...). Nevertheless, local detection seems to be well suited for redundancy by experts. So, there are several potential methods with specific advantages and drawbacks (strain gauges, thermosensitive cable, IR thermography associated with image processing, humidity distributed fiber sensor, ...).

Humidity distribution sensing fiber -for instance- based on hydrogel is a new concept[34] and experiments in hostile environments will have to be done and the difficult problem of polymer life time under high temperature to be assessed. Electrical resistance devices are inexpensive, but these devices tend to fail in under high temperature and ageing affects linearity. Moreover, strain gauges and thermo-cables need current supplies which could be limited in case of accident. IR camera is a global non-contact method which fails to detect hot spots in presence of a large amount of steam and these devices need to be frequently recalibrated to quantify the leak rate (emissivity problem). In case of an IR detection an active method with a modulated photo-thermic effect could however overcomes steam screen but the methodology is more complex. On the other hand, fibers wound around or fixed along large pipes and connected to a sensor system are able to detect strains and/or rises in temperature (hot spots) due to leaks of pressurised water or steam, and to locate them remotely (Figure 18.).

Mainly, two techniques could be used for this purpose. The first one is based on Optical Time Domain Reflectometry (OTDR) enabling distributed measurement and the second is based on FBGs enabling quasi-distributed measurement.

3.3.1 OTDR Method. OTDR method consists in an intensity measurement performed in the time domain of the light backscattered by the fiber material (silica). Two technical ways are possible, the first consists in observing Raman scattering whereas the second consists in observing Brillouin scattering.

- Raman scattering is based on Stokes and anti-Stokes band detection. The intensity of anti-Stokes line is affected by temperature while Stokes line intensity remains constant. This enables an absolute and selective measurement through the ratio of these two signals. Intensive development efforts have been done in the last decade and some products are now commercially available (resolution ~ 1 °C; spatial localisation ~ 1 m).

Brillouin scattering effect: the spectral positions of the Brillouin lines are differentially affected by temperature and strain, enabling the determination of these two parameters. Nevertheless, this elegant method called B-OTDR is less mature than the Raman-OTDR and progress will have to be done before any industrialisation.

3.3.2 In-Fiber Bragg Grating technology. The second approach uses small optical Bragg gratings written in the fiber core itself at any specific location. As for concrete monitoring, the wavelength of the backreflected energy (Bragg wavelength) of any temperature sensor (packaged to be free of strain) or strain sensor (associated with an unstrained grating for thermal compensation) is directly related to the needed parameter. Up to now, several ways have been investigated to multiplex these Bragg transducers. The CEA R&D prototype is now in operating condition[35,36] and products able to multiplex a large number of sensors will be commercially available within the next 2 years.

The main difference between these techniques is due to the distributed (OTDR) or quasi-distributed (FBG) aspects of the measurements. At first glance, a distributed measurement seems well suited to detect an unknown location event. That means a continuous monitoring with three or four fibers (120° or 90° oriented) of any pipes. Moreover, it may be possible to improve spatial resolution in some local areas by helically winding the fiber. Nevertheless, the amount of data rapidly increases in « real-time » monitoring therefore leading to a quick overflow of data. On the other hand, one could be pragmatic and choose a simpler approach (mainly from the installation point of view) able to detect 95 % of the potential events, because preliminary analysis can identify locations where stress concentrations may be expected to occur. So, a quasi-distributed measurement with sensors localised on specific zones (elbows, soldering, valves, ...) could be a good approach especially for revamping. Obviously, fibers allow a remote monitoring from the piping.

Both approaches (Raman OTDR and FBG) offer high intrinsic safety. Moreover, the remote optoelectronic systems could be time-shared between several pipes and locations, with an usual electronically driven optical switch.

3.4 Nuclear radiation detection

In nuclear industry, maybe the most important point is the protection of workers and environment. That's why, in the frame of the ALARA concept (As Low [human dose] As Reasonably Achievable) and in order to improve the dosimetry monitoring of workers and also for remote dosimetry within nuclear installations, we develop in our laboratory, specialist in nuclear sensors, a new approach based on optical fiber and luminescent materials. The system is based on the phenomenon called Optically Stimulated Luminescence (OSL).

As OSL is closely related to thermoluminescence, every OSL material is thermoluminescent; though the opposite is not true. OSL material as TLD can trap electrons on stable levels. But instead of the former, where they are released by heating the material, in OSL materials they are released by light. The OSL process can be described using as a band diagram. In the forbidden band they possess trap levels and luminescent centers. So, OSL material are able to trap electrons created by several kind of irradiation (UV, X, γ, ...). The number of trapped electrons correspond to the «data stored» which is the energy left by irradiation. Following this step, the retrieval of data can be obtained by stimulating these electrons with light. The released electrons produce a luminescence which is proportional to the data left by irradiation enabling the dose measurement.

The OSL phenomenon offers the same advantages as TLD (long data storage time, good reproducibility, large dynamic range, low level of dose achievable, ...), and the interesting possibility of a remote optical stimulation instead of post-heating.

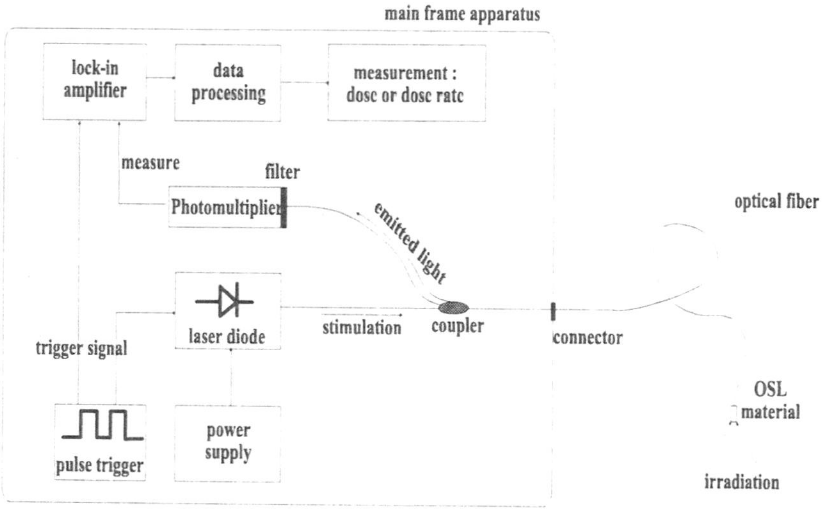

Figure 19. OFS based on OSL material for dosimetry.

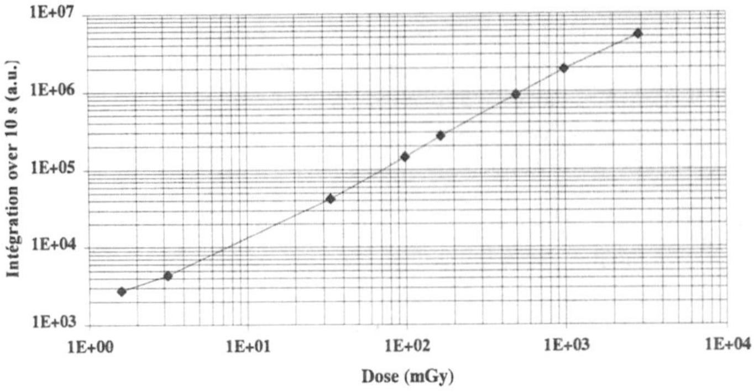

Figure 20. Integration of OFS signal versus dose for MgS:Sm.

That's why, in connection with optical fiber, it can be used for dosimetry of workers as well as for nuclear plants or waste disposal area monitoring. For dosimetry purposes, the main way of use of this phenomenon consists to connect (or glue) a small quantity of OSL material at the end of an optical fiber.[37] A prototype of such an Optical Fiber Sensor dosimeter is currently under test in our laboratory (Figure 19.).

This prototype (Figure 19.) is able to measure both dose and dose rate. By fast interrogation rate it will provide a real-time measurement very useful for workers under radiation fields or for remote dosimetry. It can also multiplex sensors. In case of electronic failure, the remote mainframe apparatus can be repaired or changed without any loss of information stored in the passive sensors. This new kind of optical fiber sensor can certainly improve the management of radiation protection for workers or nuclear plants by its real-time display of dose or dose rate. Moreover, OSL phenomenon is of great interest for dosimetry but also for other fields of research like medical imaging or optical memory.

Preliminary result show a minimum of detection of 1 mGy, a dynamic range exceeding 5 decades and cadence of interrogation of 10 seconds (Figure 20.). The perspective at short term is the improvement of optical and electronic noise reduction to lower the minimum of detection as well as reduce the interrogation time mainly due to a long erasing time.

3.5 Waste conditioning and disposal monitoring

96 % of uranium and 1 % plutonium left after operation is recycled and used again to make more nuclear fuel. Only 3 % of the fuel becomes waste after reprocessing. There are three kinds of radioactive wastes. Each is dealt with according to how radioactive it is - low, intermediate and high -. High level waste is radioactive enough to generate heat. This waste is turned into very dense solid glass blocks, vitrified and poured into stainless steel canisters. At low activity, waste is readily stored in protected surface sites for acceptable short periods, high activity waste must be kept over long terms by underground storage. In France, a law passed on 30 Dec. 1991 sets out the lines of research which, 15 years on, will enable the parliament to select the most appropriate solutions to the problem of long-lived, highly radioactive wastes. As a result of concertation programme between local authorities, led by the French MP Christian Bataille, several sites where selected at the beginning of 1994 for future underground laboratories.

Several solutions are currently under development for long-term management of long-lived waste. They could be implemented on an industrial scale within fifteen years. Figure 21. shows a cross-section of an underground repository as conceived by ANDRA, the French National

Radioactive Waste Management Agency which was created as part of the French Atomic Commission in 1979. The waste act of Dec. 1991 turned ANDRA into an independent public service company. Such future concept could consist in a network of disposal galleries interconnected *via* main galleries with a main access shaft to ground surface.

In the same way, many European repository concepts have arisen. For example, the Belgian concept initiated by SCK-CEN Mol Research Center.[38] A variety of tests involving multiple disciplines will be performed in the underground laboratories. The tests are scientific and technological in nature, and may take an «X-ray» of the sub-surface from the surface or probe the environment from the underground laboratory. Obviously, the tests must be appropriate to the nature of the site and environment that constitutes the host medium. Current research activities are led in the underground laboratory HADES.

Current research involves monitoring of pressure, temperature and dose in the local surroundings of canisters. In Germany, experiments will be under investigation in a near future in the ASSE and GORLEBEN salt mines, both with traditional and with Optical Fiber Sensing of pressure, temperature and strains. Built-in, disposable OFSs could provide the necessary information for inspection and quality insurance and therefore anticipate a potential decommissioning. Bragg grating-based sensors can ensure pressure and temperature measurements. Conversely, remote fibers may help determining distributed temperature and irradiation dose by Raman Spectrometry and OTDR of radiation-induced attenuation[39] respectively.

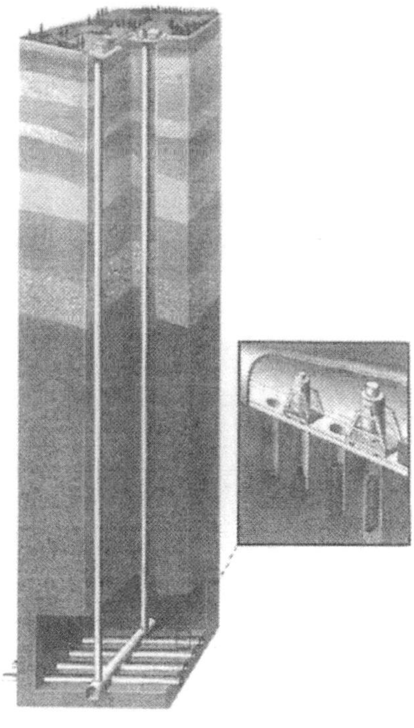

Figure 21. French concept of underground waste storage site [concept from ANDRA].

4. CONCLUSION

In a nuclear power plant, it is obviously of prime importance to avoid any failure in safety-critical components and nuclear safety can never be taken for granted and requires permanent questioning in order to determine the required actions for progress to be made. In this paper, we have reviewed exciting opportunities to improve both safety and monitoring in the nuclear industry with Optical Fiber Sensing Technology from mining to waste conditioning. US and French enquiries prove that a lot of opportunities already exist for the industrial application of Optical Fiber Sensors and related Networks within this industry and some of them are very attractive for the next generation of Nuclear Power Plants which is underway. Clearly, for Optical Fiber Sensors, two markets are emerging for the future: one involves the construction of new generation of power plants and the other deals with the renovation and maintenance of existing installations as inevitable ageing of the installed systems causing operators to renovate all or part of their supervision and control system, while increasingly seeking for accurate non-destructive control methods. Moreover, the halt in plant construction, outside Asia, pushed the nuclear power actors towards maintenance.

Several applications have been reviewed and discussed. Some of them are devoted to safety improvement (nuclear building and primary steam pipe monitoring, dosimetry, ...). Others are devoted to improve maintenance of devices and structures. The remaining are devoted to metrological purposes (current measurement *via* Faraday effect, ...). Some of the French experiments have been described in this document.

Some sensor technologies are ready for experiments and testing while others require further developments and long term R&D. As pointed out in the EPRI mentioned report: *«testing the existing options is the best method of understanding fiber-optic capabilities».*

We have demonstrated the configurations for which the use of OFS(N)(s) gives clear industrial advantages and shown that main applications are related to "structure" monitoring and safety measurements. Moreover, the good adequation of the very promising field of applications of Bragg gratings, with the requirements of some of these applications has been shown. The use of this new kind of transducers is extremely interesting because it allows to perform a lot of measurements, both in terms of sensing (temperature, strain, pressure,...) and in terms of network topology, due to an intrinsic spectral encoding. As a consequence, the Bragg grating approach looks very attractive, from the end user-point of view, because a part of the R&D costs could be shared between some different applications, including "nuclear" ones due to their good behaviour under gamma-ray exposure.

ACKNOWLEDGEMENTS. The authors wish to thank their partners from EDF and FRAMATOME for helpful discussions during the CORA 2000 workshops, from COGEMA and from ANDRA for their help and all of them for financial support.

REFERENCES

1. P. Ferdinand, Y. Denayrolles et al., The potential for distributed sensors and optical fibre sensor networks in the electric power industry, Meas. Sci. Technol., Vol. 1, 1990, pp. 908-916
2. F. Morange, P. Ferdinand, B. Jarret et al., Capteurs à Fibres Optiques et réseaux associés. Etat de l'art - tendances du marché - Propositions d'actions (Optical Fibre Sensors and related Networks. State of the art - Market trends - development proposals), Technical Report, Ref. EDF/Dir. de l'Equipement/SEPTEN/EL/92-94 (in French)
3. Research Planning Study of Fiber-Optic Sensors, Electrical Power Research Institut report, EPRI ER-6428, Project 8004-1, Final report August 1989

4. Fiber-Optic Sensing Workshop, Electrical Power Research Institut report, EPRI ER-6537, Project 8000-34, Oct. 1989
5. Research & Development Program 1991-1993, Electrical Power Research Institute report
6. Long-term Research and Development Program Plan 1991-2000, CRIEPI Report, Oct.1990
7. W.W. Morey et al., United Technology Research Center, EPRI Final Report TR-105190, September 1995
8. G. Meltz, W.W. Morey, W.H. Glenn, J.D. Farina, In fiber Bragg grating sensors, OFS-88, 27-29th January, New Orleans, USA, pp. 163-166
9. G. Meltz, W.W. Morey and W.H. Glenn, Formation of Bragg grating in optical fibers by a transverse holographic method, Opt. Lett., Vol. 14, 1989, pp. 823-825
10. E.J. Friebele, C.G. Askins, M.E. Gingerich and K.L. Long, Optical fiber waveguides in radiation environments, II Nucl. Inst. Meth. Phys. Res. B1, 1984, pp. 355-369
11. E.J. Friebele, M.A. Putnam, C.G. Askins and M.E. Gingerich, The effect of single mode fiber cladding composition on low dose rate radiations response, ECOC - IOOC '91, pp. 337-340
12. G. Champion, Y. Mallet, A. Dubail and M. Mourgues, Exposure doses inside French PWR containments under normal operating conditions, OPERA'86, Albuquerque, USA, Oct. 1986, pp. 503-508
13. F. Bouquet and J. Winslow, Radiation data for design and qualification of nuclear plant equipment, EPRI report NP-4172 M, August 1985
14. R.T. Johnson Jr, F.V. Thome and D.M. Craft, A survey of aging of electronics with application to nuclear power plant Instrumentation, IEEE Trans. on Nuclear Science, Vol. NS-30, N°6, 1983, pp. 4358-62
15. C.E. Barnes and J.J. Wiczer, Fiber optic data links optimized for radiation environments, IOOC - ECOC' 85, pp. 707-710
16. M. Kyoto, et al., Gamma-ray radiation hardened properties of pure silica core single-mode fiber and its data links system in radioactive environment, J. of Lightwave Technol., Vol. 10, n° 3, 1992, pp. 289-294
17. M. Watanabe, K. Matsui, H. Yokota, H. Kanamori, G. Tanaka and J. Isoya, The reduction of the radiation induced loss increase of pure silica core single mode fiber, IOOC - ECOC ' 85, pp. 259-262
18. G. Breuzé, Ph. Jucker, J. Serre, E. Garnero, T. Hamet and D. Colas, Fiber Optics Compatibility with Radiation Environment inside PWR Containment, IAEA Specialists' Meeting on "Improvements in Nuclear and Radiation Instrumentation for Nuclear Power Plants", 18 - 20 Oct. 1993, Saclay, France, CEA - Conf. 1644
19. K.O. Hill, Y. Fujii, D.C. Johnson and B.S. Kawasaki, Photosensitivity in optical waveguides : Application to reflection filter fabrication, Appl. Phys. Lett., Vol. 32, 1978, pp. 647-649
20. K.O. Hill, B. Malo, F. Bilodeau, D.C. Johnson, T.F. Morse et al., Photosensitivity in Eu2+ : Al2O3-doped core fiber: Preliminary results and application to Mode Converters, OFC'91, Post Pap. PD 3, pp. 14-17
21. M.M. Broer, R.L.Cone and J.R. Simpson, Ultraviolet-induced distributed-feedback gratings in Ce3+ - doped silica optical fibers, Opt. Lett., Vol. 16, 1991, pp. 1391-1393
22. L. Dong, P.J. Wells, D.P. Hand and D.N. Payne, UV-induced refractive Index change in Ce3+ - doped fibres, Tech. Dig. Conf. Lasers Electro Optics, 1991, pp. 68-70
23. K.O. Hill, B. Malo, F. Bilodeau and D.C. Johnson, Photosensitivity in optical fibers, Annu. Rev. Mater. Sci., Vol. 23, 1993, pp. 125-157
24. D.L. Williams, B.J. Ainslie, R. Kashyap, G.D. Maxwell, J.R. Armitage, R.J. Campbell and R. Wyatt, Photosensitivity index changes in germania-doped silica fibres and waveguides, SPIE, Vol. 2044, 1993, pp. 55-68
25. P.J. Lemaire, R.M. Atkins, V. Mizrahi, W.A. Reeds, High pressure H2-loading as a technique for achieving ultrahigh UV photosensitivity and thermal sensitivity in GeO2-doped optical fibres, Electron. Lett., Vol. 29, 1993, pp. 1191-1193
26. T. Erdogan, V. Mizrahi, P.J. Lemaire and D. Monroe, Decay of UV-induced Bragg gratings, OFC'94, paper TuLu, p. 50
27. P. Ferdinand, et al., Optical Fibre Bragg Sensors for Structure Monitoring within Nuclear Power Plant, Optical Fibre Sensing and Systems in Nuclear Environments, October 17th-18th, 1994, SCK-CEN Mol, Belgium
28. P. Niay, P. Bernage, M. Douay, E. Fertein, F. Lahoreau, J.F. Bayon, T. Georges, M. Monerie, P. Ferdinand, S. Rougeault and P. Cetier, Behaviour of Bragg gratings, written in germanosilicate fibres, against g-ray exposure at low dose rate, I.E.E.E. Phot. Tech. Lett., Vol. 6, n° 11, 1994, pp.1350-1352
29. R.L. McCreery, Fiber optic probe for remote Raman spectrometry, Anal. Chem., vol. 55, 1983, pp. 146-148
30. S.D. Schwab and R.L. McCreery, Versatile, efficient Raman sampling with fiber optics, Anal. Chem., Vol. 56, 1984, pp. 2199-2204
31. N.D. Dao and P. Plaza, Possibilités de mesures à distance et multi-sites par spectrométrie Raman-laser avec association de fibres optiques, Analysis, vol. 14, n° 3, 1986, p. 119-124 (in french)
32. S.M. Adler-Golden et al., Laser Raman Sensor for measurement of trace hydrogen gas, Appl. Opt., Vol. 31, n° 6, 1992, pp. 831-835
33. B. Schrader, Raman/Infrared atlas of organic compounds, VCH, 2nd Ed.

34. W.C. Michie, B. Culshaw, L. McKenzie, C. Moran, N.B. Graham, F. Santos, P.T. Gardiner, E. Bergqvist and B. Carstrom, A fibre Optic/Hydrogel Probe for Distributed Chemical Measurements, OFS10, Glasgow, 1994, pp. 130-133
35. P. Ferdinand, et al., Mine Operating Accurate STABILity Control with Optical Fiber Sensing and Bragg Grating Technology : the BRITE STABILOS Project, OFS10, Glasgow, 1994, pp. 162-166. Extended paper, J. Of Light. Techn., Vol. 13, n° 7, 1995, pp.1303-1313
36. P. Ferdinand et al. « Applications of Bragg grating sensors in Europe », International Conference on Optical Fiber Sensors OFS'97, Oct. 28-31, 1997, Williamsburg, Virginia, USA
37. O. Roy, Time Decay of the Optically Stimulated Luminescence in MgS doped with rare earths, 6th Intern. Conf. on Luminescent Materials, Paris, 1st-5th Sept. 1997
38. M. Decréton, V. Massaut, P. Borgermans, Potential benefits of fiber optics in nuclear applications. The case of the decommissioning and the waste storage activities, Optical Fibre Sensing and Systems in Nuclear Environments, October 17th-18th, 1994, SCK-CEN Mol, Belgium, idem, SPIE, Vol. 2425, 1994, pp. 2-10
39. P.B. Lyons, Review of high bandwidth fiber optics radiation sensors, SPIE, Vol. 566, 1985, pp. 166-171

CHLORINATED HYDROCARBONS TRACE DETECTION IN WATER BY SPARGING AND LASER IR GAS PHASE DETECTION

A. Lancia

TRI srl
Via Aldo Moro, 1
24020 Scanzorosciate, Italy

1. INTRODUCTION

This paper presents some of the contents of EC Project "MIRACLE - Mid Infra-Red Analysis for a Cleaner Environment" (E5V5 CT93 9339-CTEE), which addressed the development of a set of laser spectrometry based techniques for the trace level measurement of chlorinated hydrocarbons in water. The project was carried out at Dublin City University (Ireland), Ceramoptec GmbH (Germany) and TRI srl (Italy).

Chlorinated hydrocarbons are one of the most important classes of water pollutants due to their toxic properties and to their wide use in industry. For instance Dichloromethane (DCM) has been employed in paint stripping, metal cleaning, pharmaceuticals, aerosol propellants and in acetate films, Chloroform (TCM) in fluorocarbon synthesis and pharmaceuticals, 1,1,1 Trichloroethane (TCA) for metal and plastic cleaning, in adhesives and inks and as a general solvent, and many other ones are daily used within chemical plants or in manufacturing practices.[1,2]

Infrared absorption is one of the useful techniques for the quantitative and qualitative analysis of chlorinated substances in water with special reference to the strong C-Cl stretching bands which are located within 13 and 14 μm for compounds containing isolated chlorine atoms and more generally in the 12 to 17 μm range for the whole set of chlorinated organic compounds. In practice the use of conventional analytical IR methods is not suitable for a water pollution monitoring system since the direct spectroscopic measurements on water do not achieve the required sensitivity while a conventional sample treatment is a complex procedure which cannot be easily automated for field monitoring.[3,4,5]

2. GENERAL ABOUT THE SPARGING TECHNIQUE

The project investigated two alternate techniques for performing the monitoring of chlorinated hydrocarbons in water. One of such techniques is the use of polymer coated fibres for

Optical Sensors and Microsystems: New Concepts, Materials, Technologies
Edited by Martellucci et al., Kluwer Academic / Plenum Publishers, New York, 2000.

performing in situ measurements by evanescent wave spectroscopy (this subject will not be treated here). The other approach consitutes the subject of the present paper and consists in extracting the volatile chlorinated hydrocarbons with air and analysing them in the gas phase by infrared absorption spectrometry. The extraction process, often called "sparging", was performed in a few different ways and the spectroscopic measurements were made by a tunable diode laser spectrometer, by a variable wavelength CO_2 laser system and by FTIR. In general the sparging process has the basic advantage of eliminating the spectral absorption of the analytical matrix and of performing the measurement in the vapor phase where the spectral features are sharper and more selective. Moreover a careful design of the equipment and a proper data handling allow to make the method sensitive, fast and accurate.

In practice the air sparging of a volatile organic compound in air can be made in four ways corresponding to the liquid and the gas phase being stationary or mobile (Figure 1.).

The "stationary aqueous phase and stationary vapor phase" method is used in head space chromatography and in the analytical techniques based on creating the equilibrium condition and then analysing the vapor phase analyte concentration. Care must be taken in sampling the vapor phase (e.g. by drawing a small vapor amount or by having a syringe like arrangement). Of course the basic application of this technique is just giving one equilibrium measurement and in such a case the knowledge of the H constant for the anaytical matrix is required for determining the analyte concentration in water. Alternatively it is possible to repeat the measurement after stepwise dilution working out the formulas to process the concentration data in the gas phase to obtain the H constant and the analyte concentration. In general this techique is appealing for obtaining accurate equilibrium data but is generally quite time consuming and it requires discontinous operation.

The "mobile aqueous phase and (quasi)-stationary vapor phase" approach is in general the fastest way to take to equilibrium a certain amount of vapor phase. The two main uses we can envisage for this technique are the obtainment of a certain amount of equilibrium vapor phase in slow mass transfer conditions or the leveling of a time dependent concentration of the analyte in a moving aqueous phase.

The "stationary aqueous phase and mobile vapor phase" method was the main subject of our investigation and will be discussed below.

The last approach, i.e. "mobile aqueous phase and mobile vapor phase" is the analytical version of the conventional air stripping technology used at industrial scale to remove volatile pollutants from water. Its use for analytical purposes is particularly appealing for the continous monitoring of the volatile analyte concentration in a water stream where speed is more important than accuracy. In such a case the measured value could be subject to a drift associated to changes in the analytical matrix but its simplicity and speed makes it ideal for implementing a simple system for tripping a pollution alarm in the fastest way. It is moreover possible to implement this technique with variable ratio of the water and air flow to compensate for analytical matrix effects or to optimise the system precision as a function of analyte concentration.

3. PROJECT DEVELOPMENTS AT TRI LAB

The third approach was chosen at our laboratory for a deeper investigation because of its potential accuracy and speed and for the possibility to develop a compact sparging system requiring small samples.

Some previous studies already pointed out that in time-dependant sparging systems where air is traversing with a good mixing a stationary sample of water predicts that it is possible, by interpolating the logarithmic decay data of the pollutant concentration in air, to obtain the pollutant concentration in water regardless the variations in the pollutant solubility that affect the

Henry coefficient. Doyle[6,7,8,9] also developed a FTIR based analytical system based on such principle.

Due to our interest in developing a fast and sensitive sparging device we developed an analytical model taking into account the non-equilibrium mass transfer of the analyte due to the use of a relatively high flow rate of air. In this model we considered the sparging vessel as a stirred reactor where a certain volume is present of a liquid solution of the analyte. In developing the equations we assumed that the sparging gas is flown through the sparging vessel at a fixed flow rate and that the system and the stream are isothermal.

The integration of the differential equation (a symbols legenda can be found at the end of the article)

$$dm_L^{\cdot}(1+\varphi\cdot H^*) = -dt\cdot m_L^{\cdot}\cdot\frac{H^*\cdot F\cdot\varphi}{V_G} \tag{1}$$

allows to obtain the "dynamic sparging equation"

$$C_L = C_L^0\cdot e^{-\left(\frac{H^*\cdot F\cdot\varphi}{(1+\varphi\cdot H^*)\cdot V_G}\right)\cdot(t-t^\circ)} \tag{2}$$

which can be rewritten as

$$C_L = C_L^0\cdot e^{-\left(\frac{H\cdot F\cdot\varphi}{(1+H\cdot\varphi\cdot\frac{V_G}{V_L})\cdot V_L}\right)\cdot(t-t^\circ)} \tag{3}$$

in terms of H instead of H^*. For simplicity of writing we define the "sparging dynamic constant"

$$\vartheta = \frac{H\cdot F\cdot\varphi}{(1+H\cdot\varphi\cdot\frac{V_G}{V_L})\cdot V_L} \tag{4}$$

which allows to write the sparging equation as

$$C_L = C_L^0\cdot e^{-\vartheta\cdot(t-t^\circ)} \tag{5}$$

or

$$\ln(C_L) = \ln(C_L^0) - \vartheta\cdot(t-t^\circ) \tag{6}$$

The last equation is the key to the treatment of experimental data from the dynamic sparging experiments. By plotting $\ln(C_L)$ vs. t it is possible to obtain $\ln(C_L^0)$ and ϑ (by least squares linear fitting).

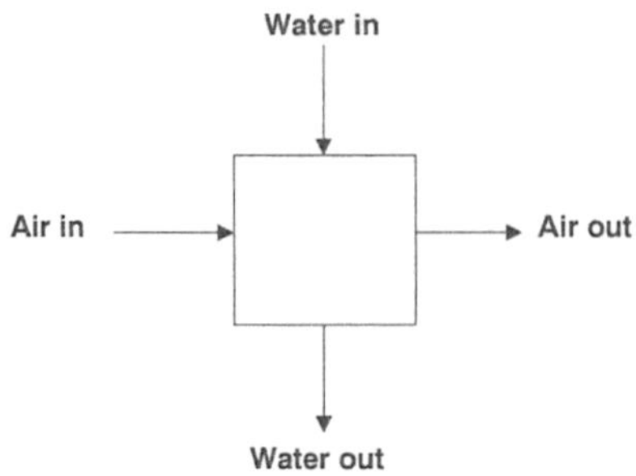

Air	*Water*
Steady	Steady
Flow	Flow
Steady	Flow
Flow	Steady

Figure 1. Sparging options.

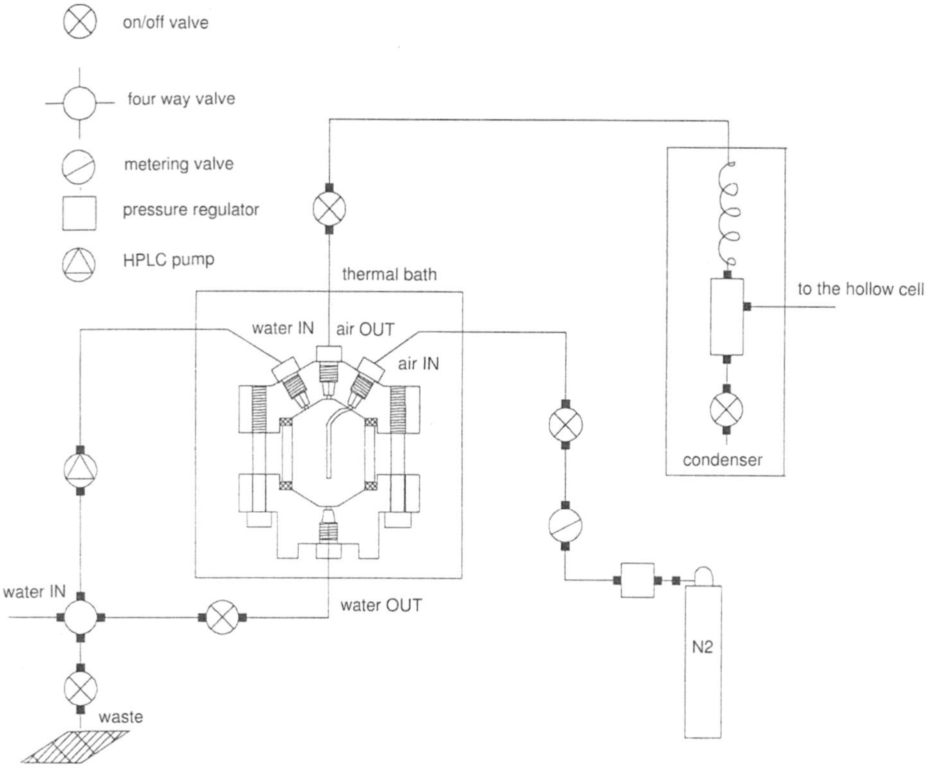

Figure 2. Scheme of the time dependant microsparger developed at TRI Lab.

In order to calculate the actual quantity of analyte in the solution sample it is necessary, if $V_G \cdot H$ is not negligible, to apply to C_L^0 a correction for taking into account the analyte which is present in the vapour phase at time t°:

$$C_L^{00} = C_L^0 + \frac{V_G \cdot C_L^0 \cdot H}{V_L} = C_L^0 \cdot \left(1 + \frac{V_G \cdot H}{V_L}\right) \qquad (7)$$

This correction is however requiring the availability of H that can be obtained by a single value of ϑ only if φ =1.

H can be obtained (if needed) by performing a series of experiments with different flow rates and then extrapolating to F=0 (equilibrium conditions).

In general H is not needed for small values of $V_G \cdot H$ and the beauty of this meaurement method is the possibility of calculating C_L^{00} without knowing H i.e. even if H changes because of the analytical matrix.

We also draw the attention on the fact that for small values of $V_G \cdot H$ the ϑ coefficient is simplified to the value:

$$\frac{H \cdot F \cdot \varphi}{V_L} \qquad (8)$$

As a matter of fact the above treatment is incorrect when φ <1 since at the beginning of the sparging a transition occours from the equilibrium state to the dynamic sparging condition. This problem can be solved by postulating that the initial concentration is such that

$$C_L^{00} = C_L^0 + \frac{\varphi \cdot V_G \cdot C_L^0 \cdot H}{V_L} \qquad (9)$$

In practice this means that we introduce an error in C_L^{00} of the order of $(1-\varphi) \cdot \frac{V_G \cdot C_L^0 \cdot H}{V_L}$.

Such error can be minimised by minimising $\frac{V_G}{V_L}$. By doing this we also minimise more generally the error related to the actual reach of equilibrium before the sparging process begins.

In practice, if the gas volume in the sparger is considerable the above error is potentially high but it can be largely reduced by the direct measurement of the initial fraction of analyte in the gas phase (actually by the gas phase concentration data collected just after the air flow is started).

An alternate treatment based on a standard "chemical engineering approach" to non-equilibrium mass transfer conducted to a set of differential equations that proved to yield by numerical integration a set of sample problems solutions equivalent to the one of the above analytical model.

Our theoretical model was first successfully tested by some preliminary non-equilibrium sparging devices using a photoionization detector to measure the gas phase concentration.

We then designed and constructed better (faster and smaller devices with a moderate departing from gas/liquid equilibrium conditions) sparging systems suitable for fully automatic operation. The best sparging device we constructed (Figure 2.) had a total volume of about 10 ml and allowed to perform a typical analysis in less than 10 minutes. Of course the small size of the device requires to use a gas analysis device with a very small equilibration time (small cell volume and fast gas turnover) but with a rather high sensitivity. The system used for this purpose comprises a hollow core guide gas flow cell and a laser spectrometer which are described in the other paper from our group in this same collection of papers.

Figure 3. shows a non-equilibium "dynamic sparging" curve (gas phase concentration vs. time) which exhibits a curvature despite the linear log dependance predicted by the model. This was mainly due to the effect of pollutant concentration on surface tension (and therefore on gas/liquid equilibration rate) and to the use of dry purge nitrogen which causes a cooling of the water due to mass transfer. The non-linearity problem was first solved by interpolating the log curve by a quadratic function and extrapolating the initial concentration value. We then eliminated the water cooling problem by saturating with water the purge gas before its immission in the mini-sparger. The surface problem effect is instead almost negligible for low pollutant concentrations in water (as it happens in the cases of practical interest).

By working with Chloroform and TCE water solutions containing different concentrations of electrolites and organic compounds we could demonstrate the insensitivity of the method to the analytical matrix and achieve a sensitivity of about 1 mg/l . Further improvements are considered possible (e.g. by using FM or TTFM TDLAS operation) and able to increase such sensitivity by almost two orders of magnitude (0.05 mg/l) for most volatile chlorocarbons.

4. CONCLUSIONS

Sparging devices provide a flexible set of solutions to monitor several VOC dissolved in water by measuring their concentration in the purge gas.[10,11,12]

The time-dependant measurement of concentration while sparging a steady water batch allows to determine a VOC concentration in water irrespectively of the knowledge of the Henry constant which is dependant on temperature and on the analytical matrix (ionic strenght and organic contaminants dissolved in water).

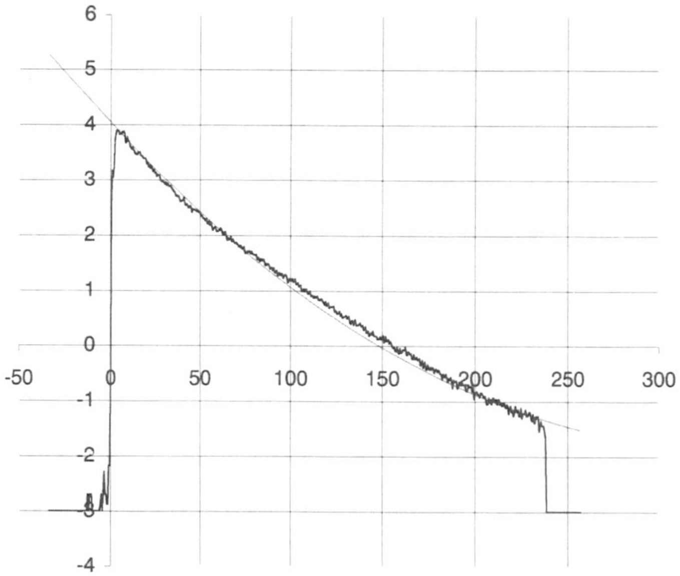

Figure 3. Logarithmic graph showing the quasi-linear concentration curve (with quadratic fitting).

An analytical model was developed of the time dependant sparging process in non equilibrium conditions which suggested the feasibility of faster analytical systems based on smaller "dynamic spargers".

A set of miniature non-equilibrium spargers were developed and the best model was successfully tested using as a gas phase VOC detector a hollow core waveguide gas cell coupled to a tunable diode laser absorption spetrometer operated in wavelegth modulation mode.[13-20]

The experimental results demonstrated the possibility of selectively detecting 1 ppm levels of chlorocarbons in water with less than 10 minutes total analysis time. A detection limit lower than 50 ppb should be obtained by more sensitive operational modes of the TDLAS system.

ACKNOWLEDGEMENTS. The author wishes to thank the European Commission, DGXII for having co-funded the "MIRACLE" project (EC contract E5V5 CT93 9339-CTEE) and all the colleagues who participated in the work. Special thanks to Patrizia Barranco and Sabrina Fumagalli (TRI, I), Brian McCraith (Dublin City University - Ireland), Slava Artushenko (Ceramoptec GmbH, D) and to the Project proposer Clive Worrell (University of Sussex, UK).

LEGENDA OF SYMBOLS

$C_G^{\cdot}$	Concentration of the analyte in the gas phase at time t
$\overline{C}_G^{\cdot}$	Equilibrium concentration of the analyte in the gas phase (at time t)
C_G^0	Concentration of the analyte in the gas phase at time t°
C_L^{00}	Concentration of the analyte in the sample before the loading of the sparger
$C_L^{\cdot}$	Concentration of the analyte in the liquid phase at time t
C_L^0	Concentration of the analyte in the liquid phase at time t°
$m_G^{\cdot}$	Mass of analyte in the gas phase at time t
$m_G^{\cdot\cdot}$	Mass of analyte in the gas phase at time t+dt
$m_L^{\cdot}$	Mass of analyte in the liquid phase at time t
$m_L^{\cdot\cdot}$	Mass of analyte in the liquid phase at time t+dt
F	Flow rate of air through the sparger
φ	Equilibration coefficient. Equal to 1 for equilibrium conditions
V_G	Volume of the gas phase in the sparger
V_L	Volume of the liquid phase in the sparger
H	Henry constant i.e. partition coefficient between the gas and the liquid phase
H^*	Partition coefficient expressed as the ratio of mass in the gas and the liquid phase
t°	time at which the sparging process begins

REFERENCES

1. N.F. Gray: Drinking Water Quality; Wiley, 1994.
2. K. Verschueren: Handbook of Environmental Data on Organic Chemicals; van Nostrand Reinhold, N.Y. 1983.
3. L.J. Bellamy: The Infrared spectra of Complex Molecules; John Wiley & Sons, N.Y. 1958; and The Infrared spectra of Complex Molecules; Vol.1, 3rd ed., Wiley/Halsted N.Y. 1975.
4. C.J. Pouchert: Aldrich Library of Infrared Spectra; 2nd ed. Milwaukee, Al.Chem.Company, Inc. 1975.
5. S.W. Fleming, B.B. Baker, Jr., B.C. McIntosh "On-line analyzer for chlorocarbons in wastewater", Am. Chem. Soc., Symposium series, vol. 508, 1992, pp.48-61.

6. W.M. Doyle, "Analysis of trace concentrations of contaminants in water by sparging FTIR", SPIE Conf. Proc., 8th International Conference on Fourier Transformation Spectroscopy, Vol. 1575, 1991, pp.199-200.
7. W.M. Doyle, "Continuous monitoring of organic pollutants in water by sparging-infrared", Process Control and Quality, Vol. 3, 1992, pp85-98.
8. B.C. McIntosh, D.W. Vidrine, W.M. Doyle, "Real time waste stream monitoring", American Laboratory, Vol.23, No.18, 1991, pp19-22.
9. W.M. Doyle "Two dimensional sparging IR analysis of trace organics in water", SPIE Conf. Proc., 9th International Conference on Fourier Transform Spectroscopy, Vol.2089, 1993, pp 456-457.
10. S. Lesage, S. Brown, "Dynamic headspace analysis of volatile organic solvents in water", Anal. Chem., Vol.66, 1994, pp572-575.
11. G. Baykut, P. Kowalski, J. Wronka, F. Laukien, "Automated spray extraction of volatile organic compounds from aqueous system", SPIE Conf. Proc., Environmental and Process Monitoring Technologies, Vol.1637, 1992, pp234-240.
12. R.L. Johnson, P.C. Johnson, D.B. McWhorter, R.E. Hinchee, I. Goodman, "An overview of in situ air sparging", Ground Water Monitoring Remediation, Vol. 13, No.4, 1993, pp127-135.
13. S.J. Saggese, J.A. Harrington, G.H. Sigel, "Hollow waveguides for sensor applications", SPIE Vol. 1368 Chemical, Biochemical and Environmental Fiber Sensors II (1990).
14. S.J. Saggese, J.A. Harrington, G.H. Sigel, "Attenuation of incoherent radiation in hollow sapphire and silica waveguides", Optics Letters, Vol.16 No.1, pp. 27-29 (1991).
15. J.P. Conzen, J. Burck, H.J. Ache, "Characterization of a Fiber-Optic Evanescent Wave Absorbance Sensor for Nonpolar Organic Compounds", Applied Spectroscopy, Vol. 47, No. 6, 1993, pp.753-763.
16. R. Krska, E. Rosenberg, K. Taga, R. Kellner, A. Messica, A. Katzir, "Polymer coated silver halide infrared fibers as sensing devices for chlorinated hydrocarbons in water", Appl. Phys. Lett., Vol. 61, 1992, pp.1778-1780.
17. J. Burck, J.P. Conzen, B. Beckhaus, H.J. Ache, "Fiber-optic evanescent wave sensor for in situ determination of non polar organic compounds in water", Sensors and A., Vol. 18-19, 1994, pp.291-295.
18. P.J. Glatkowski, M.A. Druy, W.A. Stevenson, "Detection of aromatic hydrocarbons with evanescent wave fiber optic sensors", SPIE Conf. Proc., Chemical, Biochemical and Environmental Sensors IV, Vol. 1796, 1992, pp.243-250.
19. A. Messica, A. Katzir, U. Schiessl, M. Tacke, "Liquid and gas fiberoptic evanescent wave spectroscopy by tunables lasers", SPIE Conf. Proc., Vol.1591, Infrared Fiber Optics III, 1991, pp.192-199.
20. I. Schnitzer, A. Katzir, U. Schiessl, W. Riedel, M. Tacke, "Fiber-optic-based evanescent field infrared spectroscopy using tunable diode lasers" SPIE Conf. Proc., Vol. 1048, Infrared Fiber Optics, 1989, pp.133-140.

HOLLOW CORE FIBER GUIDES AS GAS ANALYSIS CELLS FOR LASER SPECTROSCOPY

A. Lancia

TRI srl
Via Aldo Moro, 1
24020 Scanzorosciate, Italy

1. INTRODUCTION

Optical fibres have been used in the last twenty years within a number of laboratory experiments and instruments prototypes for the analysis of gases. In most such cases the fibers are used as a waveguide for connecting a remote cell (an extrinsic optical fiber sensor) to an infrared spectrometer. More rarely waveguides were used in gas analysis prototypes within fibre lasers or for the fabrication of evanescent wave probes (an instrinsic fiber sensor) where the absorption by the gas is measured by the ATR (Attenuated Total Reflection) principle.

This paper addresses a third application of infrared waveguides in gas detection i.e. the use of hollow core infrared fibres as optical absorption cells for the gas to be analysed. The original work reported here was conducted at TRI (Italy), the University of Sussex (UK) and INERIS (F) within the EC RTD project "IRGAS" (MAT1-CT94-0026). The project addressed the use of open path, evanescent wave and hollow core fibres cells coupled to a diode laser spectrometer operating in the 3-16 μm range and to a variable wavelength CO_2 laser spectrometer. In some of the tests the gas analysis cell was connected to the spectrometer by silver halide solid core infrared fibres.[1]

This paper addresses some general aspects of the IRGAS project and the specific activities carried out in TRI laboratories and particularly the development of hollow core waveguides gas cells coupled to a TDLAS (Tunable Diode Laser Absorption Spectrometer). A detailed account of the whole research program is contained in the Project technical reports while some open literature papers describe the individual work by each of the three Partners.

Hollow core infrared fibers have been the subject of investigation in Europe, USA and Japan mostly in relation to their potential use for delivering power lasers radiation for medical and industrial use.[2,3,4,5] In general hollow core infrared waveguides (HIW) are small pipes with a circular (or more rarely rectangular) hollow core whose functioning can be easily explained by the classical theory of multiple reflections or by a wave approach. In general the diameters of interest are rather large (i.e. 500-2000 μm) since they are simpler to fabricate and because the guide transmittance varies with the inverse of the core diameter cube power.[6]

HIW can be classified in two classes corresponding to refractive index higher or lower than the one of the core (actually the one of air). HIW based on materials with n>1 are called "leaky guides" and they generally consist of a metallic pipe whose inner surface is coated by a dielectric substance to decrease the excessive attenuation of bare metal guides in the infrared region. HIW made of n<1 function in the same way of solid core fibres i.e. by total reflection and they can be made of different types of dielectric materials exhibiting anomalous dispersion in the infrared wavelenghth range of interest (e.g. alumina, sapphire silica/germania glasses, heavy metals doped silica, etc.).[7,8,9,10,11] HIW have a rather small numerical aperture which may be a problem when coupling them with astigmatic sources but when properly interfaced they can reach an attenuation minimum as low as 0.1 dB/m. Additionally the absence of the solid core eliminates the reflections at the guides end facets.

A general feature which is still putting a strong limitation to the use of HIW is their rather strong bending loss which results to be linearly dependent on the fibre curvature (1/R).

The principal interest in the use of HIW as gas detection cells is related to the possibility to achieve an outstanding volume/pathlength ratio i.e. allowing the analysis of rather small volumes of gas with an optical absorption path length of some meters thus enabling to enhance the sensitivity of the measurement.[12] A further appealing feature of HIW gas cells is that they can be used in flow mode with a very short time constant for concentration equilibration due to plug flow. Another advantage of HIW for gas analysis is the possibility to fabricate cells which are rather slim and are less subject to optics misalignements in the field than coventional bulk optics cells (e.g. Herriott multipass cells).[13,14,15,16,17,18]

2. TDLAS EXPERIMENTS WITH HOLLOW WAVEGUIDES GAS CELLS (IRGAS PROJECT, TRI LAB)

The HIW gas cells work conducted in our laboratory were based on two types of fibres. The HIW by Prof. Harrington of Rutgers University (Piscataway, NJ - USA) consisted of a thin silver film deposited inside a silica glass tubing (700 μm bore diameter), lined by a thin silver-iodine layer to protect the metal and enhance reflection. This fibre exhibits a quite broadband transmission range from 2 to 20 μm. Additionally we performed measurements with a pure silica hollow core fibres (700 μm bore diameter) (supplied by Ceramoptec GmbH, Bonn - D) presenting anomalous dispersion from 7 to 9 μm.

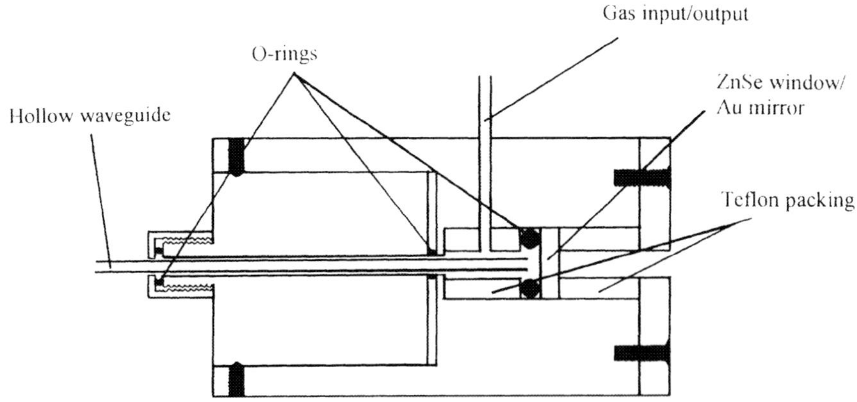

Figure 1. Section of one of the terminal parts of a hollow waveguide gas cell fabricated at TRI.

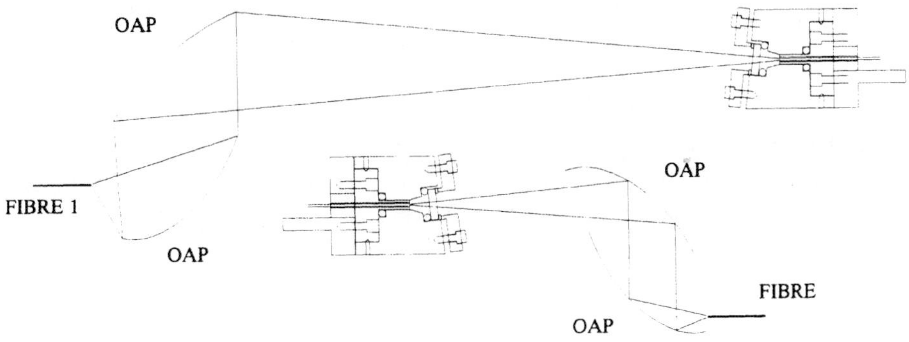

Figure 2. Scheme of the coupling of a hollow waveguide gas cell to multimode silver halide fibres for coupling to a remote TDLAS spectrometer (TRI Lab, IRGAS Project).

The cells were carefully designed in order to achieve a small coupling space of the hollow core fiber to the gas pipings to reduce the dead volume in dynamic concentration measurements. Figure 1. shows a section of one of the two cell terminal parts. We constructed a few cell units allowing a path length from 1.3 to 1.8 m and dead volumes of about 0.5 cm^3.

Due to the high attenuation of transmission in HIW when upper modes are excited the launching angle has to be minimised finding a compromise with the focused spot size (proportional to the f-number) that has to be matched to the fibre core diameter. When coupling the HIW cell to the laser diode spectrometer by multimode silver halide solid core fibers we employed couples of OAP mirrors of different f values (Figure 2.).

The laser diodes spectrometer we developed in our work was based on "lead salt" diode lasers with tuning ranges centered to about 8 and to 12.5 µm with a coupled optical power in the order of 0.1 mW. LN cooled MCT detectors were used (Figures 3. and 4.).

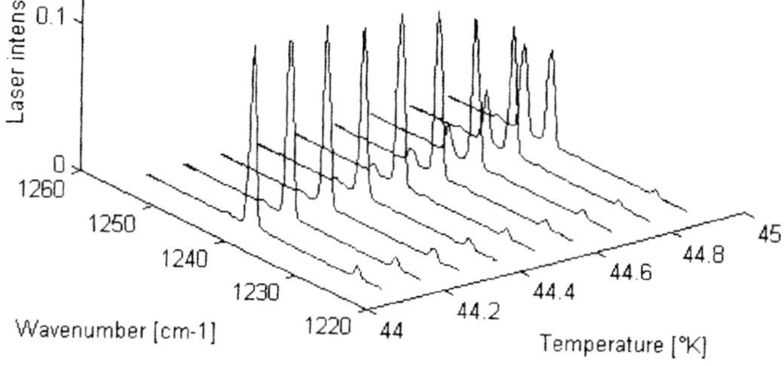

Figure 3. Sample emission profile for the 8µ lead-salt diode lasers used in the IRGAS project. TDL Multimode emmission at 8 micron acquired at temperatures running from 44°K to 44.9°K with step of 0.1°K, keeping the current constant at 386 mA.

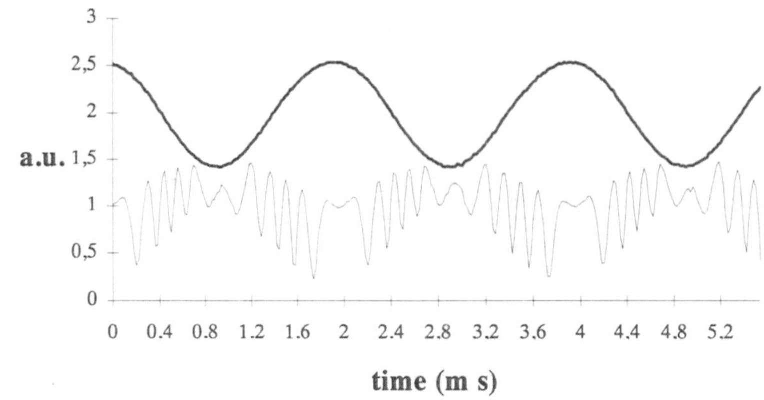

Figure 4. Example emission power and etalon signal for the 12 μ lead-salt diode lasers used in the IRGAS project (1 kHz modulation of laser diode current). Laser emission and etalon signals for the 12 μm TDL with a modulation amplitude equal to 22.7 mA. (T=50°K, frequency=1KHz).

The spectrometer was operated in low frequency wavelength modulation mode (the measurement of the 1f and 2f harmonic component in the transmitted beam allows the measurement of spectral signals which are nearly proportional to the 1st and 2nd derivative of trasmission vs. wavelenght).

The gas analysis system[19,20,21,22,23] was tested for the trace analysis of chloroform and of N-nitroso-diethylamine in air. The system performance when the HIW cell was interfaced to the spectrometer by silver halide fibres was heavily affected by the excessive total loss which makes such option practically inapplicable for very low power diode lasers.

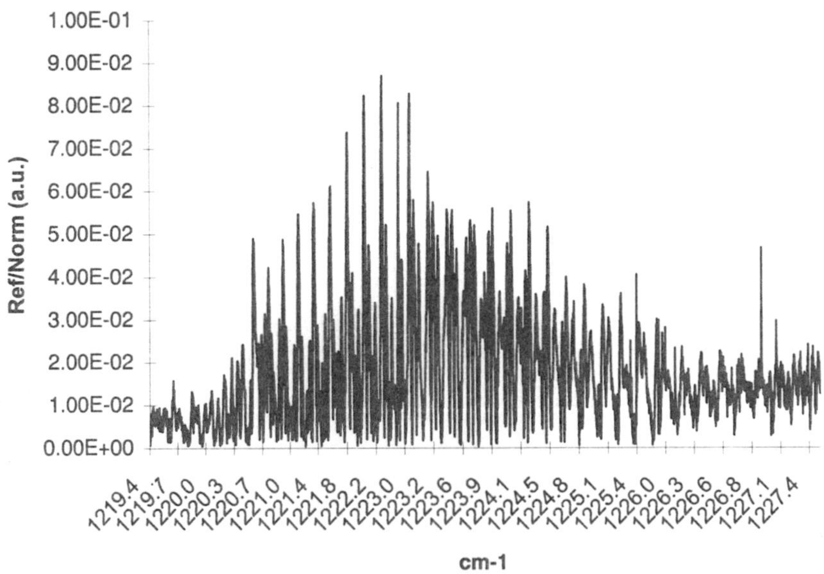

Figure 5. Sample spectrum of chloroform vapors spectrum: 2f modulus signal in the reference cell (500 Pa), current from 473 to 627 mA.

Satisfactory results were obtained when the HIW gas cells were coupled directly to the spectrometer beam achieving sensitivities of 1 ppm*m when reducing the total cell pressure to about 50 hPa. At least one order of magnitude in sensitivity would be achieved by operating the diode laser spectrometer in FM or TTFM mode (Figure 5.).

An analysis of high resolution FTIR spectra of a large set of organic pollutants used in the chemical industry showed that the peak wavelegth of certain functional groups such as the -C-Cl one allows both the identification of the compound and its quantitative measurement.

The most interesting possible applications of IR hollow core waveguides coupled with laser spectrometers are the ones which require at the same time a high sensitivity and a low sample volume. Some examples are gas chromatography, monitoring of VOC from analytical miniature spargers and thermal desorption experiments.

3. CONCLUSIONS

Hollow core infrared optical fibers can be used to fabricate gas cells with a high value of the optical path / sample volume ratio.

A new cell of this type was developed at TRI within the EC IRGAS project and gas detection experiments were conducted by a "lead salt" Tunable Diode Laser Absorption Spectrometer operated in Wavelegth Modulation around 8 and 12.5 μm bands achieving at reduced pressure ppm sensitivity for chlorinated hydrocarbons and nitrosamines.

The remotisation of the hollow waveguide gas cell from the spectrometer is possible using suitable multimode IR waveguides providing that the IR source energy is sufficiently high to compensate for the high total optical losses in the system.

Hollow core waveguides gas cells coupled to laser or FTIR spectrometers are an option to be considered when the analysis must be conducted on a small gas sample and a high sensitivity is required.

ACKNOWLEDGEMENTS. The author wishes to thank the European Commission, DGXII for having co-funded the "IRGAS" project (EC contract MAT1-CT94-0026) and all the colleagues who participated in the work. Special thanks to Patrizia Barranco and Sabrina Fumagalli (TRI, I), Clive Worrell (University of Sussex, UK) and Michel Rose (INERIS, F).

REFERENCES

1. C.A. Worrell, A. Lancia, M. Rose et al., "The IRGAS EC RTD Project", (EC-DGXII Contract MAT1-CT94-0026), Final and Summary Reports, September 1996.
2. C.A. Worrell, "Trace-level detection of gases and vapours with mid-infra-red hollow waveguides", J.Phys. D: Appl. Phys. 30 (1997) pp. 1984-1995.
3. C.A. Worrell, "Infrared optical constants for CO_2 laser waveguide materials", Journal of materials science 21 (1986), 781-787.
4. M. Saito, N. Baba, N. Sawanobori, M. Miyagi, "Hollow glass waveguides for Mid-Infrared light transmission", Jpn.J. Appl. Phys., Vol. 33, (1994).
5. M. Miyagi, A. Hongo, Y. Matsuura, "Hollow IR waveguides", SPIE Vol. 1228 Infrared Fiber Optics II, 1990.
6. C.A. Worrell, "Transmission properties of some hollow glass waveguide at 10.6 μm wavelength", Reprinted from Electronics Letters 27th April 1989, Vol. 25, n. 9, pp.570-571.
7. M. Saito, S. Sato, M. Miyagi, "Spectroscopic gas sensing with infrared hollow waveguides", SPIE Vol. 1796 (1992).
8. A.M. Scheggi, R. Falciai, R. Gironi, "Characterization of oxide glasses for hollow-core middle IR fibers", Applied Optics, Vol. 24, No. 24, pp. 4392-4394 (1985).

9. S.J. Saggese, J.A. Harrington, G.H. Sigel, "Hollow waveguides for sensor applications", SPIE Vol. 1368 Chemical, Biochemical and Environmental Fiber Sensors II (1990).
10. S.J. Saggese, J.A. Harrington, G.H. Sigel, "Attenuation of incoherent radiation in hollow sapphire and silica waveguides", Optics Letters, Vol. 16, No. 1, pp. 27-29 (1991).
11. Katsuyama & H. Matsumura, "Infrared Optical Fibers", edited by Adam Hilger, 1989.
12. W. Rudolf Seitz, "Analytical spectroscopy with mid infrared transmitting optical fiber; Spectroch", Acta Rev., vol. 15, n.6, pp.447-492, 1993.
13. N.S. Kapany, R.J. Sims, "Recent developments in infrared fibre optics", Infrared Phys. Vol.5, 1985, pp. 69-80.
14. E.M. Dianov, "Advances in IR fibers", Tech. Dig. (Los Angeles, CA:SPIE) paper 320-04.
15. S. Mitachi, S. Sakaguchi, H. Yonezawa, K.Shikano et al., Jpn. J. Appl. Phys. 24, 827-8 (1985).
16. T. Katsuyama, H. Matsumura, Appl. Phys. Lett. 49, 22-3 (1986).
17. J.P. Conzen, J. Burck, H.J. Ache, "Characterization of a Fiber-Optic Evanescent Wave Absorbance Sensor for Nonpolar Organic Compounds", Applied Spectroscopy, Vol. 47, No. 6, 1993, pp.753-763.
18. R. Krska, E. Rosenberg, K. Taga, R. Kellner, A. Messica, A. Katzir, "Polymer coated silver halide infrared fibers as sensing devices for chlorinated hydrocarbons in water", Appl. Phys. Lett., Vol. 61, 1992, pp. 1778-1780.
19. A. Messica, A. Katzir, U. Schiessl, M. Tacke, "Liquid and gas fiberoptic evanescent wave spectroscopy by tunables lasers", SPIE Conf. Proc., Vol. 1591, Infrared Fiber Optics III, 1991, pp.192-199.
20. D. Bomse, A. Stanton, J. Silver, "Frequency modulation and wavelength modulation spectroscopies: comparison of experimental methods using a lead-salt diode laser", Applied Optics, 718-731, Vol. 31, No. 6, (1992).
21. J. Silver, "Frequency modulation spectroscopy for trace species detection: theory and comparison among experimental methods", Applied Optics, 707-717, Vol. 31, No. 6 (1992)
22. G.R. Janik, C.B. Carlise, T.F. Gallagher, "Two-tone frequency-modulation spectroscopy", J. Opt. Soc. Am. B, 1070-1074, Vol. 3. No 8 (1986).
23. J. Reid, R. Sinclair, "High sensitivity detection of trace gases at atmospheric presure using tunable diode lasers", Optical and Quantum Electronics 31-39. Vol.17(1985).

CHEMILUMINESCENCE IMAGING OF PLANT ORIGIN MATERIALS

D. Slawinska

Agricultural University
Department of Physics
Wojska Polslkiego Str. 38/42
60965 Poznan, Poland

1. INTRODUCTION

Chemiluminescence (CL) is the emission of light accompanying chemical reactions:

$$A + B \xrightarrow{k} P^* \xrightarrow{k_{fp}} P + h\nu_p \qquad (1)$$

Chemiluminescent reactions produce a reaction intermediate or a product P. in an electronically excited state (P*). Subsequent radiative decay of the excited state to the ground state (P) is the source of light. Chemically produced excited state behave like radiation induced states. CL can be in UV, visible and infrared, CL is the result of two subsequent processes: chemiexcitation (ce) and radiative relaxation of the product (luminescence, 1) therefore, the quantum efficiency of CL process (Φ_{cl}) is defined as the quotient of the number of photons $N_{h\nu}$ and the number molecules that react N_r:

$$\Phi_{cl} = Nh\nu / N_r = \Phi_{co}\Phi_1 \qquad (2)$$

Analytical methods are based on proportionality between the intensity of CL (I) and the rate limiting step reaction (w):

$$I = \Phi_{cl} w \qquad (3)$$

Many weakly CL reactions can be brightened dramatically upon the addition of highly fluorescent acceptor molecules (A):

$$P.^* + A \xrightarrow{k_{et}} P. + A^* \qquad A^* \xrightarrow[k_{fa}]{} A + h\nu_a \qquad \text{when } k_{et} > k_{fp} \text{ and } \Phi_s > \Phi_{clp} \qquad (4)$$

Moreover, ultraweak CL accompanying numerous oxidative processes can be strongly enhanced after addition of chemilumonogenic probes, such as e.g. 5-amino-2,3-dihydrophthalazine-l ,4-dione (luminol) or N,N'-dimethyl-diacridine dinitrate (lucigenin). These probes reacting with reactive oxygen species (ROS) like O_2, OH, HO_2, H_2O_2, OCl^-, $^1O^*_2$ give strong CL typical of the probe used.[1] For example, luminol in the oxidative ring-splitting reaction produces 5-amino-ortho-phthalic acid (anionic form) in the first excited singlet state that emits a blue light (λ =427 nm, Φ_{cl}= 0.02):

$$luminol + H_2O_2 + OH^- \rightarrow {}^1(amino\text{-}orthophtalic\ anion)^* + N_2 + H_2O \qquad (5)$$

Interest in analytical techniques based on CL detection continues to grow.[2] One of the primary reasons for this interest is that CL methods are generally capable of the low detection limits, do not require external photoexcitation and are competitive to radioisotopic ones. They have found a wide application in biomedical and environment protection sciences, as ultraweak luminescence is emitted from all forms of living organisms. The emitted light carries vast information on the spatial, temporal and energetic characteristics of the light-producing systems.[3] The recent commercial availability of improved instrumentation have made CL techniques more attractive. Nowadays, new low-light imaging instrumentation based on the use of intensified Vidicon tube or charge coupled devices (CCD) have become commercially available. These devices are characterized by a low thermal noise, wide dynamic range, high sensitivity and spatial resolution as well as provide a quantitative relationship between photons detected and pixel addressed. Moreover, it has been previously proved that these instruments can be connected with optical microscopy reaching resolution of few micrometers. Methodological problems related to instrumental performance, data processing and the sample to be analyzed still limit a quantitative use of this principle. Critical points are the chemistry (kinetics, photochemistry) of the CL system used and the sample to be imaged in term of size, geometry thickness, aspecific signal and light scattering or reflection artefacts.[4,5]

In order to evaluate the suitability of CL imaging for quantitative monitoring of slow exergonic oxidation processes of plant origin materials, a slow scan cooled CCD camera was employed. These processes cause deterioration, ageing, corrosion and weathering of natural and artificial materials. The kinetics, emission spectra and 2-dimentional surface emission pattern of plant-origin materials were determined in model experiments simulating certain detrimental processes.

In this chapter we exemplify these studies by means of two systems/processes: 1/ the cereal food products undergoing slow autooxidation and water imbibition, and 2/ the weathering of different kinds of wood exposed to the UV-VIS radiation and humid atmosphere.

2. MATERIALS AND METHODS

2.1 Instrumentation

2.1.1 Single photon counting imaging. For the purpose of analysing 2-D emission pattern and intensity of extremely weak CL of cereal products, a single photon imaging system in the Inaba Biophoton Project Laboratory, Sendai, Japan was used.[6] It employed an intensified cooled camera (Hamamatsu Photonics, Hamamatsu, Japan) containing an imaging PMT. Photoelectrons ejected from the photocathode strike a stock of microchannel plates which provides amplification of the order of 10^8. A burst of secondary electrons emerging from the last microchannel plate then impacts a resistive anode position detector. In this way, the x-y position of the photoevent is measured and the information displayed on a monitor and stored for processing. The wavelength range of the photodetector was 350-850 nm, 512-483 pixels and dark current less than 4 e^- per pixel per second.

CL imaging of wood samples was performed using a slow scan CCD camera (a macro version) "Night Owl" Molecular Light Imager LB 981 EG&G Berthold, Wildbad, Germany. Camera records the image directly to a CCD without an image amplifier. The CCD element (thinned, back illuminated) is fixed into a hermetically sealed container to overcome condensation problems. Cooling system (4-stage Peltier element) keeps 200 K. It has the spectral sensitivity in the range of 180-1100 nm and 40% quantum efficiency at 650 nm, 578x 385 pixels, thermal noise as low as 1 e^- /1000 s and readout noise 6 e^- (Figure 1.). A powerful WinLight operating under Windows$^{(TM)}$ controls all parameters of the camera. It also ensures sophisticated processing and representation of images and analysis of digital data.

2.1.2 Spectral analysis. Conventional spectrometers are usually not possible to use for measurements of the spectral distribution of CL emitted from plant origin materials because of the weakness of the signal. Majority of spectra show broad emission bands, therefore a high resolution is not critical. In order to increase the signal-to-noise ratio (S/N), a large emitting area of the sample and of the detector has to be used. The single photon counting method combined with cut-off filters was applied. Cut-off filters have a sharp wavelength boundary of transmission. If two filters are put successively into a light beam, then the difference of current intensities or photocount rates Δi characterizes the light intensity within the region between short-wavelength boundaries of the filters:

$$\Delta i_{kl} = i_k - i_l = \gamma \int_{\lambda_k}^{\lambda_\sigma} \sigma I_\lambda (T_k - T_l) d\lambda \qquad (6)$$

where k and l are numbers of successive filters, I_λ is the light intensity, T is transmittance of filters, σ denotes spectral sensitivity of a PMT-photocathode, γ is proportionality coefficient, λ_σ is the long wavelength sensitivity of PMT and λ_k is short wavelength boundary of a filter transmission. For the majority of filters used T_{max} was the same and thence, $T_k - T_l \neq 0$ only within the region close to some λ value, being characteristic of a given pair of filters. λ corresponds to the maximum difference of transmittance for two filters. One can assume that the light intensity I does not depend upon λ in this region and is equal to:

$$I(\lambda) = \langle I \rangle (\lambda_{kl}) \qquad (7)$$

Then

$$\Delta i_{kl} = \gamma I(\lambda_{kl}) \int_{\lambda_k}^{\lambda_{kl}} \sigma (T_k - T_l) d\lambda = \gamma I(l_{kl}) C_{kl} \qquad (8)$$

From the above the average ultraweak luminescence intensity $\langle I \rangle$ expressed in arbitrary units may be calculated:

$$\langle I \rangle (\lambda_{kl}) = \Delta kl / C_{kl} \qquad (9)$$

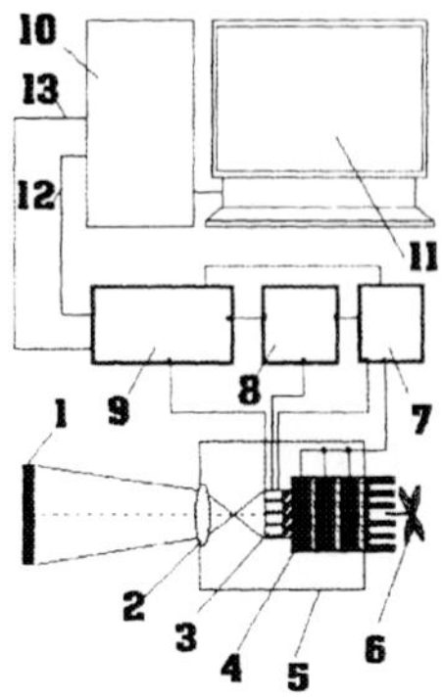

Figure 1. Scheme of the Molecular Light Imager Night Owl Luminograph LB 981 (EG&G Berthold):1-Sample, 2-Lens, 3-CCD sensor, 4-Thermoelectric (Peltier) Cooler, 5-Camera, 6-Ventilator, 7- Power Supply, 8-Thermostat, 9-Camera Control and Interfaces, 10-Computer and Software, 11-Color Screen, 12-Data of Camera Control Trasmission, 13- Image Signals Trasmission.

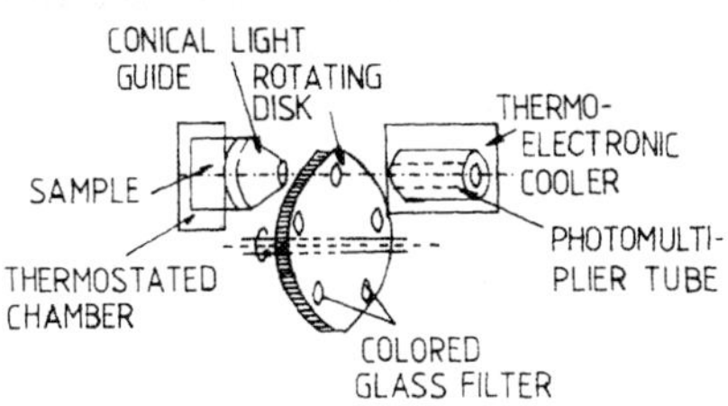

Figure 2. Schematic illustration of the ultraweak light spectral analyzer using the single photon counting method and cut-off filters.

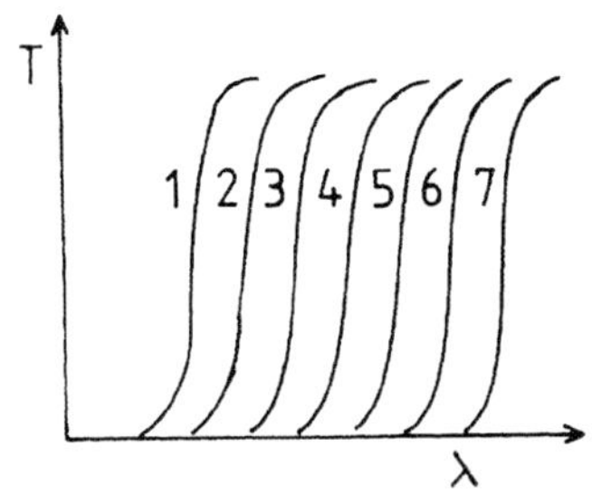

Figure 3. Schematic representation of trasmission characteristic of colored glass filters used as cut-off filters for spectral measurements of ultraweaik luminescence.

C_{kl} is the coefficient computed for the set of filters taking into account their transmittance and the relative sensitivity of the PMT (Figure 2.).

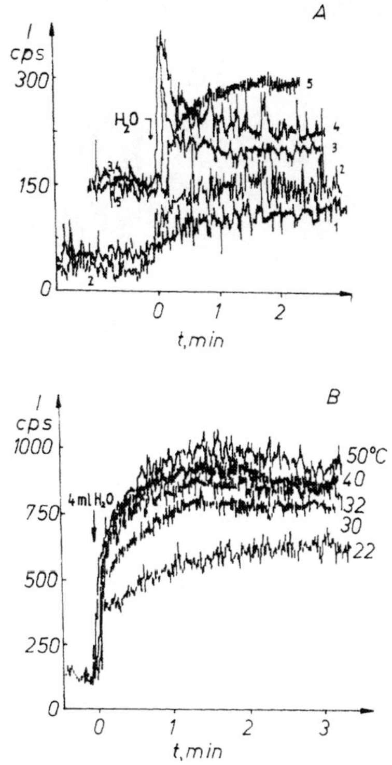

Figure 4A. Kinetic of chemiluminescence from various cereal grains upon treatment with water: 1 Lenns esculenta, lentil, 2-Triticum, wheat var "Heiduk", 3 and 4 Pisum sativum, var green pea (3), field pea (4), 5-Avena sativa, oat (the I scale decreased 2).

Figure 4B. The influence of temperature on the kinetics of chemiluminescence from wheat grains Triticum var "Heiduk" during the imbibition process.

Reproduced with the permission from John Wiley & Sons, D. Slawinska, J. Slawinski, J. Biolumin. Chemilumin. 12, 251, 254 (1997).

In the case of CL spectra of cereal food products, the measurements were performed using a red sensitive R-1333 Hamamatsu PMT and cut-off filters.[7] In our spectral analyzer employed to measurements of wood CL spectra, the set of 20 calibrated cut-off filters GOST/941 1 of Russian production was mounted in the rotating disk. The PMT was an EMI 9558QB cooled to 220 K (Figure 3.).

Kinetic measurements were performed by means of an EMI 9558 QB PMT (220 K) and also by the "Night Owl" imaging system from the set of consecutive images quantitatively elaborated.

2.2 Methods

Air-dried samples of cereal products, grains or flour were inserted in a quartz cuvette and kept for 20-30 min inside a light-tight camera and allowed to extinguish delayed luminescence. Then, the appropriate amount of water or test solution was injected into the cuvette by way of a light-tight tube. The resulting CL signal was counted as a discrete photocount-time series or imaged. Details are given elsewhere.[7,8] Air dried or water sprayed (3 mL) samples of wood were exposed for 15 min to the UV+VIS radiation from two middle pressure 60 W Hg burners Emita VP-60 at the distance of 15 cm at 58 °C. In experiments on CL imaging, all samples of wood were photographed within the CCD camera for the exposure time 0.02 s and then kept 1 h to extinguish delayed luminescence from samples and background plate.

3. RESULTS AND DISCUSSION

3.1 CL of cereal food products

The air-dried grains of 10 different varieties tested exhibit ultralow CL with the S/N ratio varying from 1 to 2 (Figure 4A.). The addition of water results in a rapid increase of CL intensity, with the S/N value reaching greater than 10.The I_{max} values and kinetic curves I = f (t) depend upon the species of cereal. The time scale and the immediate CL response from both intact and ground grains indicate that the measured emission results from physico-chemical processes, not from much slower biological ones. The effect of temperature (T) on CL of whole (intact) wheat grains is shown in Fig.4B. The Van't Hoff coefficient $Q_{10} = I_{T-10} / I_T$ and activation energy $E_a = R \ln T_2 / I_1 (T_1 - T_2)$ evaluated from the data of Figure 4B. give in average 1.44 and 25 kJ/mol, respectively. These low values suggest that the process of CL generation is controlled by weak molecular interactions rather than by biological / enzymatic ones. Analysis of the water-induced CL decay kinetics of ground grain and flours indicates on a non-exponential, hyperbolic I = f (t) dependance:

$$I = f(t) = dn/dt = I_0 t^{-\beta} = BG \qquad (10)$$

where n is the number of counts per second, and I_0 and BG are the signal amplitude and the intensity of background emission, respectively. Hyperbolic dependance is usually ascribed to charge-recombination luminescence.[2] Charged species, such as e^- and ionradicals, can be formed in cereal products because of technological processes applied during the harvesting, drying, grinding, baking, etc. In dry materials these charges are immobilized, but the motion of charges through the biopolymer matrix can be accelerated by the imbibition of water. Such charge-recombination luminescence is characterized by a small T-dependance of the recombination rate, which fits well to the experimentally found E_a= 25 kJ/mol.

Spectra of ultraweak CL accompanying auto-oxidation and hydration of cereal products cover the 380-880 nm spectral range with maxima centered around 600 nm (Figure 5.). Air-dried products emit a quasi-stationary photon flux at the rate dn/dt (n is the number of counts proportional to the number of photons emitted from a sample) of about 7-1000 cps/cm^2 and S/N of 260 (Figure 5A.). Processes underlying these Cl most probably include a very slow auto-oxidation of polyphenols and/or lipids that also emit in the same spectral region with maximum around 600 nm[9]. The water-induced CL has the same spectral distribution within the limits of experimental error (Figure 5B.). The effect of reactive oxygen species (ROS) quenching, scavenging and complexing compounds on CL was also studied. Generally these compounds do not significant change emission spectra, although they affect (predominantly decrease) the total CL intensity.[8] Pure carbohydrates exhibit much weaker CL, both in the air-dried state and hydrated. CL spectra measured only in a few cases (because of the S/N < 1) resemble spectra of cereal products: the emission covers the 450-860 nm range and about 70% of its lies beyond = 600 nm.[8] The analysis of spectral data indicates a contribution of radical reactions with the participation of excited molecular oxygen dimers $^1\Delta_g$, $^1\Sigma_g^+$ and excited carbonyls $^{1(3)}$(>C=O)*.

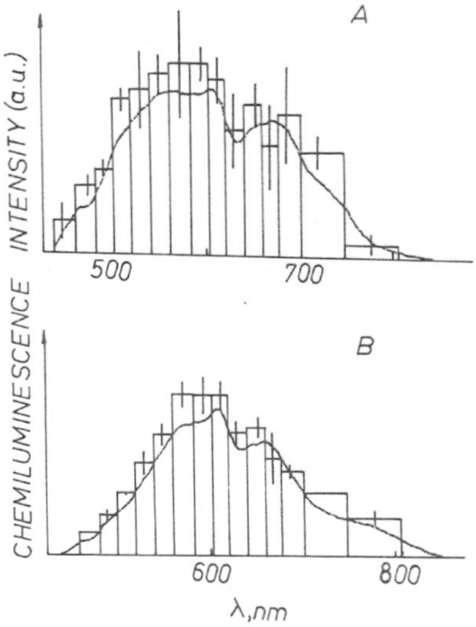

Figure 5. Emission spectra of air-dried (A) and water-treated (B) crackers. A red sensitive R-1333 PMT and 14 cut-off filters were used to measure spectra. Data average over 25 (A) and 32 (B) scans (cycles). Sampling time for each filter and background was 5 s. The width of rectangles is equal to the difference between short wavelength limits of the two consecutive filters. Error bar represent 50% confidence interval. All spectra are corrected for the trasmittance of filters, spectral sensitivity of the PMT and changes of the total emission with time. The relative intensity of spectra is expressed in arbitrary units (a. u.). Reproduced with the permission from John Wiley, & Sons D. Slawinska, J. Slawinski, J. Biolumin. Chemilumin. 13 (1998).

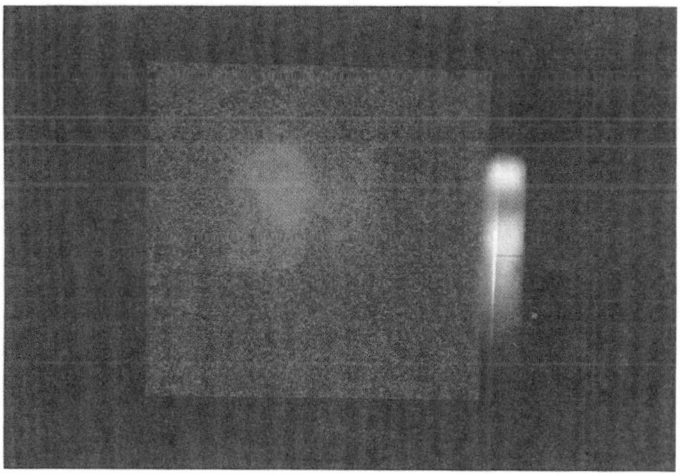

Figure 6. The two-dimensional pattern of the spatial distribution of ultraweak spontaneous emission from two crackers. At the left a wheat flour cracker with additives air-dried (lower part) and treated with water (right upper brighter part). Right: an air-dried rice flour cracker without additives.

Photon imaging or two-dimensional (2-D) spatial distribution of CL from auto-oxidizing and water -hydrated cereal products reveals the dynamics and emission pattern of complex heterogeneous chemiexcitation reactions and photon-generating processes (Figure 6.).

The effect of antioxidants and free radical scavengers, e.g. ascorbic acid (vitamin C), is clearly seen. Other compounds promoting or inhibiting radical chain reactions tested so far also gave distinct CL images.[10] These properties of CL imaging may be correlated with the supramolecular structure and energetic status of biopolymers - constituents of cereal foods. It is evident that the auto-oxidation reactions, imbibition and swelling processes which depend on hydrophillic/ hydrophobic properties of biopolymers and on water activity in the environment can be monitored and characterized by CL imaging. It seems very likely that the molecular energetic properties of cereal biopolymers, as estimated by CL imaging, are correlated with and contribute to the food quality. In this perspective, photon imaging techniques open new methodological possibilities, e.g. evaluation of grains intactness/damage which is directly correlated with the germination capacity, evaluation of the deterioration rate of food, observations of "energized" sites and locally modified or labelled biomaterials.

The problem of water-biomaterials interactions is also very interesting for theoretical reasons, since from previous observations [8] it appears that high-purity dried carbohydrates interacting with water emit ultraweak CL. Thus, the question arises whether weak molecular interactions, like hydrogen-bond formation, can accumulate excitation energy sufficient to excite electronic energy levels in the optical range.

3.2 Photoinduced Chemiluminescence of wood

The weathering process of the surface of wood exposed out of doors results in defibration loss of original colour and greying, and has serious consequences for wood technology. The loss of strip strength is associated with a light-induced depolymerisation of lignin and other cell constituents, and the subsequent breakdown of the wood microstructure. Freshly cut pine specimens did not exhibit any detectable photoinduced CL.

Changes due to outdoor exposure are collectively described as weathering. Pollutants, moisture and solar irradiation are the major weathering factors. Recently, environmental factors such an increase in the UV-B radiation connected with the ozone layer depletion may increase the rate of wood weathering. CL of wood induced by the UV and visible light (VIS) was observed for the first time and its kinetics and spectral distribution determined.[11-12]

The kinetic results of CL induced by visible light are presented in Figure 7. The observed decay of CL is best fitted by two exponential sum:

$$I(t) = a(1)\exp[-t/a(2)] + a(3)\exp[-t/a(4)] \qquad (11)$$

where I(t) is the intensity of observed CL a(1) and a(3) initial intensity value of components, a(2) and a(4) are time constant of each component, respectively. The shorter time constant a(2) is about 1.5 min whereas the longer one a(4) is about 30 min for different type of wood Such similarity of the time constants between different types of wood is suggesting a common mechanism connected with observed luminescence.

The emission spectra of photoinduced CL measured 15 min after irradiation are shown in Figure 8. All spectra have three spectral regions: 450-550 nm, 600-640 nm, and 700-720 nm, with slightly different ratios of intensities between them. A singlet oxygen quencher (sodium azide) and sensitizer (methylene blue) were used to elucidate mechanism of the measured CL. When the wood sample was sprayed with methylene blue solution prior to the irradiation, the observed spectrum was in the red region. This result clearly indicates on the singlet oxygen mechanism. Sodium azide diminishes the intensity of wood samples but does not change its spectral distribution. This result also confirms the above interpretation.

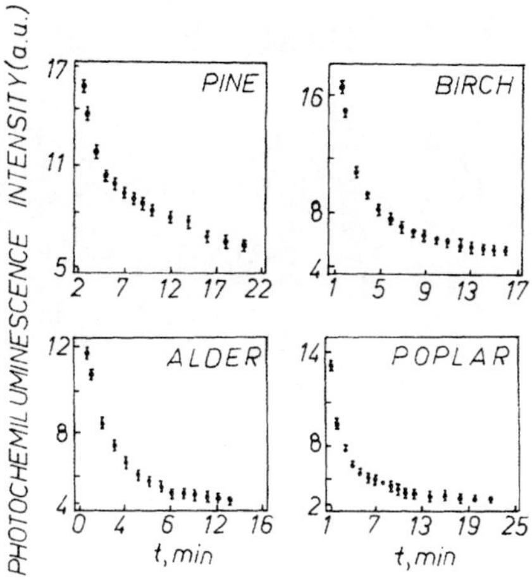

Figure 7. Chemiluminescence decay curves irradiated with visible light from xenon lamp for different wood surface.

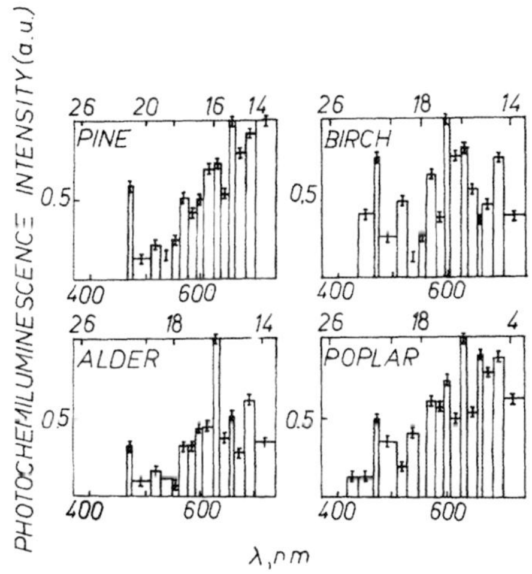

Figure 8. Chemiluminescence spectra irradiated with visible ligth for xenon lamp for different wood surface.

External components of wood, also called "impurities", such as e.g. waxes, pigments, carbonyl and quinoid structures are capable of absorbing light to raise them to excited triplet states. Thus, they can act as photosensitizers (S) in the near UV and visible part of the spectrum:

$$^1S \xrightarrow{h\nu} {^3S^*} \qquad {^3S^*} + {^3O_2} \rightarrow {^1S} + {^1O^*_2} \qquad {^1O^*_2} + A \rightarrow AO_2 \tag{12}$$

S - sensitizer (impurities), A - acceptor (substrate, a component of wood susceptible to oxidation). In this energy transfer process singlet molecular oxygen can be generated with the subsequent oxidation of wood.

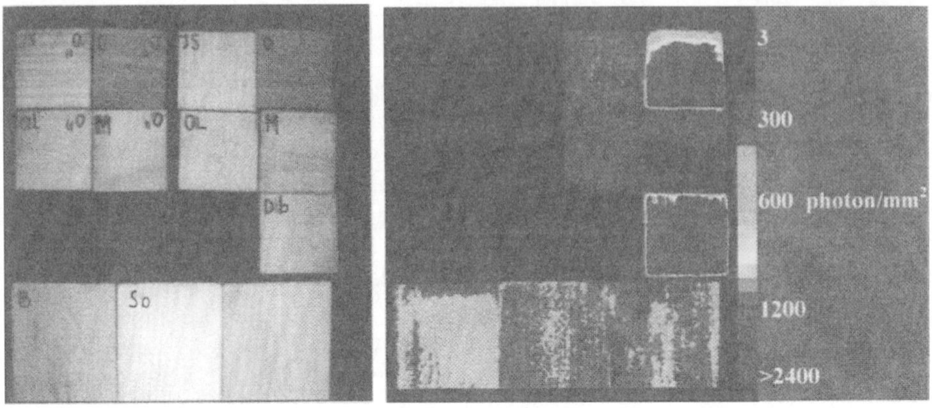

Figure 9. 2-dimensional distribution of ultraweak chemiluminescence of wood samples: IS-ash, OL-alder, M-larch, Db-oak, O-walnut, B-beech, So-pine, O-non irradiated. Left: the photograph of samples (exposure time 0.02s). Right: pseudocolored image, and the scaler (calibrated at λ = 650 nm, i.e. λ_{max} of CL). Before the irradiation procedure, samples were sprayed with 3ml water. Exposure time of luminescence imaging was 60s.

CL which occurs in the short-wavelength range 450-550 nm may be explained as originating from dioxetane mechanism and/or peroxyradical recombination. The interaction between light-induced free radicals and oxygen molecules probably leads to the formation of peroxides that are unstable towards heat and light[13]. In both cases carbonyl compound – the products of peroxides breakdown - generated in their first excited triplet states can be emitters of CL. The formation of hydroperoxides at wood surface enhances weathering degradation. The photodegradation of lignin and cellulose at wavelengths higher than 290 nm proceeds by different ways: photolysis is the main reaction for cellulose and photooxidation for lignin.[14] Our experimental data indicate that freshly cut specimens of wood do not exhibit any detectable photoinduced CL. This result correlates well with the lack of ESR signal from freshly cut pine specimens.[15]

Imaging of wood CL by means of Night Owl CCD camera is exemplified in Figure 9. As can be seen wood samples not irradiated with UV+VIS do not emit CL flux higher than about 100 photons/mm^2 s (λ = 650 nm). It cannot be detectable as an image even for 1h exposure (sampling) time. The strongest CL exhibit walnut and oak wood which are known to contain high amounts of phenolic compounds such as tannins and naphthoquinones. A satisfying spatial resolution of the CL inhomogeneity pattern and textural details are achieved. A significant role of water in generation of photoinduced CL is clearly seen: edges of samples exhibit the weakest emission since these places are drying faster than the central ones. Water adhesion to and absorption within grains (annual rings) are also manifested by differences in CL intensity. Thus, an ultrasensitive slow scan CCD camera may be an ideal tool for the investigation of kinetics and spatial patterns of interactions occuring between light, water and wood surface. The use of wood as ecological material has a wide variety of applications. There is an increasing trend towards the use of wood for exterior applications. Therefore the photon imaging of surface luminescent phenomena may be important for searching new and more effective antioxidants and retardants for protection of wood against deterioration.

4. CONCLUSIONS

Chemiluminescence imaging and auxilary measurements of CL kinetics and spectral distribution were successfully carried out with plant origin materials such as cereal grains and food products, and different kinds of wood. Using the 2-D single photon counting technique acquired by means of supersensitive imaging devices, it was possible for the first time to visualize and monitor ultraweak CL accompanying slow reactions contributing to deterioration of cereal food products and weathering of wood surface.

Photon imaging is an emerging technique ideal for the quantitative analysis of ultraweak CL from heterogeneous multiphase materials such as food products and wood. A new generation of slow scan CCD camera such as "Night Owl" Molecular Light Imager LB 981 is a powerful and sensitive tool for the investigation of energetics, kinetics and spatial patterns of interactions occuring between light, oxygen and water in complex inhomogeneous natural materials. It offers also an advantage of simultaneous CL measurements of multiple samples in the real time of the order of milliseconds to hours.

ACKNOWLEDGEMENTS. Thanks are due to Dr. Zbigniew Gorski of the Poznan University of Technology for kind help in preparing images of wood samples. This research was supported by grant number DS 508 182-2/98.

REFERENCES

1. E. H. White, Luminol Chemiluminescence, in: Chemi and Bioluminescence, J. G. Burr ed., Marcel Dekker, New York (1985).
2. G. D. Mendenhall, Chemiluminescence techniques for the characterization of materials, Angew. Chem. Int. Engl. 29, 362-373 (1990).
3. D. Slawinska and J. Slawinski, Applications of bioluminescence and low level luminescence from biological objects, in: Chemi and Bioluminescence J. G. Burr ed., Marcel Dekker, New York, 533-601 (1985).
4. B. Moeckel, J. Grand, R. Ochs, The Night owl molecular imager - a low light imaging system for bio- and chemiluminescence and fluorescence, in: Bioluminescence and Chemiluminescence - Molecular Reporting with Photons, J. W. Hastings, J. Kricka, P. E. Stanley eds., Chichester, Wiley, 539-42 (1997).
5. M. Gutekunst, M. Jahreis, R. Rein H . J. Hoeltke, The Lumi-Imager TM, a sensitive and versatile system for imaging, analyses and quantitation of chemiluminescence blots and in micro-titreplates, in: Bioluminescence and Chemiluminescence - Molecular Reporting with Photons, J. W. Hastings, L. J. Kricka, P. E. Stanley eds, Chichester, Wiley, 543-4 (1997).
6. R. Q. Scott, H. Inaba, Single photon counting imagery, J. Biolumin. Chemin. 4, 507-511 (1989).
7. D. Slawinska, J. Slawinski, Chemiluminescence of cereal products. II. Chemiluminescence spectra, J. Biolumin. Chemilumin. 12 (1988).
8. D. Slawinska, J. Slawinski, Chemiluminescence of cereal products. I. Kinetics activation energy and effect of solvents, J. Biolumin. Chemilumin. 12, 249-259 (1997).
9. D. Slawinska, Chemiluminescence and the formation of singlet oxygen in the oxidation certain polyphenols and quinones, Photochem. Photobiol. 28, 453-459 (1978).
10. D. Slawinska, J. Slawinski, Chemiluminescence of cereal products. III. Two-dimensional photocount imaging of chemiluminescence, J. Biolumin. Chemilumin. 13, (1998).
11. D. Slawinska, K. Polewski, Chemiluminescence as a new evidence for singlet oxygen participation during the weathering of wood, Polish J. Med. Phys. Engin. s 1, 249-250 (1995).
12. D.Slawinska, K. Polewski, Photoinduced chemiluminescence of wood. Symposium of Physical Organic Photochemistry, Poznan, Poland. Book of Abstracts I P-17 (1995).
13. D. N. S. Hon, Photooxidative degradation of cellulose: reactions of the cellulosic free radicals with oxygen. J. Polym. Sci. Polym. Chem. Ed. 17, 441-454 (1979).
14. J. M. Gaillard, M. L. Roux, D. Masson and X. Deglise, Analytical study of the photodegradation of wood with and without coating (alkyd resin) in: Xth JUPAC Symposium on Photochemistry, Interlaken Switzerland, 405-406 (1984).
15. D. N. S. Hon, G. Ifju, W. C. Feist, Characteristic of free radials in wood, Wood an Fiber 12, 121-130 (1980).

OPTICAL FIBER SENSORS FOR THE CULTURAL HERITAGE

A.G. Mignani, R. Falciai, and C. Trono

IROE-CNR
Via Panciatichi, 64
50127 Firenze, Italy

1. INTRODUCTION

The displaying of works of art in full respect for their artistic value involves the use of exhibition facilities that are characterized by environmental conditions which neither alter nor change the materials constituting the work of art in question. Nevertheless, many museums, especially in Europe, have been created within ancient buildings: they, too, a part of the artistic heritage. Consequently, architectonic and plant-engineering modifications such as would guarantee optimal protection to the works on display are simply not possible.

The sole precaution that can be taken in museums of this type is to introduce a network of sensors capable of identifying if and where there may exist environmental conditions that are at risk. In turn, this enables museum personnel to take the appropriate steps.

Optical fiber sensors demonstrated capable of solving many sensing problems, especially for nonstop monitoring in real time and for the implementation of sensor multiplexing over long-haul networks. For their application to the protection of the cultural heritage optical fibers offer the additional advantages of being transparent and dielectric. Transparency fulfills aesthetic requirements, while dielectricity provides intrinsic safety against dangers such as lighting, spikes and fires. In addition, fiber optic sensors offer optimized integration with fiber optic communication networks, which are being used always more often in modern museums and galleries.

The present paper presents two optical fiber sensors which have been implemented for a specific application to the proper conservation and protection of the cultural heritage.

2. OPTICAL FIBERS FOR MONITORING THE EFFECTS OF TEMPERATURE ON PICTURE VARNISHES

Transparent varnishes have been always applied to oil and tempera paintings, and have been always regarded as protective layers.[1,2] Actually, the effect of the varnish is twofold, not only on the safety of the painting, but also on the perception of the colors. In fact, the varnish

Optical Sensors and Microsystems: New Concepts, Materials, Technologies
Edited by Martellucci et al., Kluwer Academic / Plenum Publishers, New York, 2000.

substantially alters the appearance of paintings by creating a microscopically-smoother surface and by providing a refractive-index matching-medium between the air and the pigments. The incident light better penetrates the paint layer that appears glossier, and the scattered white light is reduced thus giving more saturated colors. Specialists say that varnishing paintings "bring out the colors" and "make them shiny". Because of the influence they have on the appearance of paintings, varnishes must exhibit stable or reversible optical characteristics (color, refractive index) even on ageing and in the presence of different environmental conditions.

Optical fiber technology can be used for the real-time and continuous monitoring of the temperature effects on the varnish refractive index. For this scope, a plastic-cladding silica-core (PCS) optical fiber was used. A short length of the cladding was stripped off and replaced by a layer of varnish, as shown in Figure 1. Consequently, the light intensity guided by the optical fiber was dependent on the refractive index of the varnished fiber-section, and any temperature variation affecting the varnish refractive index resulted in light-intensity modulation.

Three varnishes made of the most common natural resins used in picture restoration were tested: dammar resin, amber in linseed-oil and gum mastic, which are characterized by a refractive index of 1.539, 1.546 and 1.536, respectively.[3] Because of the high values of varnish refractive indices, the light from the quartz core of the varnished section was radiated into the varnish layer, and was totally reflected at the varnish-air interface. Polystyrene optical fibers, that have a core refractive index of 1.59, should be much more appropriate for obtaining a standing wave also in the varnished fiber section. The use of PCS optical fibers, in spite of their low value of core refractive index, was motivated by a possible implementation of the monitoring technique in a nearly-distributed network, for which low attenuation fibers are necessary.

The PCS optical fiber was a 3M-EOTec™-FP200LMT fiber with a core diameter of 200 μm. Lengths of fibers were stripped of the cladding at the center for a length of 15 mm using a razor blade; traces of plastic and grease were further removed using solvents (trichloroethylene and acetone). Each fiber was then fixed to a plastic support to prevent breakage and to be varnished and tested during thermal cycles.

The fiber optic test-unit was constituted by a LED source (@660 nm) and by two PIN detectors. The LED was connected to the common arm of a 1x2 fiber optic coupler. One arm of the coupler was directly connected to the PIN in order to check source stability (reference arm), while the other arm was connected to the varnished fiber and then to the other PIN. LED and PINs were interfaced to a PC by means of a Data Acquisition Processing (DAP) board providing LED driving current and modulation, together with signal detection and processing. The ratio between the varnished-fiber output and the reference-arm output was considered to be the sensor output.

The bare fiber-core was treated by spraying it with varnish, and the sensor output was continuously monitored for two days so as to check the varnish hardening-phase.

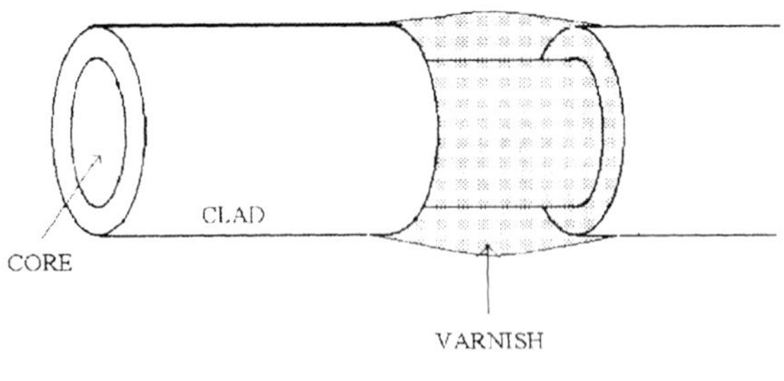

Figure 1. PCS optical fiber with varnish-replaced cladding.

Once hardened, the varnished fiber-section was heated by means of a hot plate positioned under the fiber while sensor output was being measured. The reference temperature was measured by means of a thermocouple the output of which was processed by the DAP. Several temperature cycles in the 20-65°C range were performed in order to check both thermal sensitivity and reversibility.

The temperature sensitivity of dammar and gum mastic is shown in Figures 2. and 3., respectively. All three of these varnishes exhibited temperature sensitivity to some extent: thermal behaviour of the refractive index was fully reversible for gum mastic, irreversible for dammar, slightly sensitive and poorly reversible for amber.

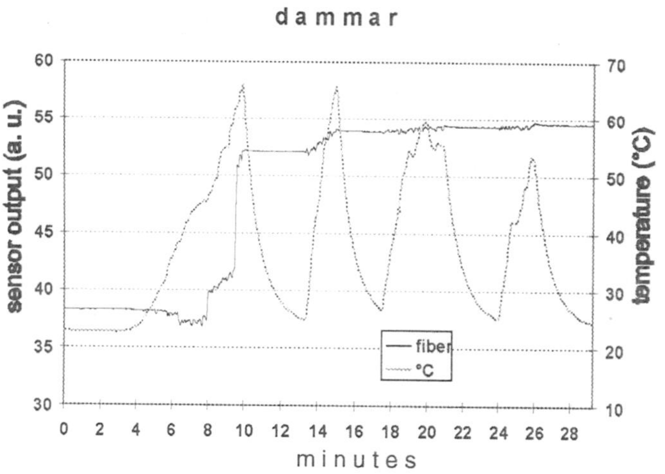

Figure 2. Optical fiber varnished using dammar: response to thermal cycles.

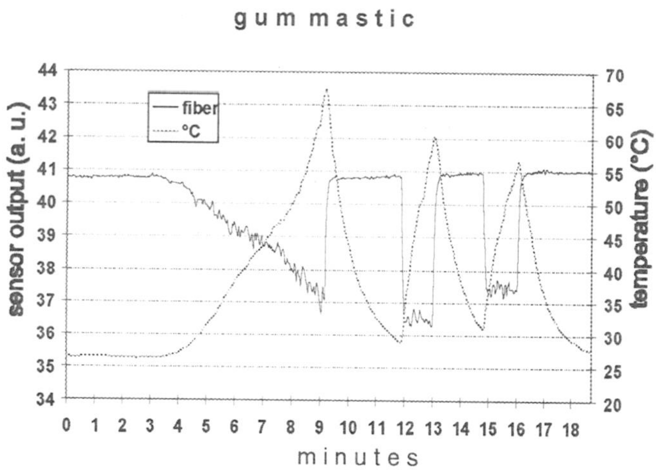

Figure 3. Optical fiber varnished using gum mastic: response to thermal cycles.

The experimental results obtained suggest two different uses for the varnishes. When a thermally-stable refractive index is required for a stable appearance of a painting, the most suitable varnish is dammar, provided that a preliminary thermal cycle is carried out. On the contrary, if the monitoring of painting thermal excursions should be performed, the gum mastic varnish is the most suitable.

3. THE MONITORING OF LIGHTING IN MUSEUM ENVIRONMENTS BY MEANS OF OPTICAL FIBERS

Light is one of the most important factors enabling visitors to fully enjoy the visual aspect of art. However, all organic materials such as paper, textiles, resins and leathers are damaged by excessive or unsuitable lighting, because of the light-induced chemical reaction. A typical reaction is photo-oxidation, which is the formation of free radicals caused by photon absorption and subsequent combination of the radicals with oxygen molecules. Photo-oxidation is a cumulative and irreversible effect; consequently, artistic material remains permanently damaged by the wrong lighting.[4] The guidelines for correct illumination suggest a maximum illuminance of 50 lux for highly delicate materials such as watercolors, paper and textiles, while many robust materials, such as tempera paintings and wood, can support an illuminance of 150 lux. As far as the fraction of UV light is concerned, all materials should be illuminated by means of 75 μW/lm at maximum.

An optical fiber sensor was implemented which is capable of measuring lighting in museum environments. Such a sensor is an extrinsic, intensity-modulated type sensor, as sketched in Figure 4. The sensitive element is a photochromic material made of fulgides[5,6] which are immobilized on a polymeric matrix. The photochromic transducer is coloured by UV-lighting, and is bleached by VIS-lighting. The temporal behaviour of fulgide colouring when exposed to both UV and VIS lighting can be expressed as

$$y(t) = -asympt + A \cdot e^{-m \cdot t} \qquad (1)$$

where A is a constant depending on the starting state of the fulgide, and asympt and m are proportional to

$$asympt \propto \frac{UV(\mu W/m^2)}{UV(\mu W/m^2) + VIS\,(lux)} \qquad m \propto UV(\mu W/m^2) + VIS(lux) \qquad (2)$$

In practice, the asymptotic value, asympt, represents the attainment of the photostationary state, while the time decay constant, m, represents the combined action of UV and VIS lighting. Consequently, measurement of both asympt and m makes it possible to obtain the fraction of UV, together with the individual UV and VIS light levels.

The probe consists of a 0.17-mm thick layer of fulgides aligned between of an optical fiber link coupled to the interrogation electro-optic unit. Multimode optical fibers 200-μm core diameter were used. The electro-optic unit makes use of commercially-available LED and PIN as source and detector. The LED spectrum is superimposed on the fulgide absorption spectra. The electronics is interfaced to a portable PC equipped with a specifically-developed software for both calibration and measurement functions.

In order to test the optical fiber sensor during normal operating conditions, both the reference photometer and the fiber sensor were exposed to natural lighting for many hours while their outputs were recorded.

The agreement between their responses was encouraging, suggesting the testing of the optical fiber sensor in a real museum environment.

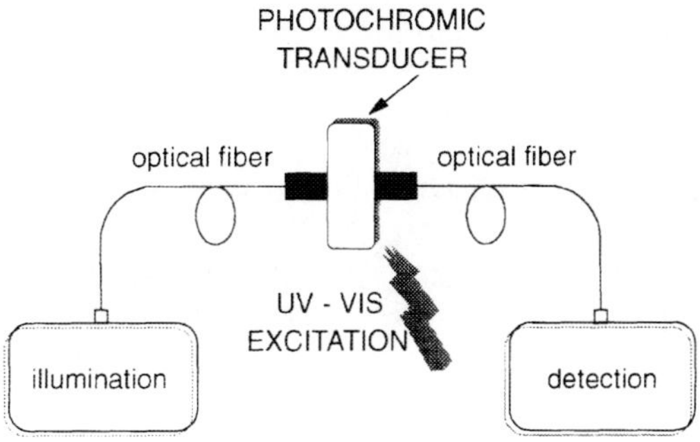

Figure 4. Sketch of the sensor for monitoring lighting conditions.

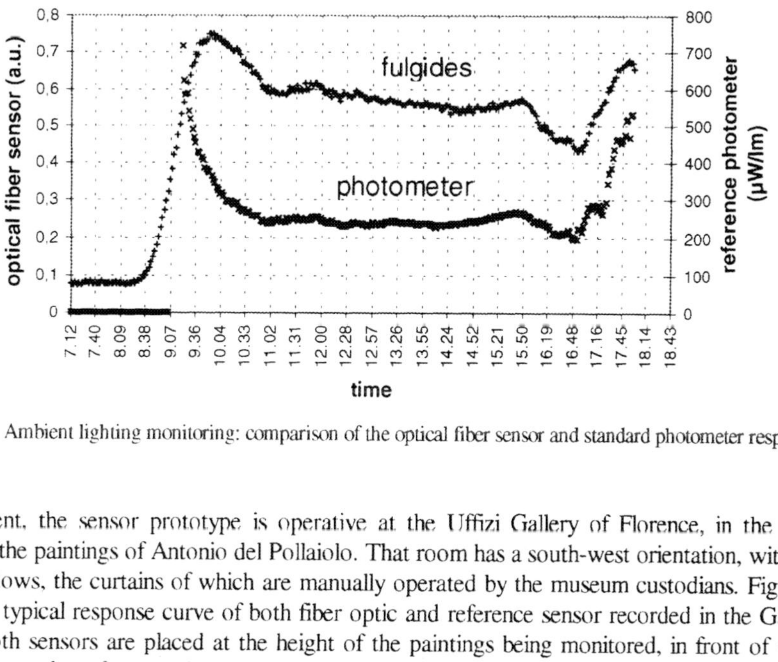

Figure 5. Ambient lighting monitoring: comparison of the optical fiber sensor and standard photometer responses.

At present, the sensor prototype is operative at the Uffizi Gallery of Florence, in the room housing the paintings of Antonio del Pollaiolo. That room has a south-west orientation, with two big windows, the curtains of which are manually operated by the museum custodians. Figure 5. shows a typical response curve of both fiber optic and reference sensor recorded in the Gallery. Since both sensors are placed at the height of the paintings being monitored, in front of which people pass, the reference photometer output, which is rather fast, is highly fluctuating. On the contrary, the long response time of the fiber optic sensor, which could be a drawback for other photometric applications, in this case gives more stable and always significant results. It is to be noted that, when the Gallery is closed to the public, emergency lighting is operative. The light level of emergency fluorescent lamps is too low to be detected by the reference photometer, while the fiber optic sensor still records the small UV fraction.

4. CONCLUSIONS

Two examples of optical fiber sensors for the protection of the cultural heritage were given. The varnished optical fiber could be used also as temperature sensor. In fact, thanks to the good temperature sensitivity and reversibility of gum mastic, it could be considered as a transducer for the implementation of a temperature sensor to be permanently inlayed in the painting. By embedding the optical fiber in the painting together with the picture varnish, for example on a corner, continuous temperature monitoring could be possible, in order to prevent risk conditions that can arise when illuminating the painting with the use of lamps, as happens during television shots.

The fiber optic sensor for monitoring of lighting in museum environments has demonstrated its functionality in real operational conditions. Another application of the sensor for safeguarding the cultural heritage, is for lighting monitoring during artwork restoration. In fact, restoring procedures are performed under high illumination, although for a short time compared to the life of the artwork. Since there are regulations also on excessive lighting for short times, continuous monitoring can indicate when the exposure is becoming too excessive. The low cost of the technology involved, indicated as a further development of the sensor an implementation as a battery-powered unit to be permanently installed where critical conditions are feared.

In addition to temperature and lighting, other parameters of interest for the monitoring of museum environmental conditions are: humidity, sound, vibrations and air pollution, especially particulates and gaseous pollutants such as sulphur dioxide, ozone, nitrogen oxides.

ACKNOWLEDGEMENTS. The work on monitoring of lighting has been partially funded by the National Research Council of Italy, under the *"Cultural Heritage"* Special Project.

REFERENCES

1. E. René de la Rie, "The influence of varnishes on the appearance of paintings", *Studies in Conservation* 32, pp. 1-13, 1987.
2. R.S. Hunter, *The Measurement of Appearance*, John Wiley & Sons, New York, 1975.
3. E. René de la Rie, "Old Master Paintings", *Anaytical Chemistry* 61, pp. 1228A-1240A, 1989.
4. G. Thomson, *The Museum Environment*, Butterworth-Heinemann Ltd., Oxford UK, 2nd Edition, 1986.
5. H.G. Heller, "Photochromic heterocyclic fulgides: part 7", *J. Chem. Soc. Perkin II*, 1992, pp. 591-608.
6. H.G. Heller, "Fulgides and related systems" in *CRC Handbook of Organic Photochemistry and Photobiology*, Boca Raton FL, 1995, pp. 174-191.

FIBER OPTICS REFLECTANCE SPECTROSCOPY: A NON-DESTRUCTIVE TECHNIQUE FOR THE ANALYSIS OF WORKS OF ART

M. Picollo, M. Bacci, A. Casini, F. Lotti, S. Porcinai, B. Radicati and L. Stefani

Istituto di Ricerca sulle Onde Elettromagnetiche "Nello Carrara" - CNR
Via Panciatichi 64
50127 Firenze, Italy

1. INTRODUCTION

The study of materials constituting works of art is of fundamental importance for conservation purposes as well as for obtaining a deeper knowledge of the techniques used by artists.

Starting from the fact that a work of art is a «unique» piece, the need to develop non-invasive methodologies is of utmost consequence. For this reason, the main objective of the research carried on by the authors at the Istituto di Ricerca sulle Onde Elettromagnetiche "Nello Carrara" (IROE) of the Italian CNR is the development and improvement of non-destructive methodologies in the visible (Vis) and infrared (IR) regions, in general, and of fiber optics reflectance spectroscopy (FORS), in particular, for the investigation of artworks.[1,2]

To the best of our knowledge, the first application of the fiber optics to the investigation of paintings was made at the National Gallery of London in the late seventies and early eighties.[3,4] This application fell within a program focused on the monitoring of light induced color changes.

In general, the use of FORS is proposed for identification of pigments,[5] by correlating the sample spectrum to a suitable data-base, for analyzing color changes on paintings,[6,7] and for monitoring the presence of alteration products[8,9] and the status-of-health of objects.

Due to its dimension and weight, FORS apparatus can be easily transported and used for collecting spectra *in situ* thus giving us the possibility to follow restoration work even in somewhat uncomfortable and difficult situations. Moreover, a large number of spectra can be recorded, offering the possibility of a statistical treatment of the data, which is useful for the identification of different compounds.[10,11] In the past years, the FORS device has also been proposed in the Near-IR for the detection of alteration products (i.e.: gypsum, calcium oxalates) on frescoes, calcareous statues and monuments.[8,9]

In the present work, the Vis and Near-IR reflectance spectra of several pigments from various paintings are reported.

Optical Sensors and Microsystems: New Concepts, Materials, Technologies
Edited by Martellucci et al., Kluwer Academic / Plenum Publishers, New York, 2000.

2. EXPERIMENTAL

FORS measurements were generally performed on pure powders, laboratory panels, and actual paintings using a spectrum analyzer UOP Guided Wave mod. 260 equipped with 316SS optical fibers. The source was a 20Watt tungsten-halogen lamp.

In the first configuration the light beam was sent to the investigated areas of the paintings by means of an optical fiber bundle ("Y" shaped) operating in the range of 400-2200nm which is constituted of 19 fibers (9 sending and 10 receiving the back-scattered light). The received radiation passed through a diffraction grating (800L/mm for the range 400-1050nm and 300L/mm for the range 1050-2200nm) and was then detected by a sensor (Silicon or PbS cell, respectively for the two intervals). The probe for the "Y" shape fiber bundle was modified by connecting the distal end of the optical fiber bundle to a drilled cylinder of blackened PVC terminated with a flat base (Figure 1.). With this configuration, since light impinges near-perpendicularly upon the surface, specular reflected light is also recorded which would alter to some extent the values of the chromaticity coordinates in the case of glossy surfaces.

The second configuration of the fibers-probe apparatus consisted of two bundles of fibers (one sending and the other –"Y" shaped- receiving the back-scattered light) operating in the range 220-1100nm constituted of 30 fibers. The probe was a dark hemisphere of a 3cm-diameter terminated with a flat base and having three apertures: one at the top of the dome (for lighting the sample) and the other two placed at 45°, symmetrically in respect to the first (Figure 1.). This set-up permitted to work in diffuse reflectance collecting the light scattered at 45° with respect to the incident light, thus avoiding specular reflected light. This arrangement was one of the configurations recommended by CIE (Commission Internationale de l'Eclairage) for evaluating the chromaticity coordinates.[12]

Figure 1. The two fibers-probe configuration used in connection with the spectrum analyzer. Left: dark hemisphere probe; right: drilled cylinder probe.

The lower surfaces of the two probes were supplied with an O-ring, in order to guarantee a reliable contact and to keep the optimum distance between the optical fibers and the painting's surface (around 3.5mm). These probes allowed us also to avoid the radiation from the environment, while giving a soft but stable contact with the surface of the painting.

Commonly, in the 400-1000nm (Vis and very Near-IR) and 1200-2200nm (Near-IR) ranges the reflectance spectra were recorded at 1nm and 2nm intervals respectively on small areas of about $0.3cm^2$ using Barium Sulfate or Spectralon® as references. For each instrumental configuration 3-5 spectra were generally collected on each spot sample and their average in the Vis region was used to calculate the chromaticity coordinates. The range 1000-1200nm is usually not considered, since the noise results roughly ten times higher than elsewhere because the two electro-optical configurations are not very suitable for detecting radiation in this range. However, sample absorption intensity for the Vis and very Near-IR configuration fitted well with the Near-IR one at 1000-1200nm.

When FORS was applied to the study of visual or color variations of paintings, a method for identifying the measured points after a given time was developed. Indeed, the positioning of the optical fibers probe exactly upon the same examined area was very important since color changes had to be determined by comparing the reflectance spectra of the cleaned samples with the spectra collected before cleaning treatments. For wall-paintings and frescoes, a weakly adhesive marks for photogrammetric recording were located on the work of art in order to accurately identify the positions of the various spots by reference to these marks. In case of canvas and panel paintings a transparent film of Mylar was fitted to the surface and fixed with adhesive tape to the border or back of the painting. Subsequently the areas to be investigated were selected and marked with a pencil, several references being drawn on the film in correspondence of clear and sharp contour. Successively, the Mylar film was removed and circular holes were made in correspondence to the marks. Moreover, white crosses were drawn on the optical fiber probes in order to align the light beam precisely on the spot to be measured. Finally, the film was fitted back to the painting and the spectra were recorded.

3. PIGMENT IDENTIFICATION

Identification of the pigments constituting the "palette" of an artist was done by comparing the spectrum of the investigated sample with the ones available in our data-base (correlation analysis or, in some cases, simply visual inspection). The data-base was prepared by collecting the reflectance spectra on a large quantity of pure pigments and its implementation has been in process for more than one decade.

The identification and subsequent discrimination among different pigments depend on the electronic structure of the chemicals constituting the pigment. Indeed, a variety of electronic (Crystal-Field, Charge-Transfer, Valence-Conduction band transitions) and vibrational processes (Overtones and Combinations transitions), falling within the Vis and Near-IR regions, may produce typical intrinsic spectral features that appear in the absorption or reflectance spectra.

Below, a set of few cases of pigment identification is reported.

The reflectance spectrum of a blue pigment collected from the wall-painting "Caduta della manna" by Allori in the old refectory of the Santa Maria Novella Church (Florence) is reported in Fig. 2. The features of the spectrum show an absorption band which is split in three sub-bands. This fact is typical of Co(II) ions in an approximate tetrahedral geometry and is a consequence of ligand field and Jahn-Teller interactions[5]. Accordingly the blue pigment was identified as smalt. In fact smalt is a potassium-silicate glass colored with cobalt oxide, then reduced to powder, where Co(II) ions are coordinated by four oxygen ions in an approximate tetrahedral geometry.

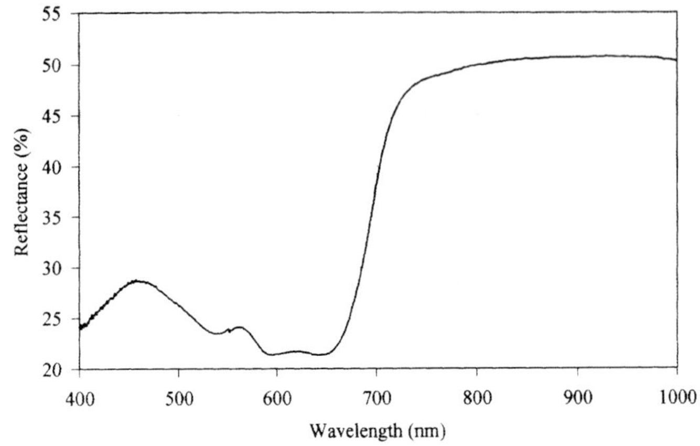

Figure 2. Vis and very Near-IR reflectance spectrum of the pigment smalt.

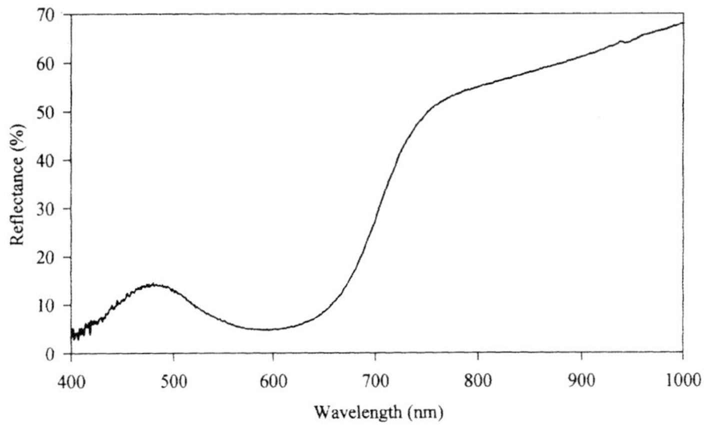

Figure 3. Vis and very Near-IR reflectance spectrum of the pigment *lapis lazuli*.

In Figure 3. the Vis and very Near-IR reflectance spectrum of a light blue spot of the mantle of the Virgin from the "Croce di Santa Maria Novella" by Giotto is also displayed. In this case, the absorption wavelength and the line-shape of the spectrum are typical of the pigment *lapis lazuli*, $((Na,Ca)_8(AlSiO_4)_6(SO_4,S,Cl)_2$. The reported *lapis lazuli* is mixed with a white pigment. Natural pigment *lapis lazuli* derives from the semi-precious stone *lapis lazuli*, which is a mixture of the blue mineral lazurite with calcite, pyrite, and other minerals. Nassau reported that the color of the pigment is due to the entrapped species S_x^- (x=2,3) *via* anion to anion charge transfer.[13] Malachite, $CuCO_3{\cdot}Cu(OH)_2$, is a basic copper carbonate and in the past was one of the most used green pigments. Its reflectance spectrum (Figure 4.) shows the presence of Copper(II) which is related to the wide absorption band from 600 to 900nm. This spectrum was collected from the fresco "La Temperanza" by Dosso Dossi in the interior of the «Sala del Camin Nero» of the Castello del Buonconsiglio (Trento).

The presence of goethite (α-FeOOH), a yellow pigment, can be revealed by its two quite well-resolved absorption edges around 650 and 900nm as well as the shoulder between 450 and 480nm (Figure 5.). Red and burgundy colors may be obtained with hematite (α-Fe_2O_3), which is spectrally similar to goethite but it has a reflectivity drop-off in the blue region (Figure 5.). The features of the reflectance spectra of these iron(III) minerals in Vis and very Near-IR regions can be interpreted on the basis of the ligand field and molecular orbital theories. Both the spectra reported in Figure 5. are from the wall-painting "Giudizio Universale" by Zuccari in the interior of the dome of Santa Maria del Fiore Cathedral (Florence).

Black and white pigments cannot be easily identified by FORS since these pigments usually do not show any typical features in Vis and Near-IR regions.

Finally, the FORS device can also be applied in the Near-IR region for studying alteration products on substrates made of carbonaceous materials (statues, frescoes, and monuments). Among these alteration products, gypsum ($CaSO_4$ $2H_2O$), and weddellite (CaC_2O_4 $2H_2O$) are the most common. Their Near-IR reflectance spectra along with the spectrum of calcite ($CaCO_3$) are reported in Figure 6.

4. CONCLUSION

From the results reported, the FORS device can be considered a useful and non-invasive tool for acquiring spectral information from paintings and wall-paintings in order to identify pigments, to analyze color changes, to monitor the status-of-health, and to detect the presence of alteration products.

Due to its dimension and weight, the FORS apparatus can be easily transported and used for collecting spectra *in situ* thus giving us the possibility to follow restoration work even in somewhat uncomfortable and difficult situations. Moreover, since a large number of spectra can be recorded, it offers the possibility of a statistical treatment of the data in order to discriminate among different compounds.

The recent advent of Mid-IR optical fibers might profitably extend in the following years the FORS domain to this spectral region (2.5-10μm) which should provide more exhaustive information.

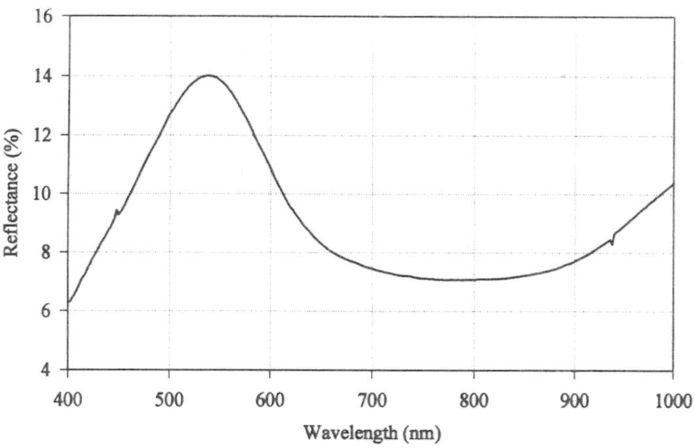

Figure 4. Vis and very Near-IR reflectance spectrum of the pigment malachite.

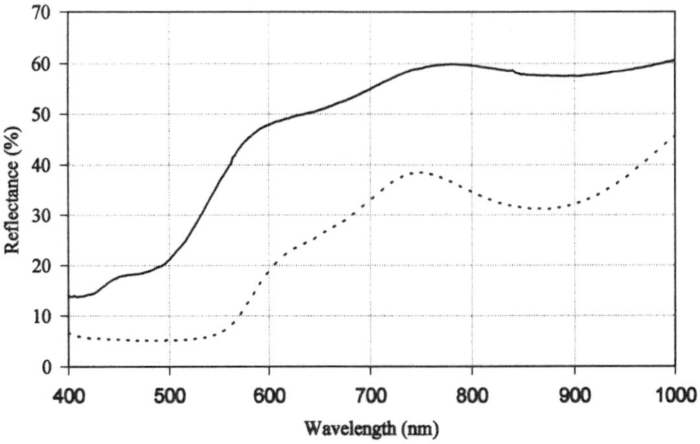

Figure 5. Vis and very Near-IR reflectance spectra of the pigments goethite (solid line) and hematite (dotted line).

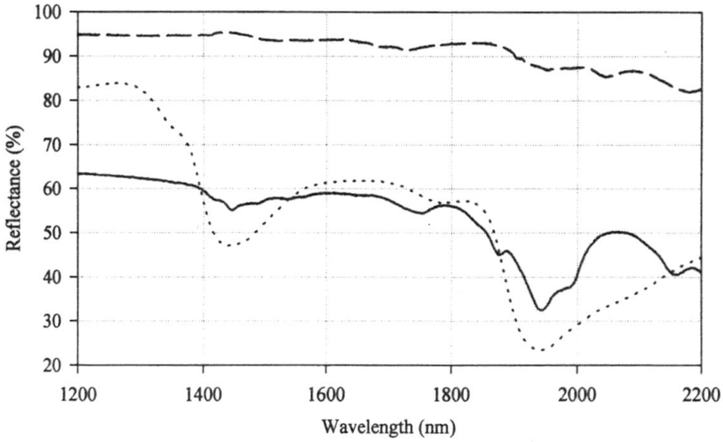

Figure 6. Near-IR reflectance spectra of calcite (dashed line), gypsum (solid line), and weddellite (dotted line).

ACKNOWLEDGMENTS. The authors thank Dr. Ornella Casazza (Sovrintendenza Beni Storici e Artistici – Florence) and Mr. Guido Botticelli (Università Internazionale dell'Arte – Florence) for allowing spectral analysis of the fresco from the old refectory of the Santa Maria Novella Church. They also thank Dr. Marco Ciatti (Painting Restoration Dept., Opificio delle Pietre Dure – Florence) and Ms. Paola Bracco (Opificio delle Pietre Dure) for providing the authors the facilities needed for recording spectra from the "Croce di Santa Maria Novella". They gratefully acknowledge Dr. Mauro Matteini (Scientific Dept., Opificio delle Pietre Dure) for giving them the opportunity to collect spectra from the wall-painting of the Santa Maria del Fiore Cathedral. Special gratitude is due to Dr. Laura Dal Prà, Ms. Francesca Raffaelli, and Mr. Roberto Perini (Ufficio Beni Storici e Artistici – Provicia Autonoma Trento) for the spectroscopic analysis and documentation of the frescoes at the Castello del Buonconsiglio.

This work was partially supported by the Progetto Finalizzato «Beni Culturali» of the National Research Council of Italy and by the EC Project Environmental Research of Art Conservation (ERA, Contract No. EV5V CT 94 0548).

REFERENCES

1. M. Bacci, S. Baronti, A. Casini, F. Lotti, M. Picollo and O. Casazza, Non-destructive spectroscopic investigations on paintings using optical fibers, in: *Mater. Res. Soc. Symp. Proc.*, **267**, 265 - 283 (1992).
2. M. Bacci, F. Baldini, R. Carlà and R. Linari, A color analysis of the Brancacci Chapel frescoes, *Appl. Spectrosc.*, **45**, 26 – 31 (1991).
3. L. Bullock, Reflectance spectrophotometry for measurement of colour change, *Natl. Gallery Tech. Bull.*, **2**, 49 – 55 (1978).
4. D. Saunders, Colour change measurement by digital image processing, , *Natl. Gallery Tech. Bull.*, **12**, 66 – 77 (1988).
5. M. Bacci and M. Picollo, Non-destructive spectroscopic detection of Cobalt(II) in paintings and glass, *Studies in Conservation*, **41**, 136 – 144 (1996).
6. M. Bacci, M. Picollo, B. Radicati and R. Bellucci, Spectroscopic imaging and non-destructive reflectance investigations using fiber optics, in: *4th Intern. Conf. Non-Destructive Testing of Works af Art Proc.*, **45.1**, 162 – 174 (1994).
7. M. Bacci, M. Picollo, S. Porcinai and B. Radicati, Non-destructive spectrophotometry and colour measurements applied to the study of the works of art, *Techne*, **5**, 28 – 33 (1997).
8. M. Bacci, S. Baronti, A. Casini, P. Castagna, R. Linari, A. Orlando, M. Picollo and B. Radicati, Detection of alteration products in artworks by non-destructive spectroscopic analysis, in: *Mater. Res. Soc. Symp. Proc.*, **352**, 153 - 159 (1995).
9. R. Chiari, M. Picollo, S. Porcinai, B. Radicati and A. Orlando, Non-destructive reflectance spectroscopy in the discrimination of two authigenic minerals: gypsum and weddellite, in: *2nd Intern. Symp. The oxalate films in the conservation of works of art Proc.*, 379 – 389 (1996).
10. A. Orlando, M. Picollo, B. Radicati, S. Baronti and A. Casini, Principal component analysis of near-infrared and visible spectra: an application to a XIIth century Italian work of art, *Appl. Spectrosc.*, **49**, 459 – 465 (1995).
11. M. Bacci, R. Chiari, S. Porcinai and B. Radicati, Principal component analysis of near-infrared spectra of alteration products in calcareous samples: An application to works of art, *Chemom. and Intell. Lab. Syst.*, **39**, 115 - 121 (1997).
12. G. Wyszecki and W.S. Stiles, *Color science: concepts and methods, quantitative data and formulae*, Wiley, New York (1982).
13. K. Nassau, *The physics and chemistry of color*, Wiley, New York (1983).

OPTICAL DIAGNOSTIC SYSTEMS AND SENSORS TO CONTROL LASER CLEANING OF ARTWORKS

R. Pini, and S. Siano

Istituto di Elettronica Quantistica - CNR
Via Panciatichi 56/30
50127 Firenze, Italy

1. INTRODUCTION

The possibility to achieve fine and selective material ablation by pulsed laser radiation has been well recognized and is routinely employed in various fields, as medical and industrial application of lasers. In the conservation of artworks, the utilization of laser-induced ablation to remove encrustation and degenerated layers from stone has been proposed since 70's by John Asmus who performed pioneering trials of laser cleaning on marble sculptures in Venice.[1,2] Still, the application of this technique to the conservation of very precious artworks has to be considered in a preliminary phase, since specific studies on the process of laser cleaning of artworks, such as stones, paintings and glasses have been reported only very recently.[3]

In this respect, it seems very important to investigate the laser cleaning mechanism with particular concern to the evaluation of the possible side effects associated with material removal. To this aim, monitoring system and sensors can be devised to provide direct control during laser operations. Moreover, considering that the interaction process of laser radiation with most of the materials involved in artworks conservation is still to be investigated, a specific effort has to be done to set up diagnostic monitoring techniques that can provide qualitative and quantitative information on the evolution of physical parameters to describe the dynamics of transient processes associated with the pulsed irradiation, as plasma formation, shock wave expansion, material ejection and so forth. These kind of measurements are possible using especially devised time-resolved imaging methods. The results of the measurements can be recorded and processed in real time, and the whole monitoring system can be arranged and engineered as an optical sensor to assist the laser cleaning operation.

2. THE CLEANING PROCESS

According to the present comprehension of the physical mechanism involved in the laser cleaning process of artworks, the main emission parameters that have to be

Optical Sensors and Microsystems: New Concepts, Materials, Technologies
Edited by Martellucci et al., Kluwer Academic / Plenum Publishers, New York, 2000.

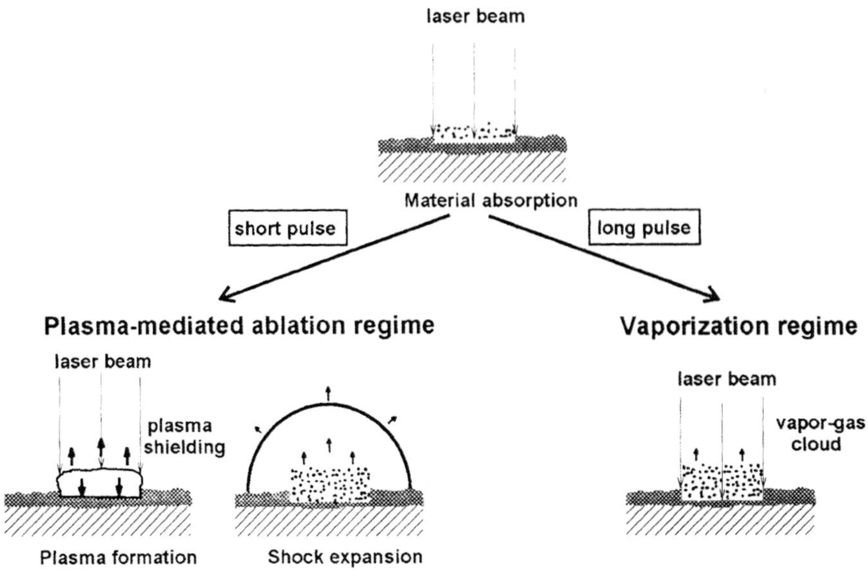

Figure 1. Scheme of laser-induced cleaning processes depending on the laser pulse duration.

controlled in order to achieve an effective cleaning action without occurrence of side-effects are the laser wavelength, the irradiation dose and the laser pulse duration. The choice of these parameters is typically related to the characteristics of the material to be cleaned, i.e. the superficial morphology, as well as physical and chemical properties of both alteration layer and substrate of the artwork. For example, since the penetration depth of laser radiation into the material depends on the laser wavelength, ultraviolet laser light is more suitable to remove thin layers of few microns or less from painting and glasses, while visible and near infrared light is more effective when thicker encrustation are concerned, up to millimeters, as it is the case of stones exposed to urban pollution. Also, the optical properties of the surface can play an important role to make the cleaning process more selective. A typical example is the possibility to obtain controlled removal of black crusts from the surface of clear stones as white marble by using visible or near-infrared laser light, since the cleaning action is driven by the higher optical absorption of the crust with respect to the stone substrate.

When the proper irradiation wavelength has been chosen, the most critical parameter for the final quality of the cleaned surface to be determined is the laser pulse duration. Given that pulsed laser radiation is more suitable than continuous laser emission for cleaning purposes since the expected result is the removal of the alteration layer rather than superficial melting, very different ablation regimes are induced by varying this parameter. Short laser pulses in the nanosecond range give rise to marked photomechanical effects as a consequence of the series of physical processes induced by material irradiation with high laser intensity. In this case the photon flux onto the target typically produces ionization, surface optical breakdown and then formation of plasma that absorbs laser light and optically shields the material surface. The rapid heating and expansion of the plasma cloud originates an intense shock wave that strongly accelerates material removal, as shown in Figure 1. However, the pressure transient associated with the shock can cause at the same time mechanical side-effects on the surface of the irradiated material, as local cracks, microfragmentation, increases porosity of the substrate. Thermal damages are usually negligible for laser pulses in the nanosecond range, because the ejection of overheated material is faster than heat diffusion into the substrate. At longer pulse duration in the microsecond range or longer, plasma formation can still occur at the stone surface, but with a lower optical density that permits direct material absorption. The removal process is more likely a fast thermal vaporisation than a plasma-mediated ablation. Considering possible side-effects, in this case photomechanical damages are less evident, whereas thermally-induced modifications assume an increased weight with longer duration pulses, up to the millisecond range where the typical effects of continuous irradiation can be recognised.

3. IMAGE DIAGNOSTICS

The phenomena described above induced during laser-material interaction can be imaged and quantitatively analyzed by means of a flash photo set up based on a pump-and-probe scheme, sketched in Figure 2. The probe beam from a Nitrogen laser (337 nm, 0.5 ns pulse duration, 50 μJ pulse energy), after spatial filtering and collimation, is sent tangentially to the sample surface to probe the air region where laser interaction and material removal occurred. The emission of the probe laser and of a Nd:YAG laser (1064 nm) used for stone cleaning are synchronized by a digital delay generator that permits a temporal scan of the whole event with a time jitter of about 10 ns in the first few microseconds. The shadowgraphic patterns produced by local refractive index perturbations and ejected particulate are detected by a CCD camera directly coupled to a frame grabber and stored in a personal computer for data analysis.

As an example of the application of this technique, we report a study on the effects of two different laser pulse durations for the cleaning of black crusts from marble samples. The laser emission under test are indicated as follows: 1) QS pulse, 6 ns duration, emitted by a Nd:YAG laser operated in Q-switching regime; 2) SFR pulse, 20 μs duration from a home made Nd:YAG laser operated in a "short pulse" free running regime, obtained by suitably shortening the flash lamp driving pulse.

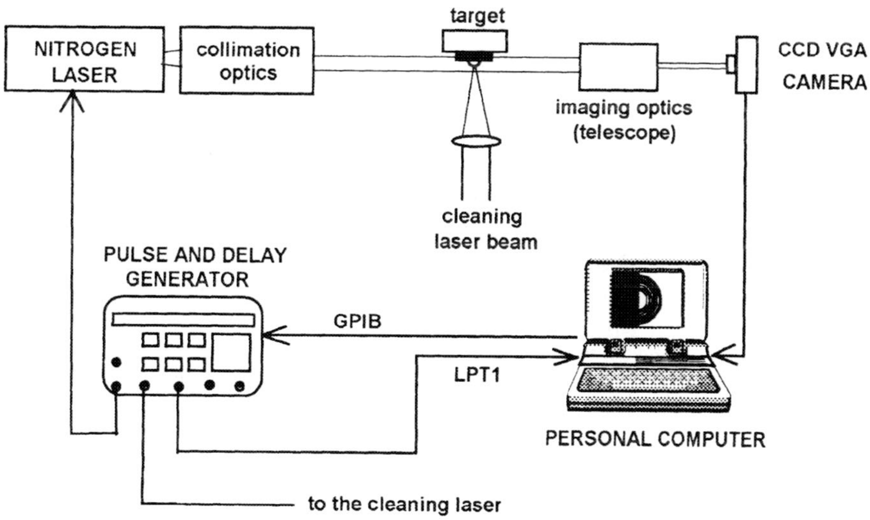

Figure 2. Flash photo set up.

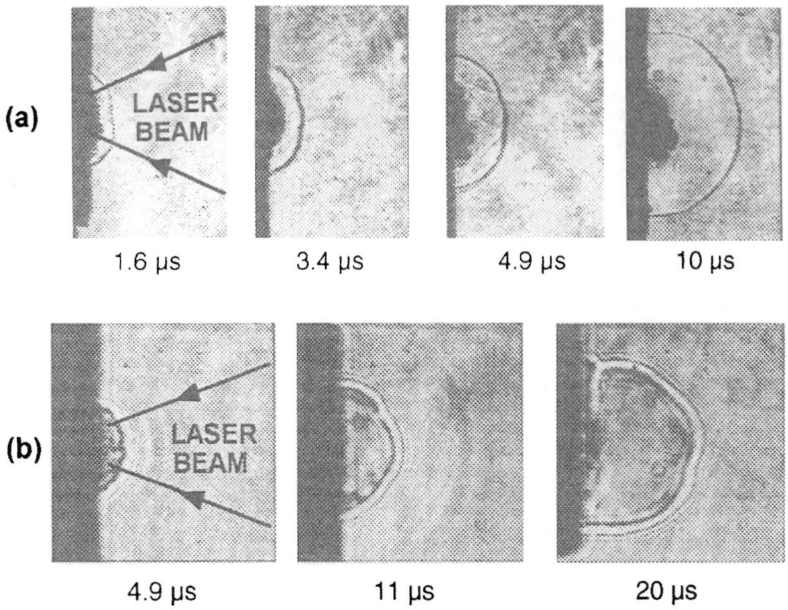

Figure 3. Sequences of shadowgraphic images showing the evolution of the ablation plume and of the associated acoustic and thermal phenomena induced by two different Nd:YAG laser pulse durations on sample of encrusted marble: (a): QS pulse, 6 ns, 1.7 J/cm^2, (b): SFR pulse, 20 μs, 17 J/cm^2.

Shadowgraphic images displaying material removal, as well as laser-induced acoustic and thermal effects developing in the air region before the target surface, are shown in Figure 3(a). and (b). for the 6 ns and 20 μs pulse duration, respectively. The QS lasers pulse generates an intense plasma, whose front expands at supersonic speed from the target surface and drives the development of a shock wave. The dark shadowgraphic ring corresponding to the shock front is typically observable at about 1 μs after laser irradiation. A dark cloud expanding at lower speed from the target surface is clearly indicative of the presence of ejected material and suggests that material removal involves mostly particulate opaque to the probe laser beam. These observations point out the relevant role of the photomechanical effect for the 6 ns pulse-induced ablation process, as it will be quantitatively discussed in the following.

Conversely, time-resolved imaging of the effects of SFR laser pulses does not show significant photomechanical effects. Only weak acoustic waves departing from the focal region are barely visible during the first microseconds of laser irradiation [see for instance the first frame of Figure 3(b).], which are related to the first intense spikes of the pulse shape. The sequence of frames shows the formation of a perturbed hot region, whose front expands at constant subsonic speed [see Figure 4(a).]. Then the front of the hot region degenerates in turbulent motions soon after the end of laser irradiation.

3.1. Analysis

3.1.1. QS pulses. Physical parameters associated to the photomechanical behavior of short QS pulses, as the pressure acting on the material surface, can be evaluated by analyzing the evolution of the shock wave expanding in air. Here the cleaning process is described by means of a model of "optical detonation". The high intensity laser irradiation is assumed to drive a

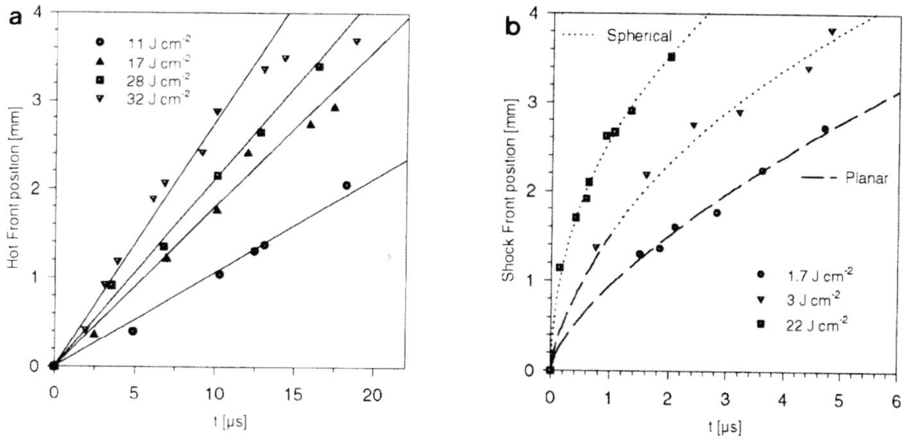

Figure 4. (a): Experimental data and linear fittings of the expansion behaviour of the front of the hot region induced by SFR pulses of increasing fluences. Front speeds of 106, 177, 211, and 276 m/s result at laser fluences of 11, 17, 28, and 32 J/cm^2, respectively. (b): Time evolution of the front of the shock wave induced by QS pulses of increasing fluences. Fittings combining planar and spherical expansion geometry have been calculated according to a model of optical detonation.

hydrodynamic regime where the strongly overcompressed and overheated gas forces a supersonic expansion of the shock front, producing the ionization of the surrounding unperturbed gas and increasing the volume of the plasma region. The speed of the shock front in planar geometry of the detonation wave can be described by:[4]

$$c_D = \left[\frac{2(\gamma^2 - 1)I_D}{\rho_0}\right]^{\frac{1}{3}} \tag{1}$$

where I_D is the detonation intensity. From Eq. (1) one can derive the whole evolution of the shock front motion supposing a planar decay at the end of the laser pulse maintained up to a propagation distance equal to d/2; after that point, the decay is assumed to be spherical. Fulfilling the continuity conditions at the border region of the three phases of planar pumping, planar decay and spherical decay, respectively, one obtains the following expression of the front motion:

$$r_f(t) = \begin{cases} c_D t & t_0 < t \le \tau_D \\ c_D\left(\frac{3}{2}\tau_D^{1/3} t^{2/3} - \frac{\tau_D}{2}\right) & \tau_D < t \le \tau_I \\ c_D\left(\frac{5}{2}\tau_D^{1/3}\tau_I^{4/15} t^{2/5} - \tau_D^{1/3}\tau_I^{2/3} - \frac{\tau_D}{2}\right) & t > \tau_I \end{cases} \tag{2}$$

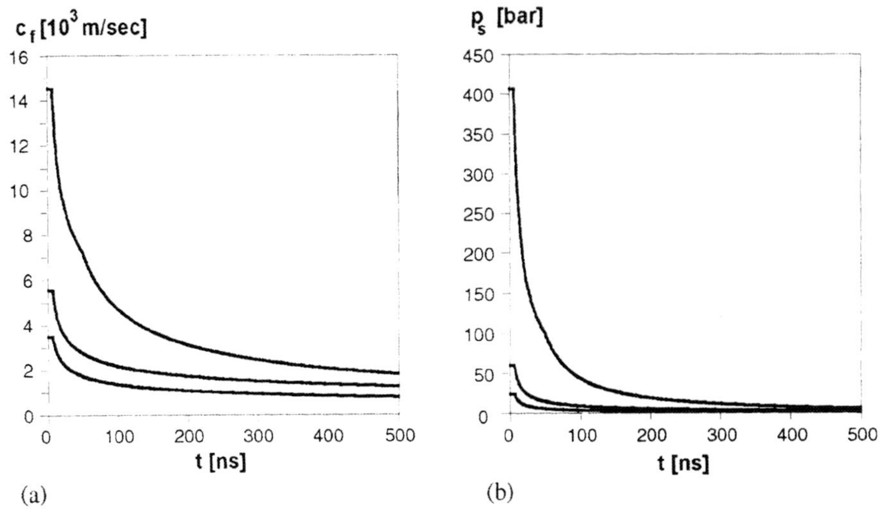

Figure 5. Calculated front speed (a), and pressure acting on the target surface (b) for QS pulses with increasing fluences of 1.7, 3 and 22 J/cm^2, respectively (from bottom to top).

where $\tau_D = \tau_L - t_0$, τ_L being the laser pulse duration, t_0 the starting time of the normal detonation regime, and τ_1 the time of switching from planar to spherical decay. Figure 4(b). reports the experimental data of the displacement of the shock front versus time for various laser fluences and the corresponding fittings, obtained from Eq. (2). Assuming $\tau_D \approx \tau_L$, the pressure acting on the target surface can be calculated as:[5]

$$p_s = \left(\frac{\gamma+1}{2\gamma}\right)^{\frac{2\gamma}{\gamma-1}} \frac{\rho_0}{\gamma+1} c_f^2 \tag{3}$$

where c_f is the speed of the shock front, given by the derivative of $r_f\,(t)$. Figures 5(a). and (b). report the front speed and the pressure at the target surface for the same laser fluence. This analysis shows that at the higher fluence value a peak pressure value as high as 410 bar can be produced on the target surface. It is worth noting that these pressure peaks can be considered as the lower limit of the real ones, since the above description does not include the recoil of material ejection due to the plasma-target energy transfer.

3.1.2. SFR pulses. The behavior observed with the SFR pulse suggests a description of the process in terms of a laser sustained combustion, where a plasma is initially generated and then it is pumped by a laser intensity much lower than that required for optical breakdown.[6] In such condition, only a plasma of low electron density can develop, characterized by a low absorption of the incident radiation and subsonic expansion of the front. The temperature inside this plasma region can be very high (in the order of 10^4 °K), but the mass density is indeed very low. In the present experimental conditions the density jump the front of the ionization zone can be obtained from:

$$\frac{\rho}{\rho_0} \approx \frac{\gamma v_f p}{(\gamma-1) I_L} \tag{4}$$

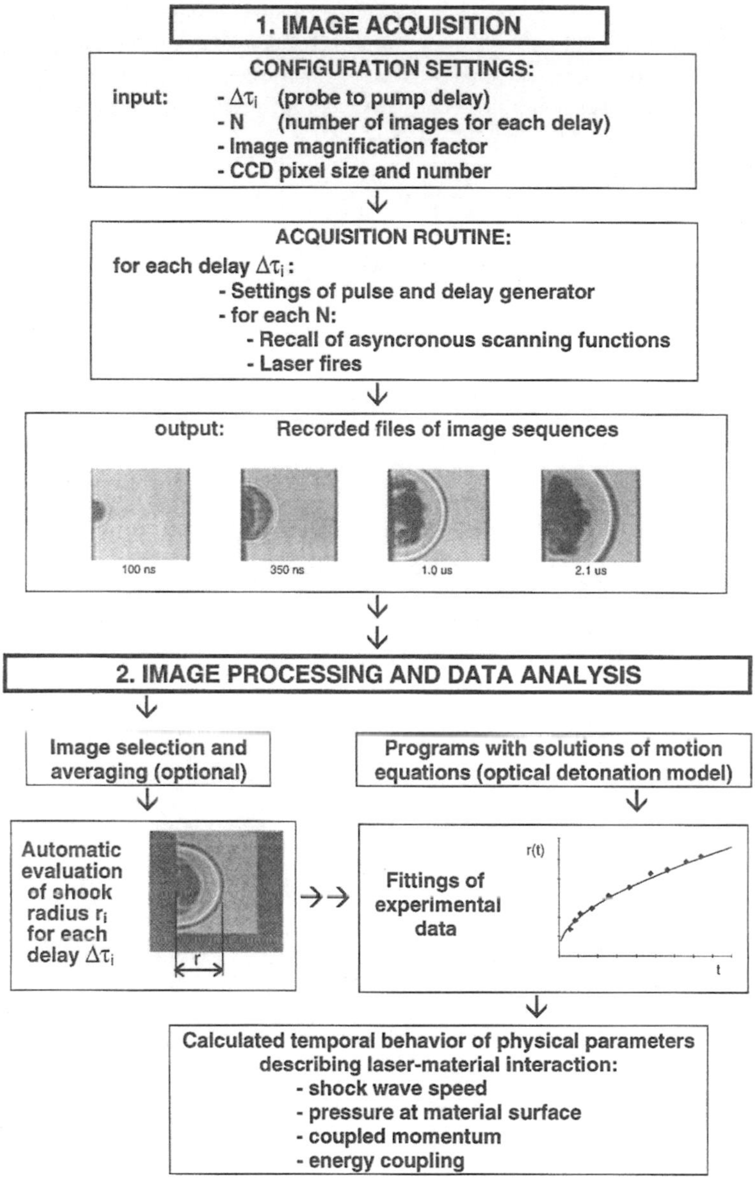

Figure 6. Block diagram of the software for automatic image acquisition, processing and analysis.

where v_f is the speed of the hot front, that was found to be linear during laser irradiation [Figure 4(a).] with estimated values in the range 100-300 m/s. Assuming $p \approx 1$ bar, Eq. (4) gives $\rho/\rho_0 \approx 0.7\text{-}2 \times 10^{-2}$ and the electron density has to be accordingly low. Such a plasma is quite transparent to the 1064 nm laser radiation, and does not play a crucial role in laser-target interaction, as in the previous QS case. Photomechanical effects as shock wave formation are thus negligible, but possible thermal side-effects have to be considered. Assuming a very simple model of quasi-continuous irradiation of material surface, a rough estimation of the penetration depth of heat into the material is given by:[7]

$$L = \sqrt{4\kappa\tau_L} \qquad (5)$$

that permits to evaluate $L \approx 5\text{-}10$ μm in our experimental conditions.

4. OPTICAL SENSOR

Based on the experimental set up scheme for image diagnostics and on the related analysis presented above, a portable diagnostic systems has been devised, performing automatic image collection and processing, and providing real-time measurements of the physical parameters by means of suitably developed computer programs.[8] The main difference with respect to the hardware shown in Figure 2. is the utilization as the probe laser of a very compact, low cost, pulsed diode laser (900 nm, 20 ns) with attached collimation optics. A dedicated software has been designed which permits fully automatic image acquisition, processing and analysis. The sequence of operations performed by this software is sketched in the block diagram reported in Figure 6. This sensor is suitable for field measurements in conservation laboratories and restoration yards as well.

ACKNOWLEDGMENT. This work has been supported by the Special Project on "Cultural Heritage" of the Italian National Research Council.

REFERENCES

1. J.F. Asmus, C.G. Murphy, W.H. Munk, Studies on the interaction of laser radiation with art artifacts, in *Developments in Laser Technology II*, Proc. SPIE Vol. 41, Bellingham, WA, pp. 19-30 (1973).
2. J.F. Asmus, L. Lazzarini, A. Martini, V. Fasina, Performance of the Venice Statue Cleaner. in: *Proc. of the Fifth Annual Meeting of the American Institute for Conservation of Historic and Artistic Works*, Boston, MA, pp. 5-11. (1977).
3. See for example.: 1st Workshop on *Lasers in the Conservation of Artworks* (LACONA), Heraklion, Greece (1995), and 2nd Int. Conf. on *Lasers in the Conservation of Artworks* (LACONA II), Liverpool, UK (1997).
4. Yu. P. Raizer, *Laser-induced discharge phenomena*, Consultants Bureau, Plenum Publication Corp., New York (1977).
5. A. N. Pirri, Theory for momentum transfer to a surface with a high-power laser, *The Physics of Fluids* 16:1435 (1973).
6. Yu. P. Raizer, Subsonic propagation of a light spark and threshold conditions for the maintenance of plasma by radiation, *Sov. Phys. JETP* 31:1148 (1970).
7. H. S. Carslaw, *Conduction of Heat in Solids*, 2nd edition, Clarendon press, Oxford, p.75 (1989).
8. Patent No. FI96A 124, deposited on 23/05/96 by the National Research Council of Italy.

ELECTRO-OPTICAL SENSORS FOR MECHANICAL APPLICATIONS

F. Docchio[1], M. Bonardi[2], S. Lazzari[2], R. Rodella[1], E. Zorzella[2]

[1]Università degli Studi di Brescia
Via Branze 38
25123 Brescia, Italy

[2]SemTec srl
Via Caselle 6
25020, Flero (BS), Italy

1. INTRODUCTION

After a considerably long incubation period, where optical technology could not compete with conventional techniques for sensing in the mechanical domain, there is no doubt that we are assisting to a period of blossoming of these techniques, which have become mature for their successful introduction in the production line. This has been made possible by a number of combined elements. Firstly, sufficient effort has been put into the transformation of optical measuring instruments into optical sensors, gaining considerably in compactness, ruggedness, cost effectiveness and reliability in the industrial framework. The availability of miniaturized components and devices, and the tremendous increase in the performances of the signal-conditioning electronics, have certainly helped this transformation. Secondly, also as a consequence of the increased reliability of the sensors, the industrial counterpart has considerably increased its acceptance of such sensors. Thirdly, the need of testing the quality of manufacts and the procedures for their production has boosted the need of sensing and measuring equipment in the various stages of production.

The present paper will deal with the state of the art and with some perspective views on optical sensors in the mechanical context. Our Laboratory has been active in the field of optical sensing for mechanics since its origin, and is committed to foster the above mentioned transformation in an active way. Thus, most of the material treated here will cover applications to industry that have been explored by the Laboratory and that have been transferred into prototypes which yield on- or off-line measurements in the industrial domain.

Optical Sensors and Microsystems: New Concepts, Materials, Technologies
Edited by Martellucci et al., Kluwer Academic / Plenum Publishers, New York, 2000.

2. E-O SENSORS IN MECHANICAL INDUSTRY: RATIONALE

Sensing and measuring are fundamental aspects in the context of modern industry. As the level of automation increases, the amount of process sensors must obviously increase to avoid malfunctioning and to guarantee full control of the production plant. In addition to this so-called "low-level sensing", two other types of events occur in the production line, which need sensors and measuring instrumentation. The first is the so-called "quality control process", which includes all the steps required to control the manufacts at the various production levels, in order to provide adequate certification of the level of their quality. Certification of the quality of a product is presently absolutely vital for its acceptability by the user and for its market success.

The second event is dimensional measurement and evaluation. A sensor, or a sensing equipment, is generally used to acquire all the necessary geometrical parameters of the manufact to be able to perform two actions:

- If the object under test is a prototype or a model, to make a computer designed replica of the object. CAD-modeling of the object will eventually make available all constructive details and process strategies.
- If the object under test is a manufact exiting the production line, to make an on- or an off-line control of the conformity of the geometry of the object with respect to the input production data. This step is in some cases vital to have the possibility of modifying the process.

In all the above sensing and measuring strategies for the mechanical industry, electro-optical sensors play a very relevant role, due not only to the intrinsically non-contact nature of the sensing action, but also to the low uncertainty achievable with most of them. Electro-optical sensors belong to different classes, and use different sensing principles. In the following we list some of them.

E-O sensors are based on the capability of the object to change one or more parameters of coherent as well as of incoherent light due to the property of the object to be measured. In this context, it is evident that coherent light is in most circumstances more usable than incoherent light. The parameters of light that are important for mechanical sensing are (i) directionality of the laser beam (which is the base for all triangulation sensors, alignment sensors, 3-D imagers based onto structured light projection, and strongly contribute to all distance/displacement sensors such as interferometers); (ii) focusability of the laser beam (fundamental for all autofocusing sensors, as well as for all sensors which gauge the object on a microscopic scale); (iii) temporal coherence or monochromaticity (all interferometric systems and all Doppler-based systems); (iv) emission of short pulses (laser rangefinding, or photoacoustic sensors); (v) sensitivity of laser light to environmental parameters such as temperature, humidity, pressure (laser-based optical thermometers, hygrometers, pressure sensors, both in vacuum and in fibers), (vi) spatial coherence (holography, ESPI).

Given the above general statements, the most common e-o sensors used to monitor a process, to control the quality of the manufacts, or to gauge their geometrical or physical parameters, are:

- Triangulation sensors. In triangulation sensors, the object is illuminated by a focused laser beam. The light diffused by the object is imaged onto one or more detectors placed at suitable angle with respect to the illumination axis. The position of the image with respect to the center of the detector is related to the position of the object.
- Autofocusing sensors. A laser beam is tightly focused onto the target. Preliminary settings result in ideal focusing if the object is in the zero position. The light is reflected back to a detector. Any departure of the object from this position in either direction will result in a different amount of light reaching the detector, due to an increased spotsize. Detection of the reflected light can be accomplished by various methods. Autofocus sensors can be passive or active, the latter including piezo-actuators able to shift the

focusing optics in response to an error signal derived from the measured position, in order to maintain the beam at focus on the target.

- Interferometric sensors. Based on the principle of interferometry, they are used to gauge a vast number of object parameters (dimensions, displacement, velocity, vibration, humidity, stress, etc.), both in free-propagation, fibre, or integrated optics configurations. Recently the progress in low-coherence interferometry made it possible to develop such interferometers for absolute reflectivity measurements, such as, for instance, the measurement of the thickness of transparent layers.
- 3-D imaging sensors. Extension of the principles of triangulation to three dimensions yield the so-called 3-D imaging sensors. Here, the deformation of an originally straight laser line, or of a set of black and white stripes, contains the information of the position of the points of a surface with respect to a reference plane. These sensors are used to measure planarity, microtopography, profiles, waviness, both statically and dinamically.
- Color sensors. Small color sensors suitable for the industrial environment are normally based on sets of calibrated emitters in different spectral ranges, and on photodetectors which detect the reflected light.
- Temperature sensors. They are normally based on radiometric principles (IR sensors or imagers), or on the properties of the material into which the light propagates (e.g., a fibre, or a grating inscribed in a fibre). The latter is also used to monitor stress in a structure in which the sensor is embedded.
- Light scattering sensors. Based on the dependence of the properties of scattered light onto the roughness of the surface, they are used to make in- or off- process evaluation, in a non-contact way, of the properties of the surface.
- Polarimetric sensors. Polarization of laser light can be altered by a number of material properties. Thus, polarimetric sensors sense the modifications of a beam initially having a known polarization state, to monitor the property of the material (e.g. stress- or temperature-altered birefringence of a film).

3. EXAMPLES OF APPLICATIONS OF ELECTROOPTIC SENSORS TO MECHANICAL MEASUREMENTS

3.1 Dimensional control

In our Laboratory there is an ongoing theoretical and experimental investigation of interferometric systems and of related instrumentation for dimensional gauging. Various interferometric principles have been explored (fringe-counting incremental interferometry, dual- or multiple wavelength interferometry, frequency-modulated interferometry, etc), in bulk optics as well as in fibre- and integrated optics. The applications have been numerous, ranging from measurement of diameters, vibration analysis, pressure measurement in industrial environment. We will not enter in detail in high-coherence interferometric measurements, as this is the subject of another paper in this book.

Low-coherence interferometry[1-4] is presently a subject of interest in the Laboratory. Several configurations of LC interferometers have been developed. One of these configurations is an LC interferometer with no external moving parts, with a stepping-motor driven reference mirror collinear to the measurement mirror[5] (Figure 1.). The detection of the fringes is performed by re-injection of the interfering light into the superluminescencent diode, equipped with an in-built photodiode. A typical application of this device is the measurement of the thickness of transparent surfaces (e.g. windscreens in car industry).

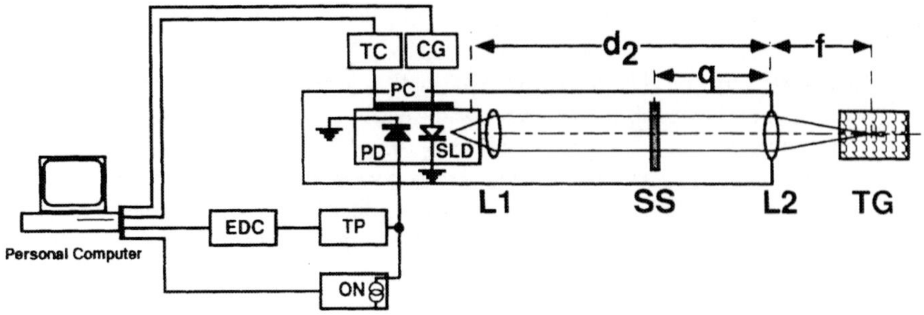

Figure 1. Low coherence interferometer with retroinjection. Internal mirror SS and target TG form the two interferometric arms.

A typical interferometric scan is shown in Figure 2. This setup can be very compact and suitable for the mechanical environment, and yields a resolution of 10 μm and working range of about 2.5 cm.

An alternative LC scheme[6] is shown in Figure 3. Here, all longitudinal modes in a multimode-emitting laser diode are made to interfere after dispersion by a grating. Any pathlength variation due to a shift of the measuring mirror results in amplitude modulation of the mode envelope. An example of such situation is given in Figure 4. Accuracy in the order of 200 nm over a range of 1.5 mm have been obtained.

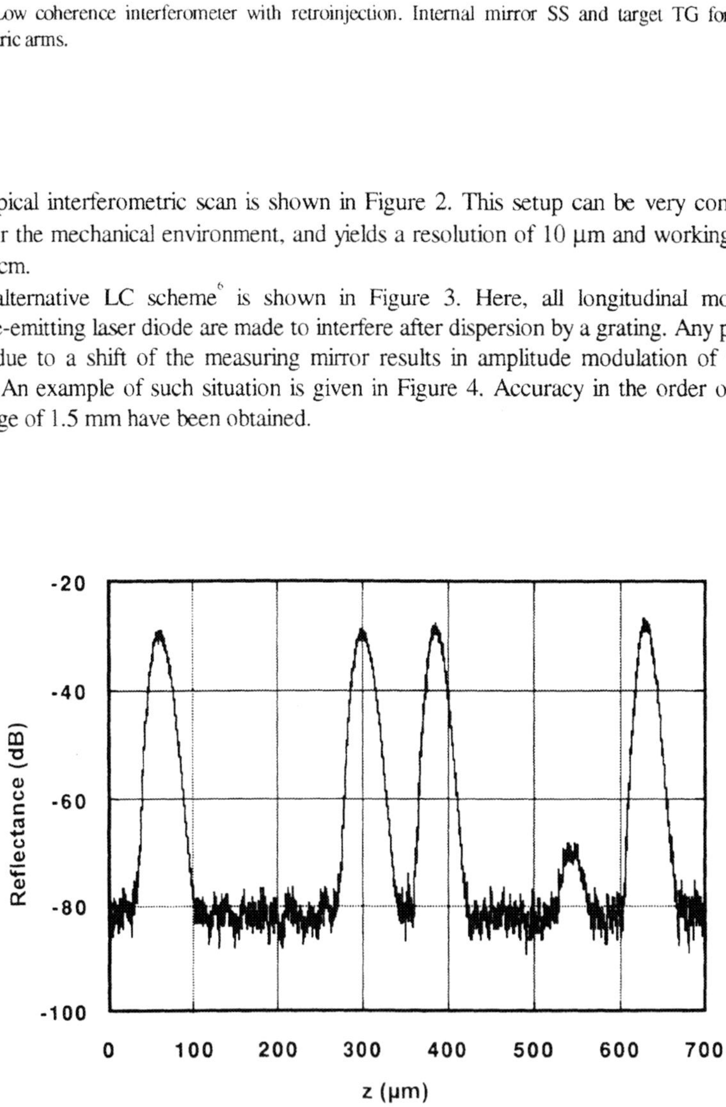

Figure 2. Typical interferometric scan produced by a pair of glass slides separated by a gap.

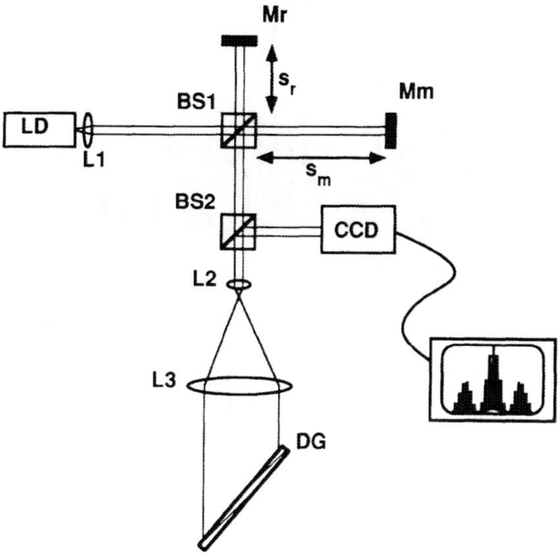

Figure 3. Block diagram of a dispersive white-light LC interferometer.

3.2 Tests of surface properties of industrial manufacts

A manufactured mechanical surface is affected by surface defects such as roughness, waviness, microcracks. Roughness is generally discriminated against waviness by means of the different spatial period, rather than of peak amplitude (Figure 5.).[7]

In most metallic and non-metallic surfaces, for the quality control of the surface a contact roughness measuring tool is not preferred, as it might damage the surface. Optical measurement system prove to be superior in this respect.[7-9] A measuring device, based upon triangulation sensors, has been recently developed to monitor surfaces on a non-contact basis. The triangulation sensor has a resolution of 0.3 μm on a range of a few mm. A typical example of measurement of a test sample is shown in Figure 6., where both the profile and its Fourier Transform, useful to extract the surface parameters, are shown.

3.3. 3-D Macro- and Microprofilometry

Macro- and microprofilometry systems[10-15] have been extensively investigated in our Laboratory. Recently, the companion company SemTec is involved in the exploitation of the amount of know how accumulated over the years. Macroprofilometry has been approached by (i) imaging combined with projection of structured light, and (ii) point measurement combined to scanning systems. The first approach has been preferred, since it seems more versatile. Different projection strategies have been developed over the years, and recently combined in a unique system[12,13]. The projecting unit is based on a Liquid Crystal Projector. Correction of all possible measuring errors in different cases has been operated.

The system, shown in Figure 7., has features that depend on the dimensions and shape of the target, but the majority of objects of industrial interest remains well below the tenth of millimeter. As examples, Figures 8.a, b, and c, show 3-D profiles of different objects, as obtained by our macroprofilometer.

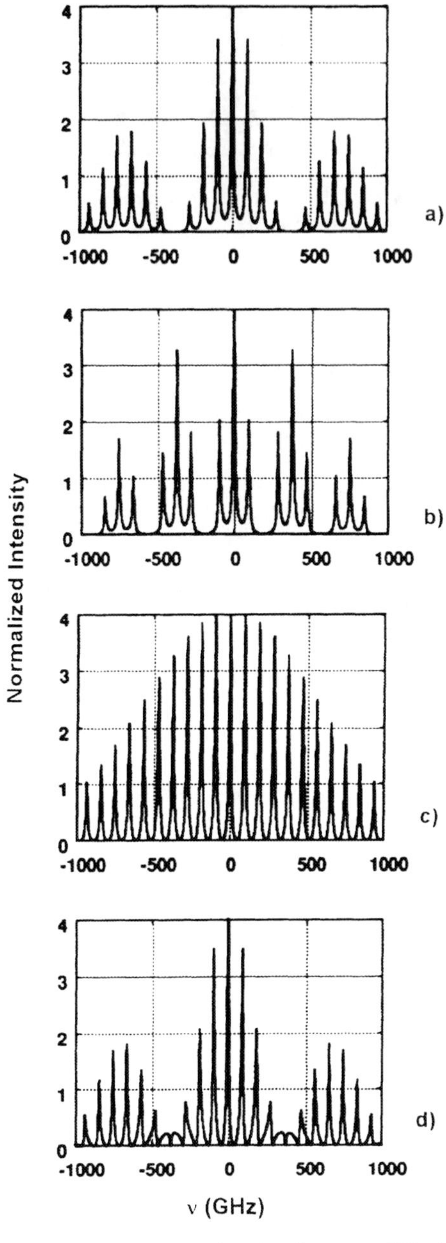

Figure 4. Amplitude modulation of the mode envelope from interferometer of Fig. 3 for distances of 0.2 mm (a), 0.175 mm (b), 0.8 mm (c) and 1.8 mm (d).

The microprofilometer setup is schematically shown in Figure 9. It makes use of either autofocusing or triangulation sensors, depending on the range and on the resolution required by the measurement. It is composed by the sensing head mounting the sensor of interest, and of a motorized x-y scanning unit.

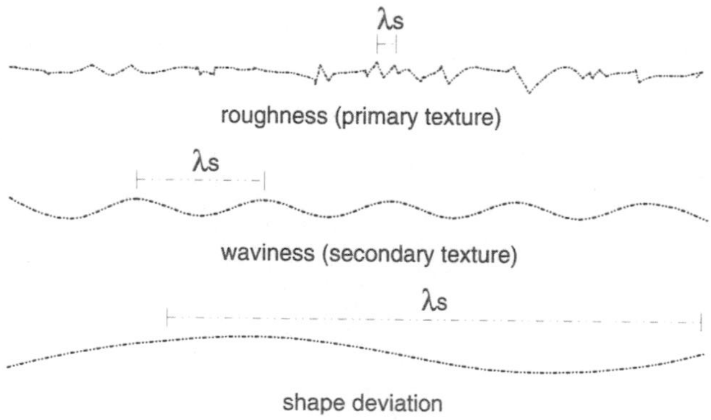

Figure 5. Roughness, waviness, and shape factor of a surface. λs is in the order of 0.1-1 mm, 1-30 mm, 30 mm and greater respectively.

As examples, Figures 10.a, b, and c, show 3-D profiles of different objects, as obtained by our microprofilometer, the first obtained by the autofocusing sensor, the second and third ones obtained using a triangulation sensor.

Among others, the microprofilometer has been used to monitor the wear of mechanical tools. Tool wear characterization is indeed crucial in mechanical industry, since the prediction and diagnosis of wear of a working tool limits the number of manufacts that must be discarded due to tool malfunctioning, and increases the safety of the production, as it is well known that tool malfunctioning may result in hazard for the operator.

3-D microprofilometry proved to be advantageous in this respect. Figures 11.a and b show an example of tool gauging: Figure 11.a shows a microphotograph of the tool. Figure 11.b show a section of the 3-D scan. In a study presently carried out by our groups the presence of defects is being correlated to the aspect of the shaving, to establish a database for the prediction of tool wear.

3.4. Color measurements

A color measuring head has been developed in the Laboratory, to perform colorimetric measurements on surfaces of industrial as well as of biotechnical interest. A layout of the colorimetric head is shown in Figure 12. The head is composed of a number of LEDs with known spectral emissivity, and of a photodiode to detect the reflected light. An example of the characterization of ceramics is shown in Figure 13., where the RGB characteristics of samples of different spectral reflectivity are shown.

4. PERSPECTIVES

Electrooptical sensors and sensing instruments seem to finally have a bright future in the industrial framework. A winning factor in this context is given by a close cooperation between laboratories and industries, devoted to know how transfer and exploitation. As the choice of electrooptical sensors with improved performances, lower cost, higher reliability increases, great attention must be paid by the system integrators to carefully match the sensor to the application, and to combine e-o sensors to sensors of other types. Ruggedness and reliability of the operating hardware and software are other key factors to ensure e-o sensors and sensing equipment the success that they deserve in mechanical industry.

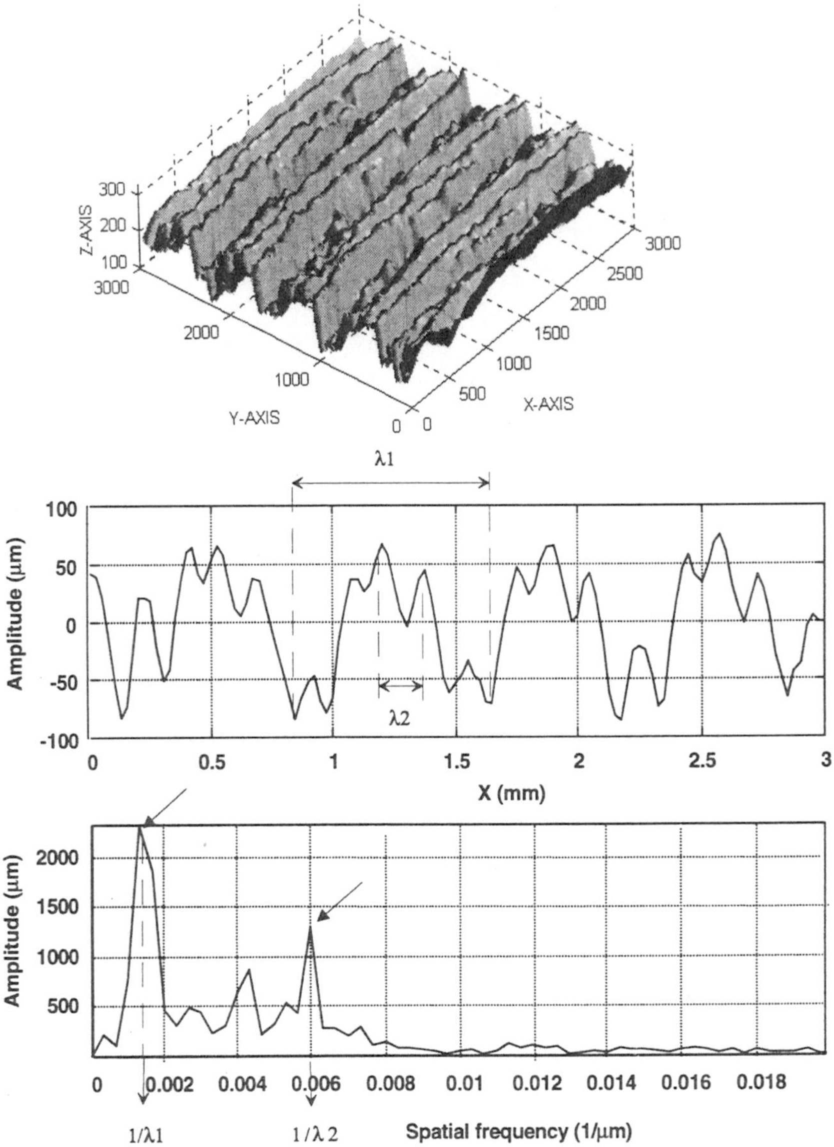

Figure 6. 3-D reconstruction of a milling surface (a), profile section and its Fourier Trasform (b).

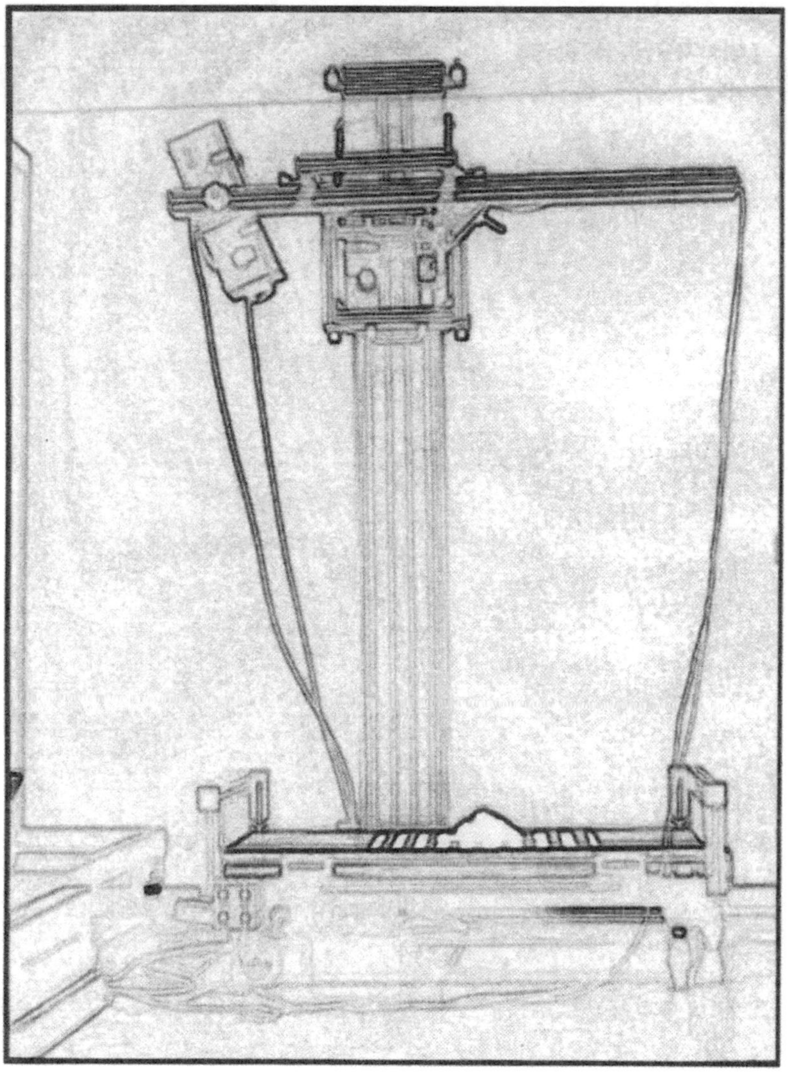

Figure 7. Layout of 3D-profilometer.

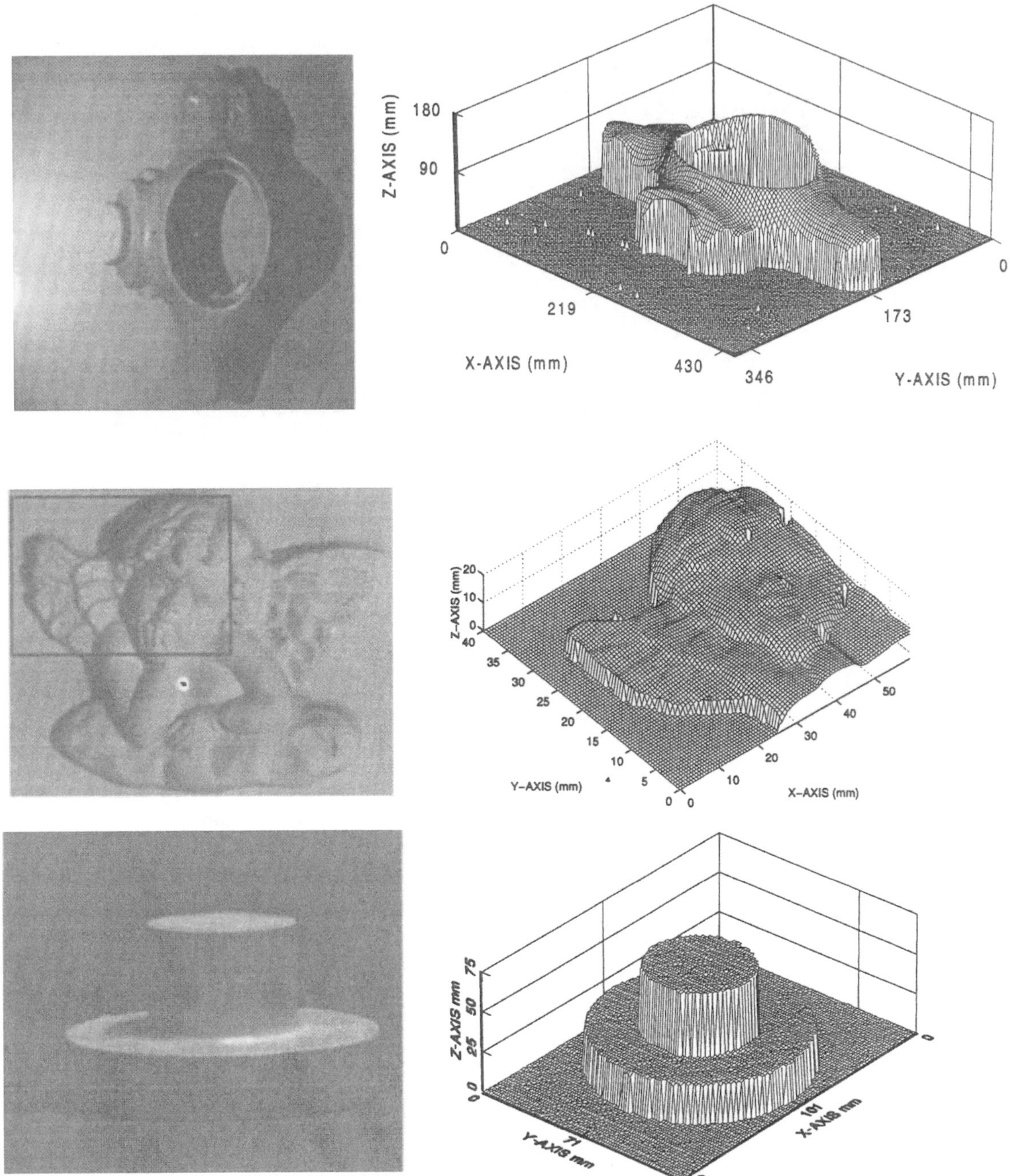

Figure 8. Photograph (left) and 3-D profile (right) of a motoscooter body (a), a chalk bas-relief (b) and a mechanical flange (c).

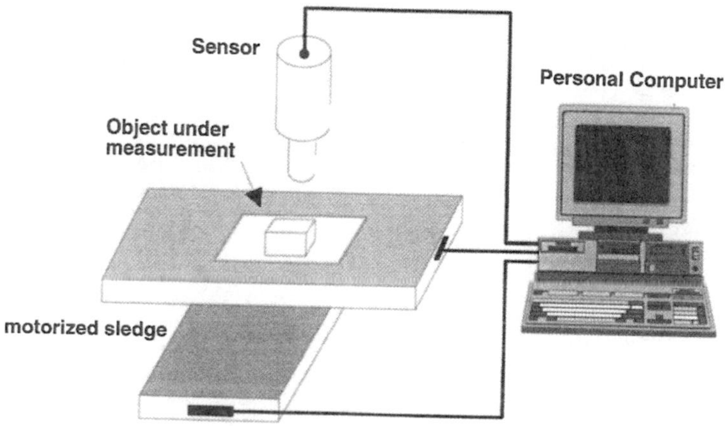

Figure 9. Setup of the microprofilometer.

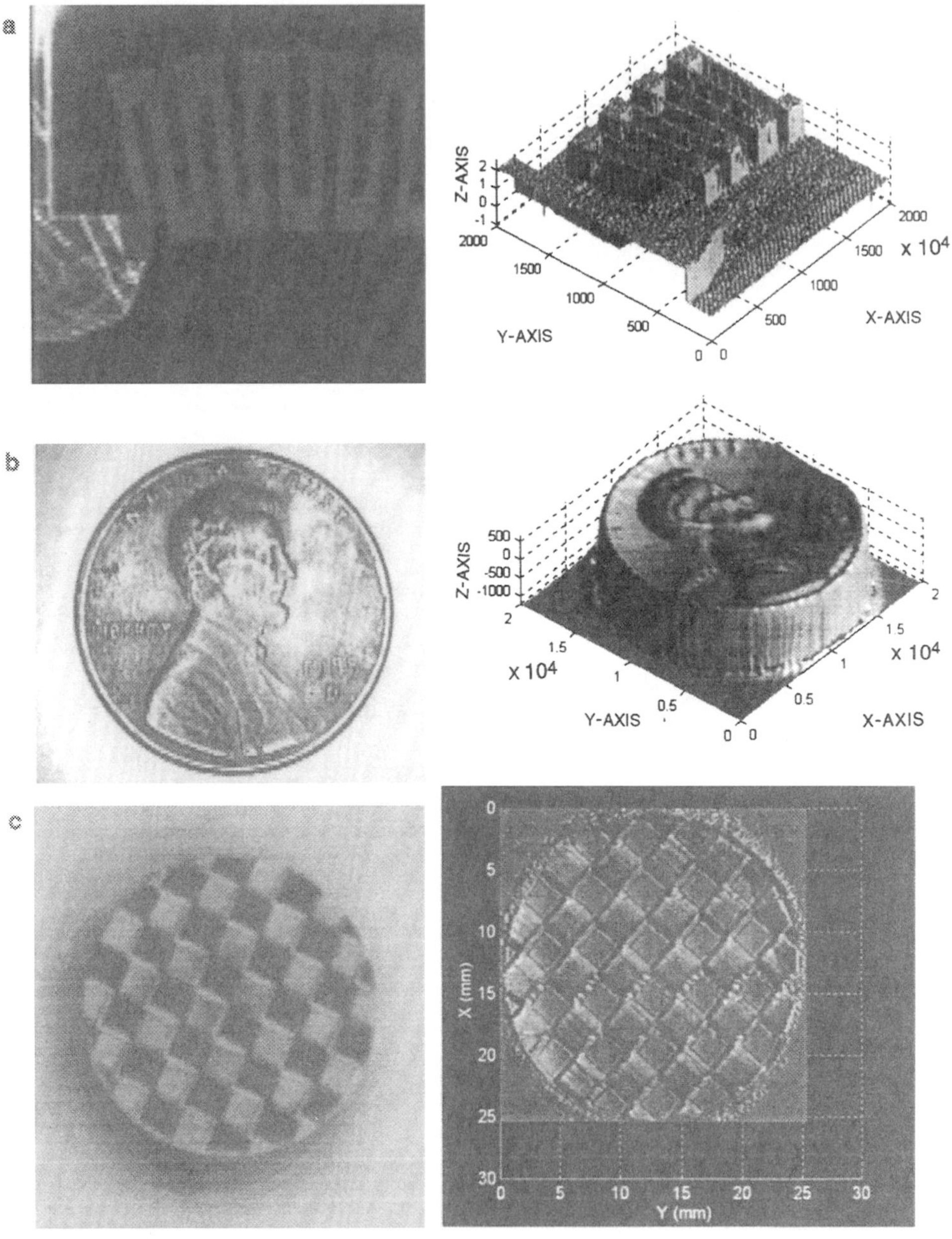

Figure 10. Photograph (left) and 3-D profile (right) of strain gauge pressure sensor (a), a US coin (b) and a button (c). Profile (a) has been obtained using the autofocus sensor, profiles (b) and (c) using the triangulator.

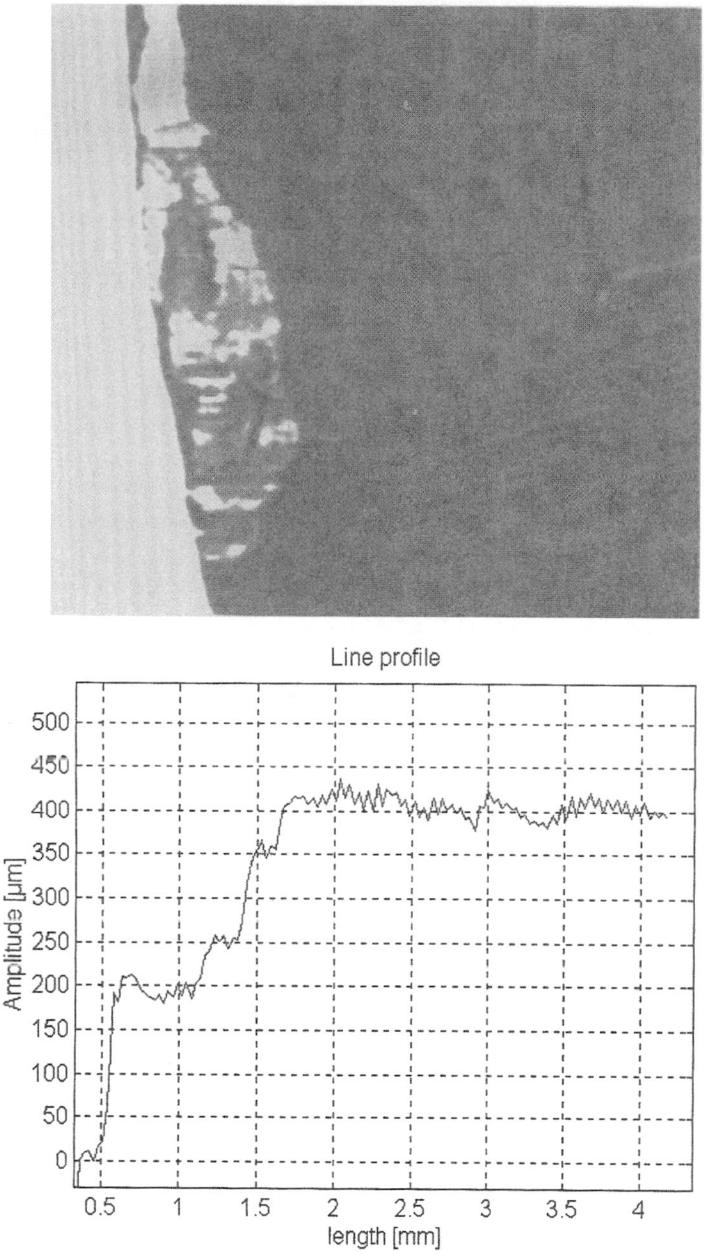

Figure 11. Photograph (a) and section profile (b) of a mill tool.

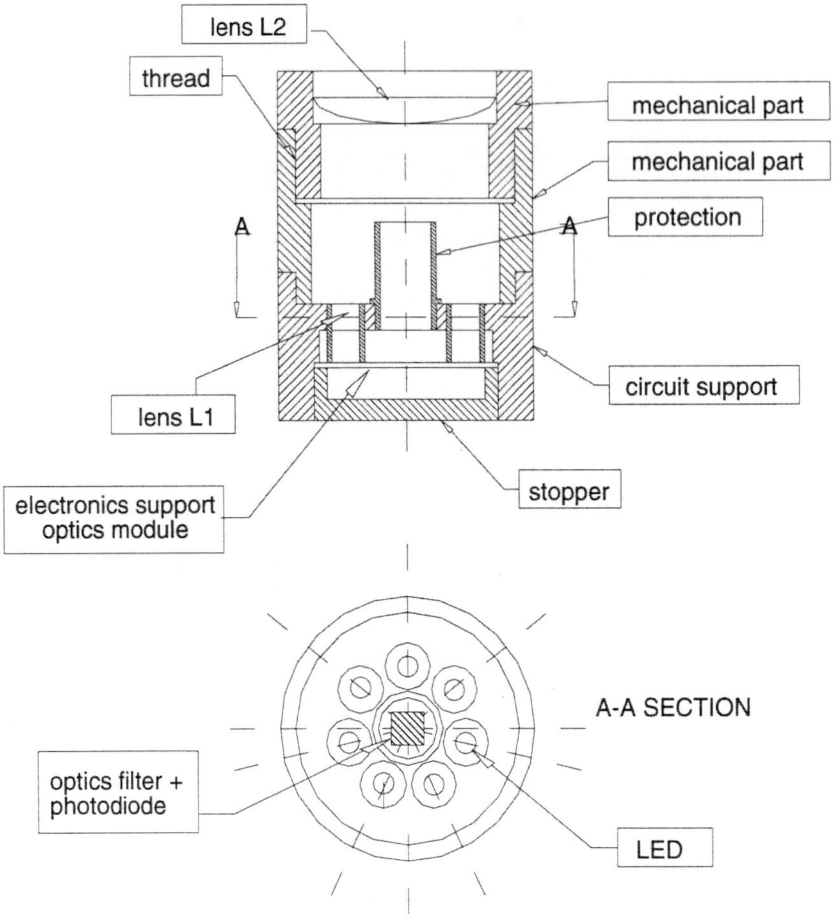

Figure 12. Details of the miniature colorimetric head.

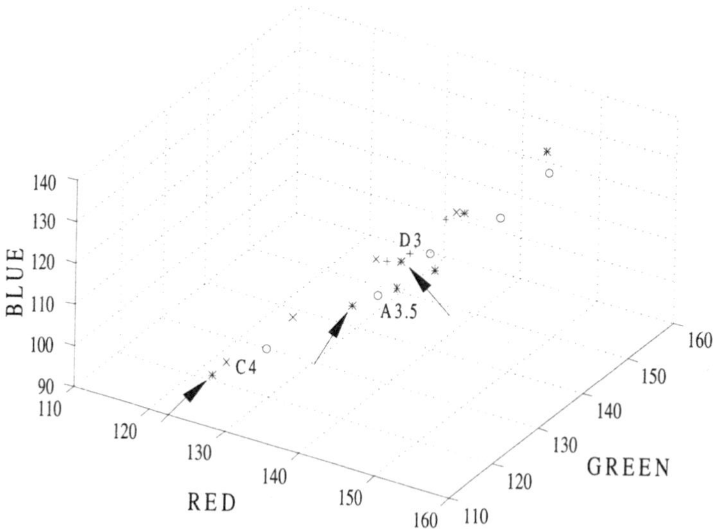

Figure 13. RGB characterization of ceramics. Different samples are represented as points in RGB 3-D space.

ACKNOWLEDGMENTS. The authors gratefully acknowledge the contribution of all the staff of the Laboratory of Optoelectronics, with particular reference to Prof. U. Minoni, Prof. G. Sansoni and Dr. L. Rovati, and are indebted with all the industries which have directly or indirectly stimulated research in the fields covered by this paper.

REFERENCES

1. D. Leslie, P. de Groot, High speed noncontact profiler based on scanning white-light inteferometry, *Appl. Opt.* 33:7334 (1994).
2. U. Schnell, R. Dändlinker, S. Gray, Dispersive white-light inteferometry for absolute distance measurement with dielectric multilayer system on the target, *Opt. Lett.* 21:528 (1996).
3. U. Schnell, E. Zimmermann, R. Dändliker, Absolute distance measurement with synchronously sampled white-light channeled spectrum interferometry, *Pure Appl. Opt.* 4:643 (1994).
4. S. Chen, A.W. Palmer, K.T.V. Grattan, B.T. Meggitt, S. Martin, Study of electronically-scanned optical-fibre white-light Fizeau interferometer, *Elect. Lett.* 27:1032 (1991).
5. L. Rovati, F. Docchio, Low-coherence interferometry using a self-mixing super-luminescent diode, *IEEE Phot. Tech. Lett.* 10:123 (1998).
6. L. Rovati, U. Minoni, F. Docchio, Dispersive white light combined with frequency-modulated continuous-wave interferometer for high-resolution absolute measurement of distance, *Opt. Lett.* 22:850 (1997).
7. ANSI, Surface texture: surface roughness, waviness and lay, American Standard ANSI B.46.1 (1985).
8. J.H. Zhang, B.H. Zhuang, New optical stylus sensor, *Proc. of SPIE* 2349:148 (1994).
9. J.M. Bennett, D.K. Burge, J.P. Rahn, H.E. Bennett, Standards for optical surface quality using total integrated scattering, *Proc. of SPIE* 181:124 (1979).
10. S. Tang, Y.Y. Hung, Fast profilometer for the automatic measurement of 3-D object shapes, *Appl. Opt.* 29:3012 (1990).
11. K. Sato, S. Inokuchi, Three-dimensional surface measurement by space encoding range imaging, *Journal of Robotic Systems*, 2:27 (1985).
12. G. Sansoni, S. Corini, S. Lazzari, R. Rodella, F. Docchio, 3-D Imaging based on Gray code light projection: characterization of the measuring algorithm and development of a measuring system for industrial applications, *Appl. Opt.* 36:4463 (1997).
13. G. Sansoni, L. Biancardi, U. Minoni, F. Docchio, A novel, adaptive system for 3-D optical profilometry using a liquid crystal light projector, *IEEE Trans. Instrum. Meas.* 43:558 (1994).
14. E. Mainsah, W. Dong, K.J. Stout, Three dimensional imaging of engineering surfaces, *Proc. of SPIE* 2599:141 (1995).
15. K.J. Stout, Three dimensional surface topography: measurement, interpretation and application - A survey and bibliography, *Penton Press* (1994).

OPTICAL FIBRES AND THEIR ROLE IN SMART STRUCTURES

B. Culshaw

University of Strathclyde, Electronic & Electrical Engineering
204 George Street
Glasgow G1 1XW, United Kingdom

1. INTRODUCTION

The term "smart structures" crept into the technical vocabulary more than a decade ago.[1-3] Since then a significant amount of philosophical debate as to what the term might mean has occupied the more academic amongst us wondering whether or not our smart structure should seek to emulate a biological organism or - whatever.

Yet others will point out that smart structures are nothing more than a sensible approach to integrated engineering (Figure 1.). The debate over the benefits and disadvantages of incorporating "smartness" into structural entities will continue. There is little doubt that the smart structure will be more costly at the outset than its less intellectually endowed ancestors. However whether this smartness costs more in the long term remains open for debate and frequently depends upon how the accountants do the accounting and who is spending from which budget and when. Equally there is little doubt that major infrastructure investments such as bridges, highways, pipelines, ships, trains, motor vehicles and all the rest could be significantly improved by incorporating as needed rather than panic measure maintenance using structural smartness as the trigger. All the better if the structure could react and adjust or heal itself.

Further all agree that the first stage in the smart structure is to find out how it feels -which means sensors. Indeed it means usually a very large array of sensors since to know how our structure feels we really need inputs from its most personal parts. This requirement for sensor arrays imposes new demands on not only sensing systems but also on the capability to process the data which is produced by such systems.

Sensor technologies are many and range from observing ground movements through satellite radar speckle interferometry, to precision GPS, to nanoprobes addressing cellular chemistry, to tunnelling microscopes for detecting individual atoms. Sensing is a niche oriented activity, the domain where the small specialised company can compete effectively with lumbering multinationals.

The niche for sensing in smart structures has its own well defined characteristics. First and foremost the sensors must be compatible with the structural environment, which means in particular for advanced structural materials such as fibre reinforced composites the sensor must

Optical Sensors and Microsystems: New Concepts, Materials, Technologies
Edited by Martellucci et al., Kluwer Academic / Plenum Publishers, New York, 2000.

be able to withstand the mechanical and thermal excursions to which the structure may be subjected during its operational life. The sensors should also be capable of addressing the whole structural volume which in most cases is very substantial and they must be able to produce a map of the parameter which they are measuring throughout this operational volume. There is then the whole issue of electromagnetic interference, reliability and stability in operation, ease of installation especially for retrofits and the usual sensor criteria concerned with cross sensitivities especially to temperature and measurement reliability.

There is already a whole range of sensor technologies for structural monitoring available and established in the field. Traditionally the strain gauge and the thermocouple have been the mainstay. However they have their limitations, the most important of which is that each thermocouple and/or each strain gauge must be individually wired and coupled to either a data bus or directly into some form of instrumentation recorder. The wiring harness rapidly becomes complicated. Its susceptibility to corrosion and its ability to pick up stray electromagnetic signals have inhibited the application of traditional sensors in structures other than in the test laboratory environment. Very few have emerged into life cycle monitoring in a practical environment. There are a few trials on-going worldwide of which the Kingston Bridge here in Glasgow is an example. This currently has around 1000 sensors attached to various parts of its structure, but all effectively on external surfaces. This particular bridge has had some significant structural refurbishment which has necessitated continuous monitoring.[4] However basing the monitoring on conventional sensors has highlighted the very substantial investment required in installation and maintenance. The extremely expensive refurbishment processes have also highlighted that an early warning of impending distress could have saved a very substantial investment. So the need for sensing is most definitely there. However it has to be made available in the appropriate technological package.

This package will probably encompass a number of different approaches. It is now well recognised that fibre optics will be a major contributor to the package. The principal reason for this is the ability of fibre optics to undertake distributed measurements from which a map of the measurement parameter of interest against position may be readily derived. Couple that to the mechanical strength, large linear strain range, temperature tolerance and measurement adaptability of optical fibres and the result is a strong case for the exploitation of optical fibre techniques[5,6] within the first measurement phase in the realisation of smart structures.

In the remainder of this paper we shall examine just some of the technologies which have emerged and are already beginning to make significant useful contributions.

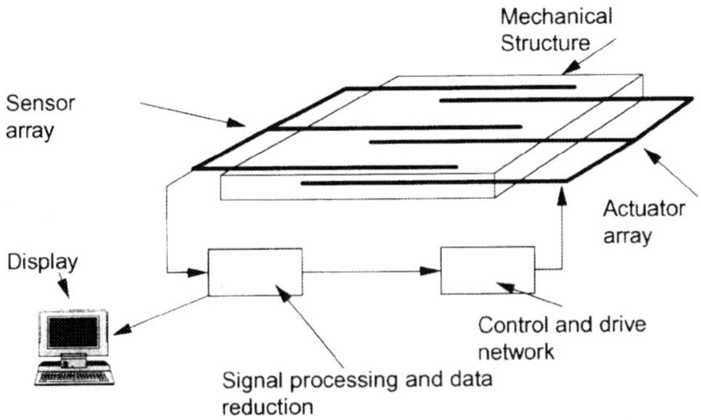

Figure 1. The smart structure concept - but is integrating these functions nothing more than good engineering practice ?

We shall see that the capacity for distributed measurement (Figure 2.), or alternatively the ability to string a succession of multiplexed sensors on a single optical fibre link, figures strongly in these applications. These capabilities are unique to optical fibre sensing technologies.

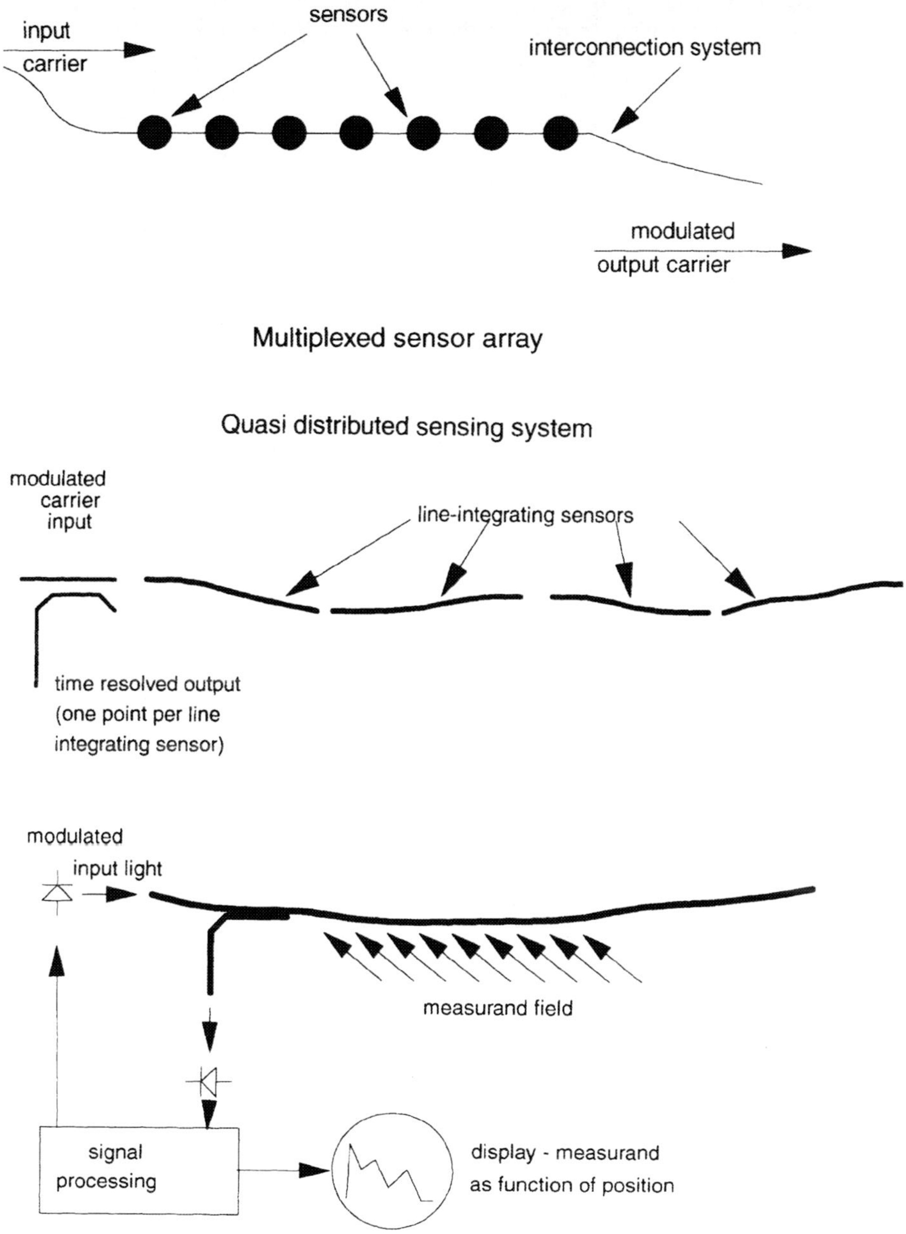

Figure 2. Fibre optic sensors in distributed, quasi distributed and multiplexed architectures.

2. THE MEASUREMENT REQUIREMENTS - SENSING ON A LARGE SCALE

The ultimate aim of any smart structure is to enhance its performance in terms of either its response to damage and deterioration, thereby enhancing overall structural safety or incorporating the ability to cope with anomalous loads. The principal issues are then safety and, associated with this, maintenance and repair and conservation of materials and resources.

Reducing these general issues to a definition of measurement requirements is straightforward in principle but fraught by the general observation that for many of the measurement requirements suitable systems to perform such measurements are currently unavailable. We can divide the types of measurement into three distinct areas - damage detection, deterioration sensitivity and adaptivity to load. There is obviously some linkage -for example any capacity to adapt to changes in load could also be configured to compensate for damage and deterioration. It is useful to review the general features of each of these measurement functions:

Damage usually occurs over a relatively short period of time, for example the result of a collision or a spillage from corrosive chemicals. Most damage events are mechanical in origin and are a consequence of impact, vandalism or natural events such as earthquakes and landslides.

Structural deterioration is a gradual process. It may be mechanical in origin for example due to foundations slipping beneath a bridge or building. More likely it is chemical in origin and a direct consequence of corrosion, scouring, erosion or similar processes.

Structural adaptability requires energy sources and reserves of physical strength to cope with anomalous loads. It implies a feedback mechanism through which a structural response may be defined and in all cases the inherent structural fabric must be capable of withstanding the anomalous loads. So in practice adaptability must be more about minimising the impact of anomalous loads for example by ensuring that the cross section of the structure to wind or wave can be readily minimised.

The measurement requirements for mechanical parameters for each of these requirements are broadly similar. A strain resolution of the order of $10\mu\varepsilon$ coupled to a temperature resolution of the order of 1°C is usually adequate. The temperature and strain fields should be sampled at a sufficiently high spatial frequency to ensure that the structural behaviour of interest is highlighted. Typically this gauge length will be some fraction - perhaps 10% - of the characteristic mechanical dimensions of the structure which determine its natural resonant frequencies. Temporal bandwidth should be about 10 times this fundamental resonant frequency.

There is considerable flexibility in implementation. The inherent assumption in this very rough estimate of spatial and temporal bandwidth is that the structure is to be sampled at frequencies comparable to its natural resonance. This in turn implies that the monitoring system logs the response of the structure to either its natural loading cycles or its base band impulse response. There are innumerable other checking techniques of which the most well known uses high frequency ultrasonic probes usually launched in a pulsed mode into the structure. This approach has the advantage of defining an excitation function and ensuring that this excitation function is well separated from any natural excitation which may occur. In this regard ultrasonic signature analysis may well prove to be effective as a first stage diagnostic probe[7] rather than the more complete but more usual ultrasonic imaging.

Most so called "smart" structures must also operate through a complete environmental specification so that whatever measurement system is installed must be capable of withstanding temperature excursions over ranges broadly from -40° to +60°C and normal vibrational and temperature cycling stresses. In some cases, for example for aircraft and space craft, the environmental specification will have to be extended perhaps significantly.

As a final comment we should acknowledge the need for efficient and effective data processing in smart structures. The capacity of these monitoring systems to create strings of data is enormous. It is not uncommon for a structure to be instrumented with 1,000 sensors. If each of these produces a 12 bit word every 5 seconds then the output is running at 4.3 Mb/hour or about

40 Gb/annum, most of which is simply a statement that all is well. Consequently the data collected from any smart structure must be continuously filtered since storing such vast volumes is not only confusing but also totally unnecessary. We shall not dwell upon the data processing issue in this paper, suffice it to say that an appropriate data handling strategy is critical to the application of any structural monitoring system.

3. THE MEASUREMENT PROCESS - POINT DISTRIBUTED AND QUASI DISTRIBUTED SENSING

Fibre optics is an unusual measurement technology. It enables a unique range of measurement functions. Conventional measurements based on established instrumentation (for example thermocouples, strain gauges, etc) almost always measure a parameter at a particular sampling point within the structure. The spatial and temporal sampling frequencies must then be sufficient to reconstruct the data field of interest. Fibre optics is of course capable to being configured as a point sensor technology with specified individual locations defined as monitoring points. This architecture we shall refer to as "point sensing".

Fibre optics however does enable other measurements, and of these measurement functions the capacity for distributed and quasi distributed measurement is extremely important. In quasi distributed measurement the measurement system determines the average value of a particular parameter (for example strain) between two predetermined points. In distributed measurements the measuring system determines the value of the data field of interest as a function of linear position along the fibre usually convolved with an interrogation window. Figure 3. illustrates these measurement functions.

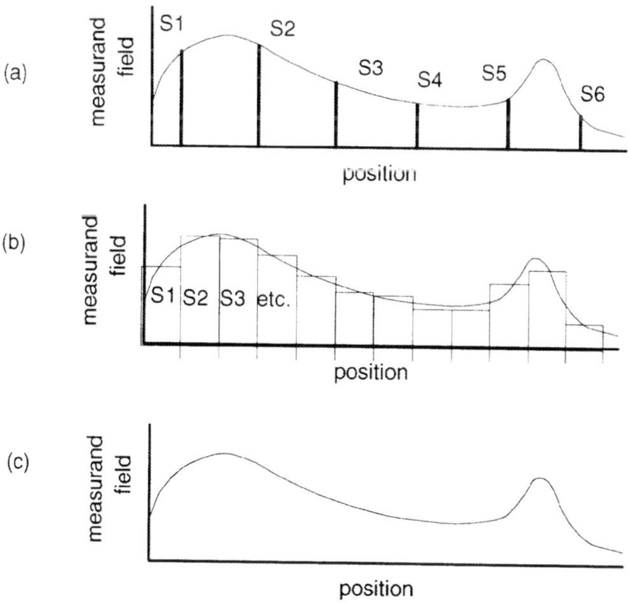

Figure 3. Illustrating the outputs for (a) multiplexed point sensor (b) quasi-distributed array and (c) distributed sensor with short pulse width.

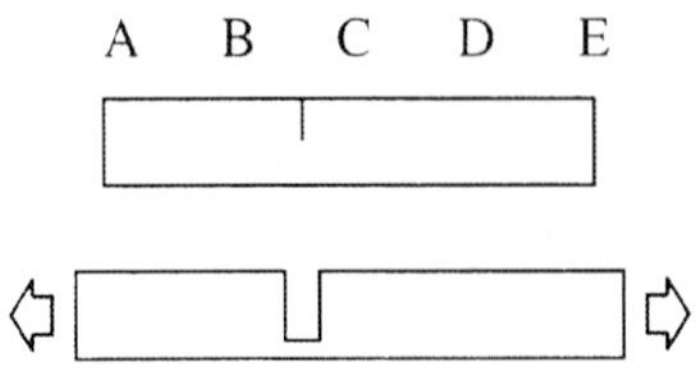

Figure 4. A simple example of the distinction between point and quasi distributed architectures and the data they produce. Strain sensors at A, B, C, D will miss the crack under load – but will show the same stress as if the crack were absent.

There is a particularly important distinction between the measurement capability of distributed and quasi distributed systems and that of point systems. Figure 4. highlights this distinction by considering a simple bar with a crack through one point in the middle. Strain gauges at points A, B, C etc along the bar will behave quite normally when the bar is subjected to tensile stress. However a quasi distributed network measuring average strain between points A-B, B-C, C-D etc will detect anomalous elongation between the points surrounding the crack. This simple observation serves to highlight the need to consider the measurement requirements carefully and relate these to the measurement technologies which are available.

4. WHY USE FIBRE OPTICS

We have already emphasised that fibre optics in structural monitoring should take its place beside the many other sensing technologies. Fibre optics does however have a number of beneficial features:

- The capability for operation over extended gauge lengths (up to kilometres) in either integrated or distributed format. This is unique to optical fibres.
- Total immunity to electromagnetic interference and zero electromagnetic radiation signature.
- Mechanical strength and chemical passivity both of which enable a very simple packaging and installation process.
- Access to a wide range of measurands sometimes through a common optical fibre sensor interface.

Consequently optical fibre sensors can be readily utilised in large structures or over entire site installations. They can be embedded in concrete[8] and in glass and carbon fibre composite materials[9] thereby ensuring intimate contact with the structure itself and they can withstand the mechanical, chemical and thermal environment.

In the remainder of this paper we shall briefly describe some of the sensor technologies which are currently being exploited in structural monitoring and some of the emerging concepts which are making their way into this new range of applications.

5. PHYSICAL MEASUREMENTS - STRAIN AND TEMPERATURE FIELDS

Optical fibre sensors offer unique measurement capabilities for both temperature and strain fields and may be realised as point distributed and quasi distributed architectures.

A temperature measurement, the Raman distributed temperature probe (Figure 5.) is a simple self calibrating system which enables distributed temperature measurements over distances to 10s of kilometres with resolutions of the order of °C in interrogation lengths of the order of metres[10]. The Raman DTS has been available as a commercial system for around a decade and whilst it is undeniably an expensive option its functional flexibility more than compensates for this expense in applications where this is required. Its uses include logging temperature fields in oil wells, as a distributed heat sensor for fire alarm systems especially in confined spaces such as tunnels and as a continuous monitor in long process ovens[11]. It could also be used for example to measure temperature fields along oil pipelines or to monitor the curing process in large volumes of material such as concrete foundations. The unique feature of the optical fibre DTS using Raman backscatter is that the return signal is a function of temperature and temperature alone. In most other optical fibre based temperature measuring systems there is also a strain sensitivity which must be taken into account in sensor design.

The Brillouin distributed sensing system[12] is probably the most important of the remaining distributed sensing technologies especially in terms of its ability to operate over very long distances (even up to 100 km range). The Brillouin system, like the Raman DTS a non-linear process, relies upon the production of stimulated Brillouin scatter which occurs at a frequency off-set from the interrogating laser frequency by an amount which depends only on the acoustic velocity within the optical fibre. This in turn depends upon the material from which the fibre is fabricated and, more important, on local temperature (through which the bulk modulus is changed) and strain. Typical values of the Brillouin off-set frequency are in the range of 10 to 15 GHz and some precise measurements of these frequencies as a function of temperature and strain for a particular fibre laser combination[13] are shown in Figure 6. From these graphs we see that the extent of the crosstalk is such that 1°C produces approximately the same off-set as a strain of 20$\mu\varepsilon$. There have been some attempts to monitor the Brillouin off-set frequency and the temperature field separately in order to apply a correction. The temperature field can of course be interrogated through a separate Raman DTS which is one possible approach.

Alternatively the spontaneous Brillouin scatter can also be interrogated exactly analogous to the Raman system to determine the temperature. Brillouin backscatter measurement systems are currently extremely expensive with the first commercial system introduced about five years ago for monitoring optical fibre communication cable installations to determine the location and extent of any ground shifting. To date there has been very little experience with the Brillouin system in structural monitoring partly due to its cost and complexity.

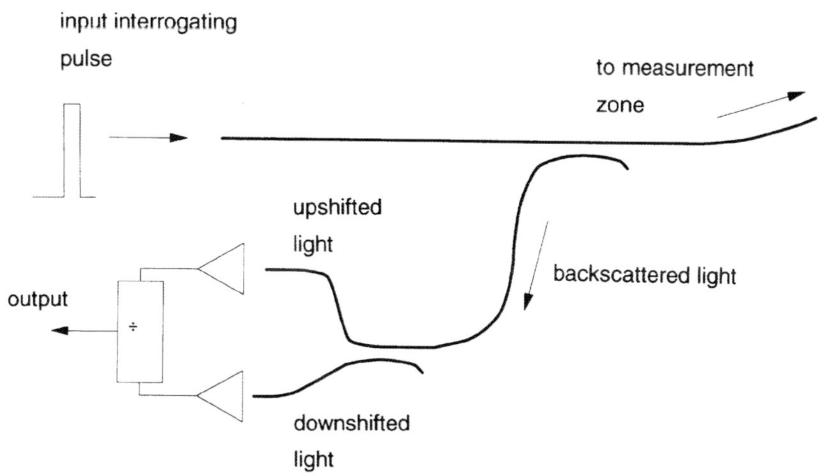

Figure 5. Raman distributed temperature probe: basic schematic.

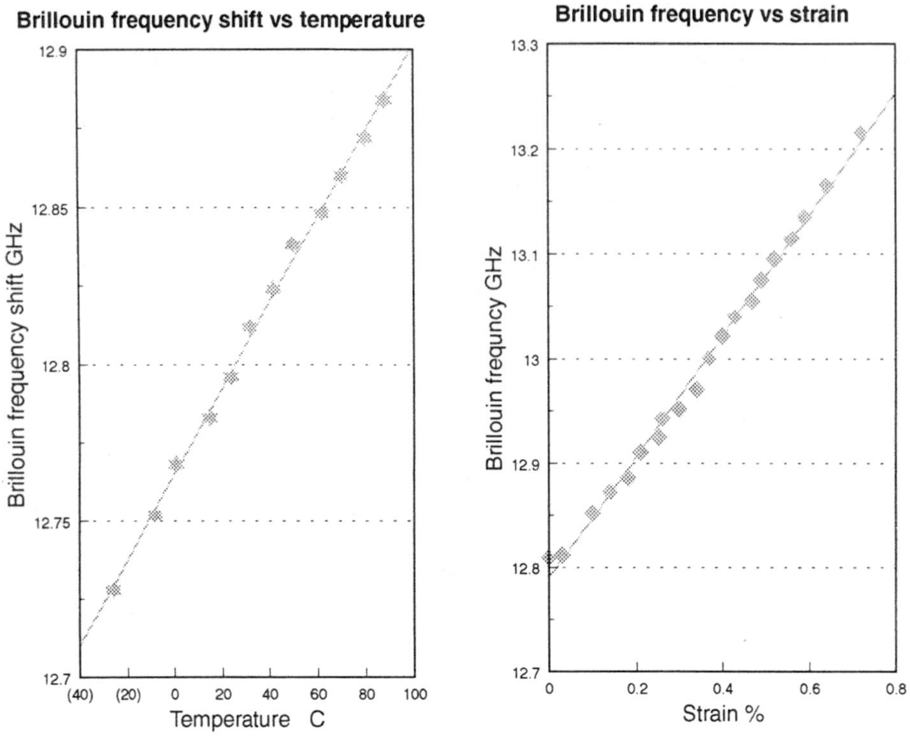

Figure 6. Strain and temperature variations in Brillouin frequency (after ref 13).

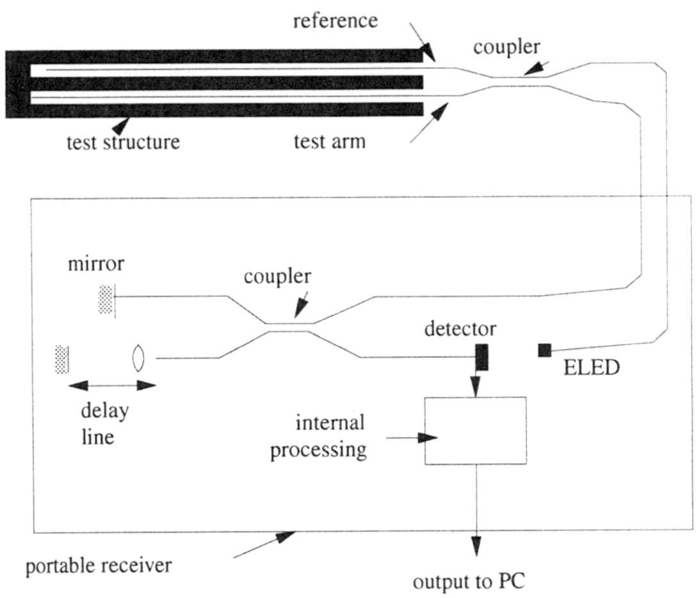

Figure 7. Basic features of the SOFO system.

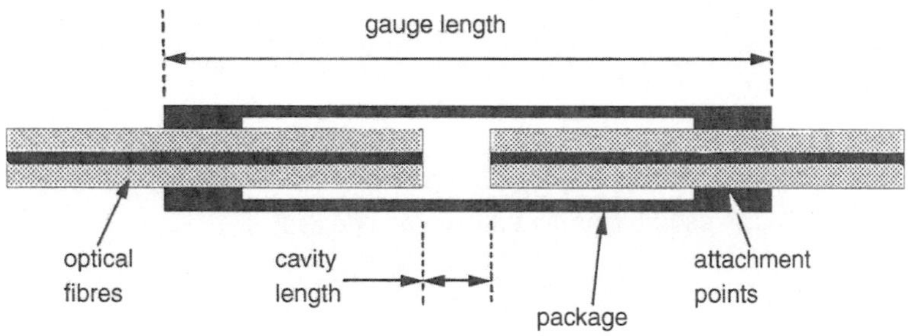

Figure 8. Schematic of the EFPI (extrinsic Fabry Perot interferometer) optical fibre displacement sensor.

However more economically designed architectures are becoming available and some initial evaluations are underway. The system offers resolutions of a few 10s of microstrain over gauge lengths of a few metres with ranges up to several 10s of kilometres though interrogation times are relatively long (seconds to minutes). As such it could find applications in intruder alarms, security systems for pipelines and other areas where very long monitoring systems are required.

Numerous quasi distributed and integrating architectures have been evaluated including both polarimetric/mode:mode differential systems[14] and direct measurement architectures.[15] Of these the SOFO system (Figure 7.) which uses white light interferometer to monitor the differential between a strained and a reference fibre mounted within a protective tube has emerged as a commercial entity. SOFO sensors can be either interrogated individually or in strains of up to about half a dozen units.[8] The principal application has been in measuring point to point displacements over gauge lengths from 1 to 50 metres in large concrete structures such as bridges and dams. The displacement measurement accuracy and resolution is of the order of 5 μm and long term stability over several years has already been demonstrated. To achieve this stability in resolution requires both careful attention to the design of the sensor units and an efficient detection system which integrates over time periods of typically 10 seconds per sensor. The reference fibre provides temperature compensation for the sensor though of course the effect of temperature on the distances between measurement points within a structure under normal operating conditions should also be recognised.

The principal point sensor architectures which have been used in temperature and strain measurement are the extrinsic Fabry Perot interferometer (Figure 8.) and the fibre Bragg grating (Figure 9.) both of which perform identical measurement functions to conventional strain gauges/thermocouples but with the benefit of simple multiplexing into strings of sensors on to a single fibre. The EFPI architecture is somewhat constrained by the relatively high losses of the individual sensor elements (several dB) so that relatively few have been demonstrated together as a single string (perhaps four). In contrast the FBG is, in practice, only lossy at the wavelength at which the FBG reflects and is essentially transparent elsewhere. Consequently, a sufficiently broadband source can be used to illuminate as many FBG sensors as the spectral bandwidth of the source permits. In practice up to a dozen have been run together on a single string. In both cases bandwidths of the order of 1Hz are typical though a few FBG sensors have been reported operating in the low hundreds of Hz. Both systems have the advantage of wavelength selectivity which can be used to tag a particular sensor element.

Both systems also exhibit some temperature:strain crosstalk and some form of temperature correction or compensation is essential. In the EFPI a strain sensitivity is dictated by the total interaction length within the EFPI package and so, for a given cavity length can be varied over perhaps an order of magnitude through different package designs.

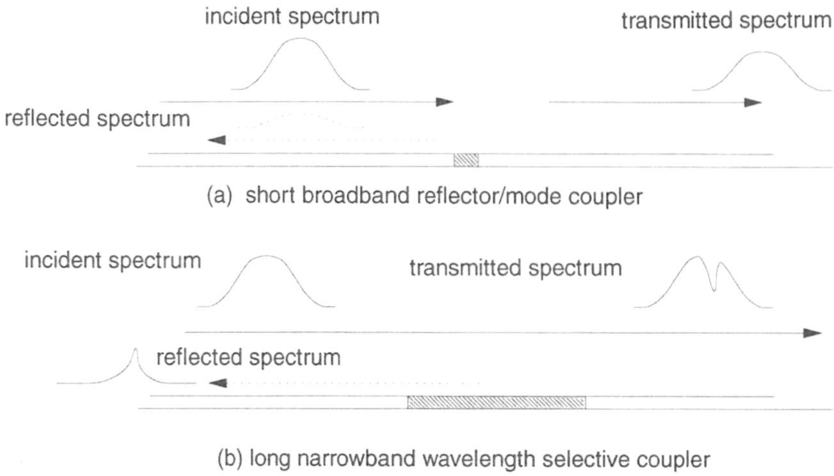

Figure 9. Functions performed by in fibre Bragg gratings.

Its temperature sensitivity is determined by the characteristics of the mounting tube within which the optical fibres are located and the differential temperature coefficient between this mounting tube and the silica fibres themselves. The temperature change in refractive index of the silica fibre has no effect whatsoever on the temperature characteristics of the sensor. Again, the sensor response is strongly influenced by the packaging.

The FBG in contrast retains the light which is used to interrogate it within the fibre. The return signal is a wavelength determined by the Bragg grating periodicity which depends directly upon matching the propagation constant of the oncoming light to the grating period. The return signal from the FBG is therefore completely dependent upon the optical frequency propagation constant within the fibre - that is the refractive index and the applied strain. A shift in wavelength corresponding to 1^{0}C change is approximately equivalent to that introduced by a 10μstrain strain excursion.

Physical sensing for structural monitoring using optical fibres has almost always been based upon one or other of the techniques mentioned here. In all cases temperature effects must be recognised and corrected for during the measurement process. Conversely for temperature measurements, interferences due to strain can also occur except for the Raman DTS. Data interpretation is critical and the measurements must interface with a suitable structural model in order to determine whether or not a fault condition has occurred and also to enable the measurement system to disentangle the effects of load induced strain from those of strain induced through temperature variations. In the structural monitoring context the ability for measurements to be made within the structural fabric and flexibility to realise point distributed or quasi distributed architectures have made optical fibre sensor systems an attractive option. There is still considerable proving through field trials to be completed before fibre sensors become universally acceptable.

6. CHEMICAL MEASUREMENTS IN STRUCTURAL MONITORING – SAFETY AND CORROSION

Chemical measurements in engineering structures are far less straightforward to design and implement. For physical measurement the interpretation of a strain and temperature field through a structural model gives information concerning structural damage or deterioration and/or overloading. Chemical measurements are almost always associated with corrosion detection or with issues concerned with operational safety. The objective with any corrosion monitoring system is to be able to nip the process in the bud to avoid immense cost and inconvenience during major repairs. Safety monitoring minimises the chances of damage to people or property. For physical measurements it is practical to define a strain or temperature, spatial and temporal sampling procedure which can be supported conceptually into the modelling process. The situation in chemical sensing is more uncertain. Both corrosion and safety hazards can, in principle, strike anywhere so that measurements should be made throughout the structural volume - in practice there is always some compromise.

Water is a major contributor to most if not all corrosion processes and can also threaten structural fabric through flooding. Consequently water detection is an important enabler within the monitoring for smart structures envelope. There are a number of water sensing systems currently available usually based upon a cable structure which short circuits whilst wet. These usually only respond whilst wet and are slow to dry out with considerable hysteresis. They can, however, be used as flood alarms. In the safety sector there is a well established technology using pellistors which is extensively used in mines, chemical plant oil exploration systems etc. Yet again, the system is limited in operational flexibility and requires persistent maintenance and checking.

Chemical sensing using fibre optics is one of the more promising applications. Many chemical measurements are spectroscopic so the fibre optic transmission medium is well suited to conveying signals to and from a remote measurement location. However, the emphasis in chemical sensing using optical fibres has been primarily in biomedical instrumentation exploiting the small size and electrical passivity which fibre optics has to offer. Structural measurement presents an entirely different set of applications criteria and in this domain as we have already noted several times, the distributed and multiplexed capability of large area systems using fibre optic interconnect is the principal benefit. There is relatively little ongoing activity in applying chemical sensing techniques using fibre optics in this particular domain. There are, however, a number of interesting possibilities which are emerging.

Figure 10. shows a longitudinal cross section of a distributed moisture sensitive cable based upon distributed microbend sensing.[16] The principle is simple - the thin hydrogel coating swells when in contact with water and in doing so forces the fibre against the Kevlar wrap and hence introduces microbend. The gel responds to high humidity in addition to total immersion so the sensor can be used as a monitor for the onset of high humidity conditions. The sensor has been used to monitor drying conditions within civil engineering construction processes and to provide a very long distributed water leakage detector which can extend up to 10km. Some typical results[17] are shown in Figure 11. This indicates the drying process within a grout filled tube simulating the processes which are used in the construction of post tensioned concrete bridges.

The sensor has also been used to monitor moisture dispersion in soils and as an indicator of relative humidity levels in the region exceeding 80%. The basic principle can be modified by modifying the hydrogel material to respond to other parameters such as pH and the presence or absence of liquid hydrocarbons. This sensing technology will, therefore, evolve into a structural sensing system for monitoring the onset of conditions under which corrosion may occur and the safety system for flood alert and pipeline leakage.

Multiplexed sensor systems also show some promise and Figure 12. shows a schematic diagram of a multiplexed system which could be used to measure methane (or similar) gas concentrations in, for example, natural gas distribution systems and oil exploration rigs. This again is a safety monitoring system and designed such that each sensing point (all addressed by the same laser optoelectronic unit) is capable of detecting about 1% of the lower explosive limit of methane gas and air corresponding to about 5% by volume. A DFB laser tuned to the absorption line in the near infra-red and providing about 1mW of output can be used in the architecture shown in Figure 12. to energise an array of up to 128 sensing points.[18] The DFB together with the frequency stabilisation and modulation optoelectronic unit is a relatively expensive item so it is essential to share this facility among as many detection points as possible.

The system configuration indicated in the diagram shows more than adequate performance in a relatively simple format requiring some care and special techniques in the frequency stabilisation optoelectronic unit, the detection optoelectronics and the design of the detection cell which is essentially a micro-lensed fibre unit (Figure 13.) with a few centimetres open path interaction length.[19]

When compared to the conventional pellistor system, the all fibre system offers intrinsic safety (no power supply required), much lower cost of ownership since the pellistor units need continuous scheduled replacement at intervals of a few months and the facility to operate at very long distances from the monitoring and control room - 25km would be no problem.

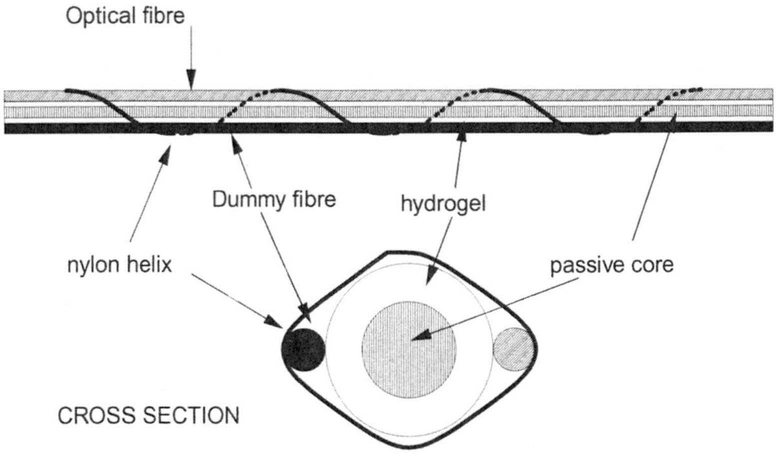

Figure 10. Fibre optic distributed moisture ingress or pH sensor (microbend sensor).

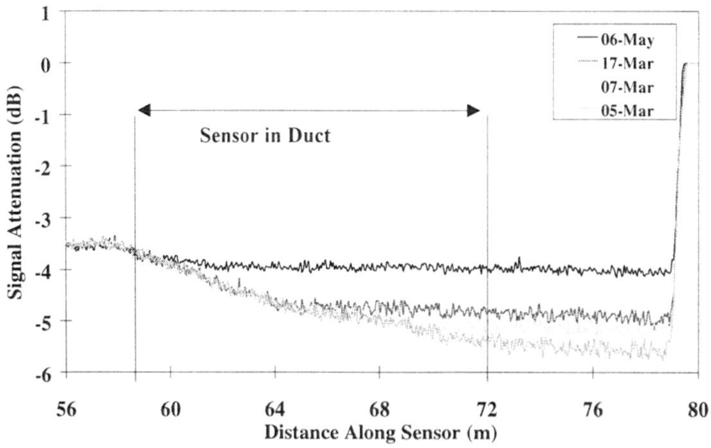

Figure 11. Distributed water ingress sensor used to monitor progressive drying of grout in duct. The high slope sections of the OTDR trace indicate wet regions.

There are numerous other safety and security systems using fibre optics including, e.g. a linear heat detector (based on Raman DTS) for fire monitoring in long structures such as tunnels and - as one of the original optical fibre sensors - a bilge water monitoring system for detecting oil suspended in the water using scatter.[20] Fibre optic systems tend to be relatively expensive and so must find their place in niche applications - but of course, the same is true for everything else.

7. CONCLUSIONS

The concept of smart structures will be an important generic enabler for mechanical and civil engineering construction techniques through the next century. The first stage in smart structures is the realisation of an effective sensory mechanism from which decisions concerning corrective action and reactions may be determined. The sensor system must be physically and chemically compatible with the environment and must also be capable of addressing large areas and/or volumes within the structural fabric. Fibre optic technology can fulfil both these requirements and as such may be viewed as a potentially strong contributor to the evolution of smart structures.

This paper has discussed some of the ongoing approaches to structural instrumentation using fibre optics. Temperature and strain measurement systems are probably the most advanced and have begun to make a contribution. Fibre optic measurement of chemical parameters also holds promise in the structural instrumentation concept and the examples mentioned here of hazardous gas monitoring and leak detection will find a niche in the instrumentation market place.

We discussed - very briefly - at the beginning of the paper the need for data processing and in particular data reduction. This is a very important issue which must be effectively addressed in any sensor system. Certainly most of the signal processing functions will be realised after the sensor data has been detected. However, fibre optics has another unique feature which is the ability to perform antenna array processing on certain types of parameter fields and in particular on acoustic and vibration fields.[21] This pre-processing and pre-filtering could be put to good effect in the longer term realisation of smart structures. Fibre optics will, of course, have another role in the evolution of advanced structural instrumentation - namely all the data that the fibre optic sensors obtain concerning structural health will undoubtedly find its way on to a fibre optic communication network for further attention. There is obviously a lot more to come.

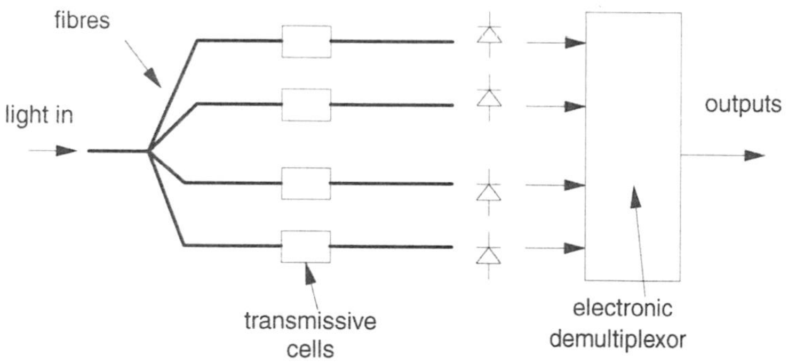

Figure 12. Multiple cell system configuration for remote gas sensing using absorption spectroscopy: trasmissive star system with electronic demultiplexor.

Figure 13. Engineered micro-optic open path absorption cell for industrial methane gas measurements.

REFERENCES

1. B. Culshaw, *Smart Structures & Materials*, Artech House, Norwood MA (1996).
2. SPIE, Bellingham, Washington have published many conference proceedings on Smart Structures & Materials technology. Full listings can be obtained from SPIE, covering European Conferences on Smart Structures and Materials and the SPIE US Annual Symposium.
3. M.V. Ghandi and B. S. Thompson, *Smart Materials and Structures*, Chapman & Hall, London (1992).
4. J. Telford *The monitoring of the Kingston Bridge Glasgow*, IMechE Seminar on Smart Materials & Systems, London (10 April 1997).
5. E. Udd, *Fibre Optic Smart Structures*, Wiley, New York (1993).
6. B. Culshaw & J.P. Dakin *Optical Fibre Sensors Vol I* (1988), *Vol II* (1998), *Vol III* (1997), *Vol IV* (1997), Artech House, Norwood MA.
7. S.G. Pierce, W.J. Staszewski, A. Gachagan, I.R. James, W.R. Philp, K. Worden, B. Culshaw, A. McNab, G.R. Tomlinson and G. Hayward, Utrasonic condition monitoring of composite structures using a low profile acoustic source and an embedded optical fibre sensor, *Proc SPIE* 3041:437 (1997).
8. D. Inaudi, Field testing and application of fibre optic displacement sensors in civil structures, in *12th Optical Fiber Sensors Conference*, OSA Technical Digest Vol. 16, 596 (1997).
9. S.S.J. Roberts *Fibre Optic Based Smart Materials*, PhD thesis, University of Strathclyde, Glasgow Scotland (1992).
10. J.P. Dakin, D.J. Pratt, G.W. Bibby and J.N. Ross, Distributed optical fibre Raman temperature sensor using a semiconductor light source and detector, *Electron. Lett.* 21:569 (1985).
11. J.W. Berthold, Sensors in industrial systems, in *Optical Fibre Sensors Vol. IV*, B. Culshaw & J.P. Dakin eds., Artech House, Norwood MA (1997).
12. T. Horiguchi, Brillouin scattering for measuring strain and temperature, in *Optical Fibre Sensors Vol. IV*, B. Culshaw & J.P. Dakin eds., Artech House, Norwood MA (1997).
13. M. Nikles, L. Thevenaz and P.A. Robert, Brillouin gain spectrum characterization in single mode optical fibres, *IEEE-J. Light. Technol.* 15:1842 (1997).
14. G. Thursby, W.C. Michie, D. Walsh, and B. Culshaw, Simultaneous recovery of strain and temperature fields by the use of two moded polarimetry with an in-line mode/splitter analyser, *Opt. Lett.* 20:1919 (1995).

15. B. Noharet, M. Turpin, J. Chazelas, P. Bonniau, D. Walsh, C. Michie, and B. Culshaw, Microwave subcarrier optical fibre strain sensor, *Proc. SPIE*, Vol. 2361;236 (1994).
16. W C Michie, B. Culshaw, M. Konstantaki, I. McKenzie, S. Kelly, N.B. Graham and C. Moran, Distributed pH and water detection using fibre optics sensors and hydrogels *IEEE-J. Light. Technol.* 13:1415 (1995).
17. W.C. Michie, G. Thursby, A. McLean, B. Culshaw, B. Verwilghen and M. Voet, Fibre optic sensor for distributed water ingress detection and humidity measurement in *12th Optical Fiber Sensors Conference*, OSA Technical Digest Vol. 16, 634 (1997).
18. G. Stewart, A. Mencaglia, W. Philp and W. Jin, Interferometric signals in fibre optic methane sensors with wavelength modulation of the DFB laser, *IEEE-Journal of Lightwave Technology,* 16: 43 (1998).
19. B. Culshaw, *Fibre optic techniques for spectroscopic methane gas detection in the near infrared - from detection concept to system realisation*, Proc. Europt(r)ode IV, Munster, Germany, (March 29-April 1, 1998)
20. G D Pitt, Private communication 1982.
21. See, for example, ref. 1, p. 140.

ALL OPTICAL FIBER ULTRASONIC SOURCES FOR NON DESTRUCTIVE TESTING AND CLINICAL DIAGNOSIS

E. Biagi, L. Masotti, and M. Pieraccini

Università degli Studi di Firenze
Dipartimento di Ingegneria Elettronica
Via Santa Marta, 3
50139 Firenze, Italy

1. INTRODUCTION

The laser generation of ultrasonic waves has been a well-established technique for Non-Destructive Evaluation (NDE) since the pioneering work of White in 1963.[1] Unique advantages of this technique are high frequency, large bandwidth, and high spatial resolution. The employment of fiber optics as waveguides for laser pulses adds further advantages: very high miniaturization and geometrical versatility, electromagnetic compatibility, ionizing radiation safety, the possibility of keeping the optoelectronic unit away from the test object.

Standard facilities for this kind of ultrasonic NDE make it possible to irradiate a test object through a pulsed laser.[2] Photoelastic waves are generated by thermal expansion due to the absorption of optical energy of a laser pulse either on the surface or inside the sample. Unfortunately, the efficiency of ultrasonic generation by direct exposure of the material under test is very low and highly dependent on the conditions of the material. Moreover, the directivity pattern of the elastic wave is very broad with respect to that of the elastic waves generated by conventional transducers (in this regard, see Figure 1.). Efficiency and directivity can be improved by a suitable photoacoustic conversion system. For example, it is possible to shape the acoustic beam in terms of directivity and propagation direction by placing appropriate fluid layers on the sample surface. In this regard, Von Gutfeld[3] and Oksanen[4] proposed different multilayered structures in which the laser radiation is absorbed by a metallic film deposited on a glass slab.

According to the theoretical support reported in the following section, innovative photoacoustic sources for ultrasonic NDE, smart structure, and clinical diagnosis are proposed (International patent pending n. FI96A 216). The working principle is based on thermal conversion of laser pulses into a metallic film evaporated directly onto the tip of a fiber optic. Efficiency and directivity are comparable to that of devices proposed by Gutfeld and Oksanen. Unique features of the proposed transducers are very high miniaturization and potential easy embedding in smart structure. Additional advantages are high ultrasonic frequency, large and flat bandwidth. All these characteristics make the proposed device an ideal source of ultrasound not

Optical Sensors and Microsystems: New Concepts, Materials, Technologies
Edited by Martellucci et al., Kluwer Academic / Plenum Publishers, New York, 2000.

only for ultrasonic imaging systems but also to calibrate any ultrasonic transducer. Moreover thanks to its dimensions, the proposed device can be used to calibrate ultrasonic transducer arrays as well.

2. THEORETICAL BACKGROUND

The authors have elaborated a theoretical formulation starting from the White theory of elastic wave generation by the direct irradiation of a test object. Temperature distribution (T) can be obtained by the usual one-dimensional heat diffusion equation

$$\hat{k}\frac{\partial^2 T}{\partial x^2} = \frac{\partial T}{\partial t} \tag{1}$$

where $\hat{k}$ is the thermal diffusivity. If we consider the optical power absorbed only at the surface x=0, the thermal boundary condition is:

$$-K\left.\frac{\partial T}{\partial x}\right|_{x=0} = \alpha I \tag{2}$$

where K is the thermal conductivity, α is the fraction of light absorbed, I is the light intensity. The Laplace transform of heat diffusion equation has the solution:

$$\overline{T} = \frac{\alpha \overline{I}}{K p} e^{-px} \quad with \quad p = \sqrt{\frac{s}{\hat{k}}} \tag{3}$$

where $\overline{T}$ and $\overline{I}$ are the Laplace transforms, respectively, of temperature and light intensity, s (Hz) is the Laplace transform variable.

Letting v be the sound velocity, we can write the equation of the u displacement field as:

$$\frac{\partial^2 u}{\partial x^2} - \frac{1}{v^2}\frac{\partial^2 u}{\partial t^2} = \beta \frac{\partial T}{\partial x} \tag{4}$$

By substituting Eq. (3) in the Laplace transform of Eq. (4), we obtain the u displacement field as sum of an acoustic wave and a thermal wave of amplitudes B and C respectively, i.e.

$$u = Be^{-kx} + Ce^{-px} \quad with \ k = s / v \tag{5}$$

where:

$$C = -\frac{\alpha I}{Kp}\beta\frac{p}{p^2 - k^2} \approx -\frac{\alpha I}{Kp^2}\beta\left(1 + \left(\frac{k}{p}\right)^2\right) \tag{6}$$

The Taylor development halted at the second order is justified by the fact that k<<p. It still remains to determine B imposing a suitable boundary condition. Since x=0 is a free surface, we can consider it stress-free, that is:

$$\left(\frac{\partial u}{\partial x} - \beta T\right)_{x=0} = 0 \tag{7}$$

By substituting Eq. (5) in Eq. (7), we obtain

$$(kB)_{stress\,free} = \frac{\alpha I}{Kp}\beta\left(\frac{k}{p}\right)^2 \tag{8}$$

Now we consider a different experimental situation. A material, which is transparent in the laser light with mechanical characteristics very similar to the irradiated test object, is affixed to the surface of the test object in order to constitute a mechanical constraint. If, in a first approximation, we consider the test object and the constraint as two semi-spaces separated by an infinite flat surface for reasons of symmetry, the local displacement is null on the surface. In this hypothesis, the condition in Eq. (7) becomes $u(0) = 0$, from which $kB = -kC/2$, and since:

$$(kB)_{withconstr.} = \frac{\alpha\beta Ik}{2Kp^2} \tag{9}$$

It is to be noted that, in this case, it is no longer necessary to develop the expression of C up to the second order and therefore it can be obtained that the amplitude of the ultrasonic waves is increased by $(k/p)^{-1}/2$; the factor 1/2 is due to the fact that, because of symmetry, ultrasounds are generated in both directions of x-axis. In the case of an ideal constraint the $(k/p)^{-1}$ factor for ultrasound at 5MHz is of an order of 60dB on the amplitude value, obviously in a real case this extreme improvement in photoacoustic conversion is not realistic. In fact experimental works report enhancement of an order of 40dB.[5] Our calculation of the conversion efficiency η gives as result:

$$\eta = \frac{I^*_{acoust}}{I^*} = \frac{\rho v^2}{2}\left(\frac{\partial u}{\partial x}\right)\left(\frac{\partial u}{\partial x}\right)^* \approx \frac{\rho v^3}{2}(kB)(kB)^* \tag{10}$$

and by substituting the Eq. (9) in Eq. (10), we obtain:[6]

$$\eta = \frac{\alpha^2\beta^2 vI}{8\rho c^2} \tag{11}$$

that gives $\eta = 4.7\times10^{-9}$, for numerical values relative to water. By assuming $I = 10^{10}$ W/cm^2 which is the maximum optical intensity which does not cause material vaporization, an acoustic intensity equal to 47 W/cm^2 is obtained.

Conversion efficiency and, consequently, acoustical intensity can be improved in the following ways: 1) by employing a film of more absorbing material (e.g. chromium oxide); 2) by employing a liquid with a higher thermal expansion coefficient and a lower thermal capacity.

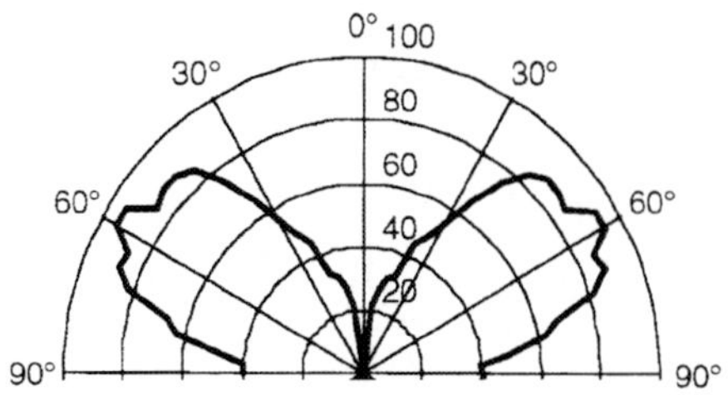

Figure 1. Ultrasonic amplitude pattern obtained by direct exposure of a metallic test object.

By adopting both these solutions, it is possible to improve the conversion efficiency of more than two orders of magnitude, obtaining acoustic intensity notably higher than conventional transducers operating in high frequency range (over 5 MHz).

The reported theory give important predictions about conversion efficiency, but it can not predict the radiation lobe shape. About this subject we can make same qualitative consideration. When a high power laser pulse is directed at the surface of a metal, the absorption takes place within a skin depth typically of a few nanometers, giving rise to transient thermoelastic stresses and strains in the surface layer as the metal tries to expand. Provided the optical pulses are very short thermal diffusion effects are relatively small so that the thermal source is localized to a very thin surface (a few micrometers) layer. If the surface of the metal is free, then the principal stresses are parallel to the surface. This generates an unusual far-field source radiation pattern: longitudinal wave amplitude pattern exhibit a cylindrical symmetric lobe at 60° with respect to the surface normal.

An experimental confirmation of the radiation lobe shape was performed by the authors by direct exposure of a metallic test object. The lobe, obtained with a Nd-YAG laser beam of 1mm diameter and peak power 5MW, is shown in Figure 1.

3. FIBER OPTIC ULTRASONIC SOURCE DESIGN

According to theoretical consideration, ultrasonic sources of good efficiency can be designed as follows. The light beam of a laser operating in a pulse mode was coupled to a fiber optic.

In order to realize the simpler kind of ultrasonic source, 'cut fiber ultrasonic source' in Figure 2a. a metallic film of chromium was evaporated directly onto the tip face of the fiber optic. The tip of the fiber optic is immersed in a liquid where the ultrasounds are generated.[7,8] The innovative feature of this very simple structure is the double function of the fiber optic that acts both as optical waveguide and as mechanical constraint for the elastic wave generated in the liquid. Furthermore, the liquid acts both as coupling medium and as high thermal expansion material for amplitude enhancement of produced ultrasounds.

For the second kind of photoacoustic source, as appears in Figure 2b., the fiber was uncladded in a short stretch, subsequently the silica core was heated at 1200 °C and with a suitable mechanical guide the fiber was U-bent. Finally, a metallic film of chromium was evaporated onto the extrados of the bent fiber optic transducers. Because of the U shape a considerable fraction of optical pulses spreads over metallic film, generating elastic wave by the thermal shocks in the material surrounding the fiber.

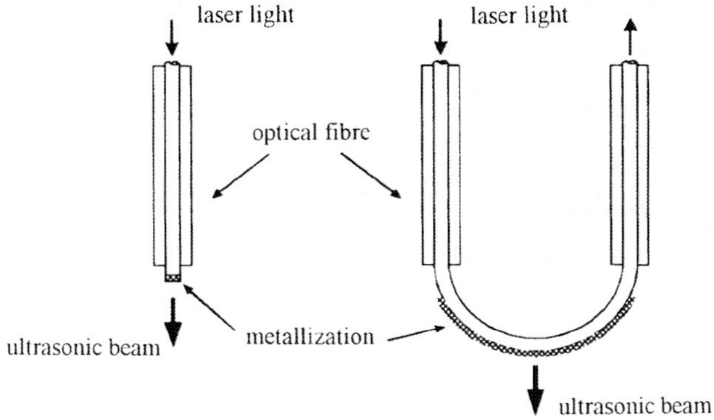

Figure 2. a) cut fiber optic ultrasonic source, b) U-bent fiber optic ultrasonic source.

Of note, the light that is not converted into heat by metallic film is again guided by the fiber and can be newly used for ultrasound generation by a second U-shaped structure sequentially introduced into the same fiber and so on. This feature is of great interest because the largest part of optical energy is reflected and not converted into heat by the metallic film. Furthermore, the U-bent fiber transducer makes it possible to realize a series of ultrasonic sources spatially modulated over the same fiber, i.e. an ultrasonic array.

4. EXPERIMENTAL SETUP AND RESULTS

A complete characterization of proposed photoacoustic source in its more simple configuration, Figure 2a., is performed in terms of ultrasonic spectral response and of ultrasonic beam spatial distribution.

The measurement setup is reported in Figure 3. Q-Nd:YAG lasers are employed as light sources. The optical pulses are coupled to 600 μm core diameter PCS fiber, through a focusing lens. An optical attenuator is interposed between laser and lens in order to control the optical power inside in fiber.

A 20 MHz V316 Panametric probe positioned in front of the all-fiber optic ultrasonic source was employed as ultrasound detector. The fiber and the ultrasonic probe are immersed in a water tank. The ultrasonic signal, whose typical shape is reported in Figure 4., was amplified by a commercial receiver (5052PR Panametrics) and acquired through a digital oscilloscope.

The employed laser pulse was characterized in terms of duration 270 ns, repetition frequency 2.00 kHz, and pulse energy 570 μJ. Spectral distribution of the obtained ultrasonic signal is plotted in Figure 5., and labeled *a.*

A second ND-YAG, with shorter optical pulses (100-150 ns), was used in the same experimental condition. Spectral distribution of the obtained ultrasonic signal is plotted in Figure 5., and labeled *b.* The obtained spectral distributions exhibit a direct correspondence with the spectral distribution of the laser pulse, consequently the spectrum of the generated ultrasound is completely controlled by the laser pulse duration.

To evaluate the performance obtained through the metallization of the fiber tip for clinical purposes, we compared the results obtained using a non-metallized 600 μm core diameter PCS fiber. In this case the light absorption occurred in the liquid surrounding the fiber tip.

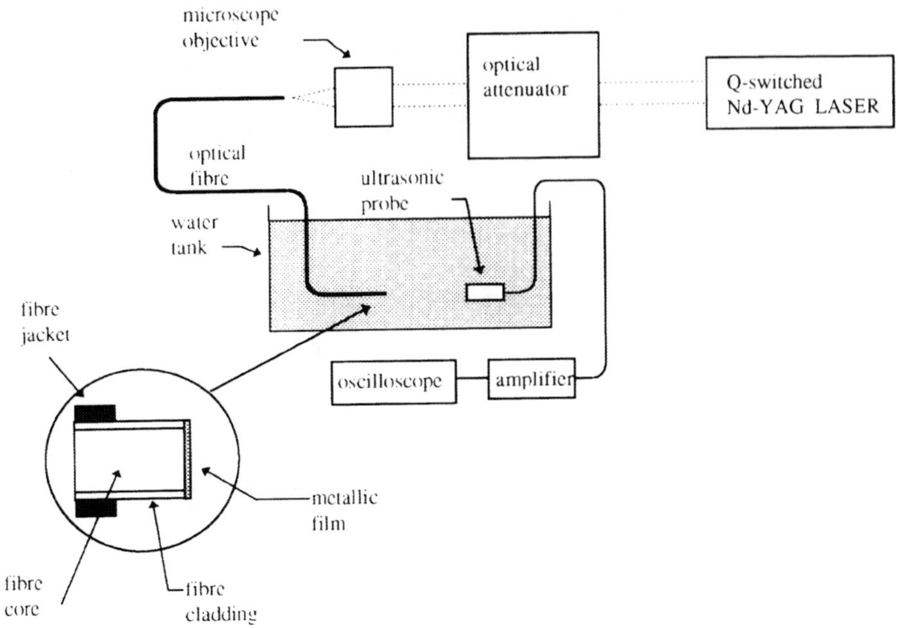

Figure 3. Experimental setup.

Keeping in mind the low absorption coefficient of pure water, we could not obtain an easily detectable ultrasonic pulse. However, biological liquid or tissue can be much more absorbent, even if they are constituted principally of water.

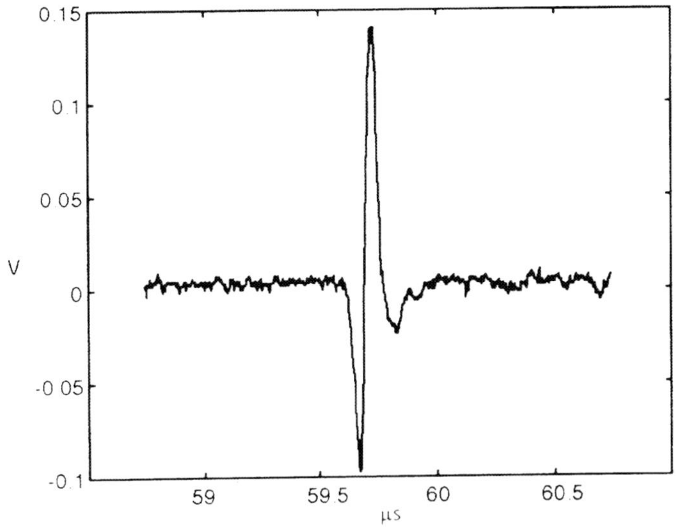

Figure 4. Ultrasonic pulse generated by the optical fiber ultrasonic source collected by a 20 MHz ultrasonic transducer.

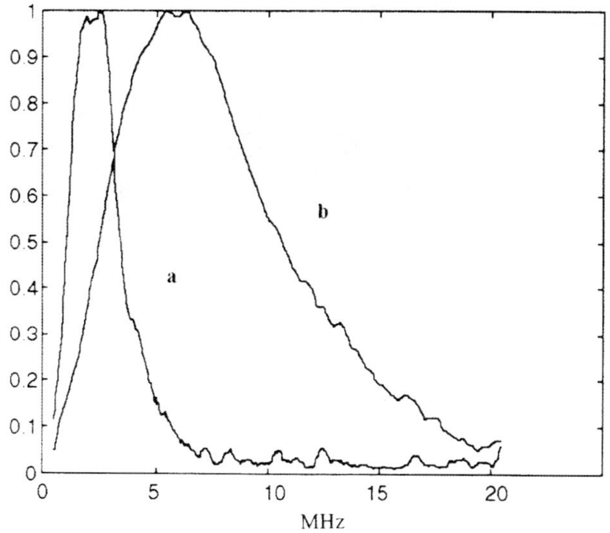

Figure 5. Normalized spectral amplitudes obtained for long (a) and short (b) optical pulses.

To increase the optical absorption and to have an experimental situation comparable with the irradiation of biological tissue, we contaminated water with China ink. Figure 6a. shows the ultrasonic signal detected by the 20MHz ultrasonic probe in the inked water while the spectrum labelled b, which is the the same of Figure 5., is related to the metallized fiber. As can be noted by comparing the spectra in Figure 6a. and 6b., the metallization dramatically increases the high frequency content of the ultrasonic spectrum.

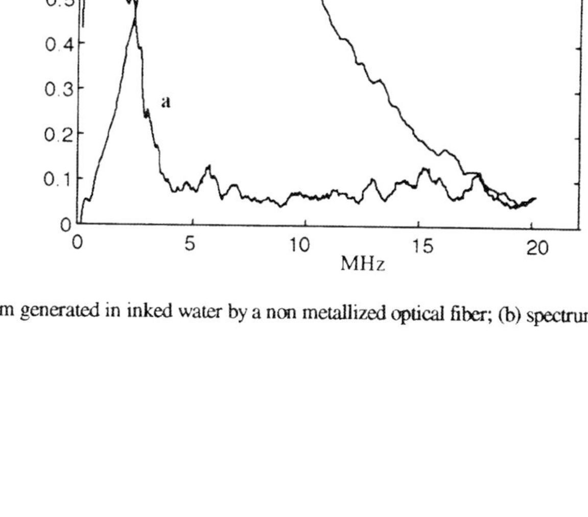

Figure 6. (a) Spectrum generated in inked water by a non metallized optical fiber; (b) spectrum of metallized fiber.

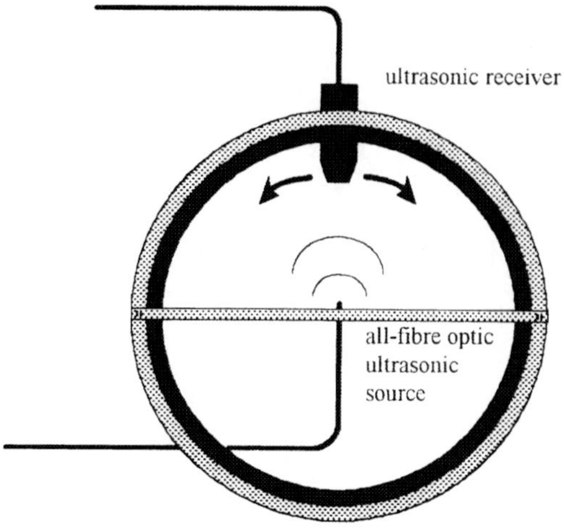

Figure 7. Experimental system for rotating the ultrasonic probe around the fiber.

This effect is closely related to the optical absorbing depth. In fact, with the metallization the optical energy is absorbed in a very thin metallic layer of the order of approximately one hundred nanometers. On the contrary, by illuminating opaque liquid with non metallized fiber, the optical absorption involves a liquid thickness of several mm. We also characterized the proposed ultrasonic source in terms of directivity pattern. A mechanical device for rotating the ultrasonic probe with respect to the fiber is shown in Figure 7.

The measured ultrasonic amplitude radiation lobe relative to the shorter optical pulses is shown in Figure 8. A -6dB spread angle of 10° can be observed. The beam divergence of a generic ultrasonic transducer can obtained by using the following equation :

$$sin(\vartheta) = 0.514 \times v \,/\, (f\ D) \qquad (12)$$

where θ is the -6dB spread angle, f is the peak frequency of the ultrasonic spectrum, v is the sound speed in water, and D the transducer diameter. The spread angle value calculated by Eq. (9) was 12.1°, which is in good agreement with the measured value by considering the measurement accuracy.

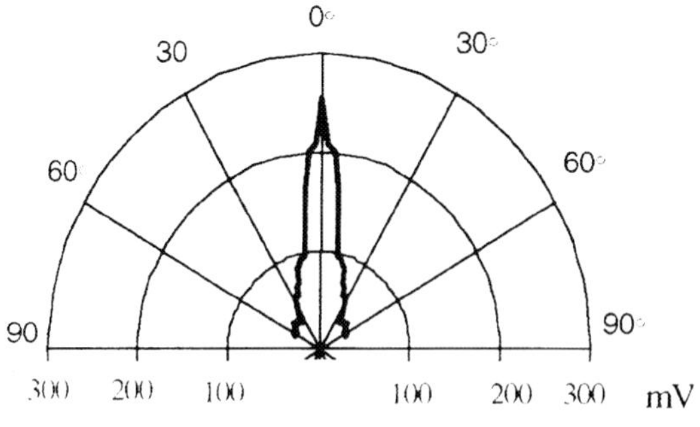

Figure 8. Ultrasonic pattern for a cut-fiber optic source.

Furthermore, we verified the photoacoustic generation through the U-bent fiber optic source. Employing the shorter pulse Nd-YAG laser, clearly detectable ultrasonic pulses are obtained. However, the spectra exhibit a complex shape probably due to multiple acoustic paths inside the fiber optic source.

5. CONCLUSION

The innovative feature of our photoacoustic source consists in the double function of the optical fibre that acts both as optical waveguide and as mechanical constraint for the elastic wave generated in the liquid.

It was experimentally demonstrated that the bandwidth and the maximum frequency of the ultrasonic signal can be controlled by the laser pulse duration. Consequently, high frequency ultrasonic transducer with high bandwidth can be realized with good miniaturization.

A peculiar characteristic of the proposed source is the high value of the generated ultrasonic intensity with respect to ultrasonic piezoelectric transducers. In fact, the acoustic intensity emitted from the photoacoustic source is indipendent from the frequency while the piezoelectric transducer intensity strongly decreases when the frequency rises in value. Afterwards the metallized fiber appears to be a source with high efficiency for frequency higher then several MHz, even if it is not competitive in the kHz range.

As mentioned above, the proposed all-optical fibre transducers are characterized by high thermo-acoustic efficiency, but the enhancement of opto-thermal conversion is ignored in this work and further efforts are still need. An array of the U-bent transducers fed by the same optical pulse was demostrated to be a praticable way to solve the efficiency problem.

REFERENCES

1. R.M. White, "Generation of elastic waves by transient surface heating", Journal of Applied Physics, vol. 34, no. 2, pp. 3559-3567, 1963.
2. B Scruby and L.E. Drain, "Laser Ultrasonic", Adam Hailer Bristol, 1990.
3. R.J. von Gutfeld and H.F. Budd, "Laser-generated MHz elastic waves from metallic-liquid surface", vol. 34, no. 10, Applied. Physics Letter, pp. 617-619, 1979.
4. M. Oksanen and J. Wu, "Prediction of the temporal shape in a photoacoustic sensing application", Ultrasonics, vol. 32, no. 1, pp. 43-46, 1994.
5. D.A. Hutchins, R.J. Dewhurst and S.B. Palmer, "Laser generated ultrasound at modified surfaces",Ultrasonics, May, pp. 103-108, 1981.
6. E. Biagi, M. Brenci, S. Fontani, L. Masotti, M. Pieraccini, "Photoacoustic generation: optical fibre ultrasonic sources for NDE and clinical diagnosis", Optical review 1997, n.4,pp.481-483.
7. E. Biagi, S. Fontani, F. Francini, L. Masotti, M. Pieraccini, "Photoacoustic generation: all-optical fibre transducers"1996 IEEE International Ultrasonic Symposium, pp. 921-924.
8. Italian Patent "Trasduttore a fibra ottica per la generazione di ultrasuoni e fibra ottica per la sua realizzazione" n. FI96A 216.

INDEX